Parametrizations in Control, Estimation and Filtering Problems: Accuracy Aspects
___

Communications and Control Engineering Series
Editors: B.W. Dickinson · A. Fettweis · J.L. Massey · J.W. Modestino
E.D. Sontag · M. Thoma

CCES published titles include:

Sampled-Data Control Systems
J. Ackermann

Interactive System Identification
T. Bohlin

The Riccatti Equation
S. Bittanti, A.J. Laub and J.C. Willems (Eds.)

Nonlinear Control Systems
A. Isidori

Analysis and Design of Stream Ciphers
R.A. Rueppel

Sliding Modes in Control Optimization
V.I. Utkin

Fundamentals of Robotics
M. Vukobratović

Michel Gevers and Gang Li

Parametrizations in Control, Estimation and Filtering Problems: Accuracy Aspects

With 60 Figures

Springer-Verlag
London Berlin Heidelberg New York
Paris Tokyo Hong Kong
Barcelona Budapest

Michel Gevers, PhD
Centre for Systems Engineering and Applied Mechanics (CESAME),
Université Catholique de Louvain, Bâtiment Euler,
B-1348 Louvain-la-Neuve, Belgium

Gang Li, PhD
School of Electrical and Electronic Engineering, Nanyang Technological
University, Nanyang Avenue, Singapore 2263

ISBN 3-540-19821-0 Springer-Verlag Berlin Heidelberg New York
ISBN 0-387-19821-0 Springer-Verlag New York Berlin Heidelberg

British Library Cataloguing in Publication Data
A catalogue record for this book is available from the British Library

Library of Congress Cataloging-in-Publication Data
A catalog record for this book is available from the Library of Congress

Apart from any fair dealing for the purposes of research or private study, or criticism or review, as permitted under the Copyright, Designs and Patents Act 1988, this publication may only be reproduced, stored or transmitted, in any form or by any means, with the prior permission in writing of the publishers, or in the case of reprographic reproduction in accordance with the terms of licences issued by the Copyright Licensing Agency. Enquiries concerning reproduction outside those terms should be sent to the publishers.

© Springer-Verlag London Limited 1993
Printed in Germany

The publisher makes no representation, express or implied, with regard to the accuracy of the information contained in this book and cannot accept any legal responsibility or liability for any errors or omissions that may be made.

Typesetting: Camera ready by authors
69/3830-543210 Printed on acid-free paper

To my parents
— *Michel Gevers*

To Qian-Rong
— *Gang Li*

Contents

Preface   xi

1 Introduction   1
    1.1 Motivation and general statement of objectives   1
    1.2 A historical view and some motivating examples   4
    1.3 Outline of the book.   17

2 Finite Word Length errors and computations   23
    2.1 Introduction   23
    2.2 Representations of binary numbers   25
    2.3 Overflow and quantization errors   29
    2.4 Arithmetic computations and roundoff errors   33
    2.5 Dynamic range and scaling   39
    2.6 Conclusions   43

3 Parametrizations in digital system design   45
    3.1 Introduction   45
    3.2 State space realization set   49
    3.3 Sensitivity measure of a state space realization.   50
    3.4 Optimal realizations with respect to a sensitivity measure   55
    3.5 Roundoff noise gain of state space realizations   60
    3.6 Minimal roundoff noise gain realizations   63
    3.7 Relationship between sensitivity measure and roundoff
        noise gain.   66
    3.8 Examples and simulations.   68
    3.9 Alternative approaches   76
    3.10 Conclusions   81
    Appendix 3: Proof of Theorem 3.2   82

4 Frequency weighted optimal design   85
    4.1 Introduction   85
    4.2 Minimization of a frequency weighted sensitivity measure   87
        4.2.1 Weighted sensitivity measure of a realization . .   87

viii   Contents

    4.2.2   Optimal FWL realizations 90
    4.2.3   Existence and uniqueness 91
  4.3 Computation of the optimal realization set 95
  4.4 Numerical example 99
  4.5 Conclusions 103
  Appendix 4.A: Proof of existence of a minimum 104
  Appendix 4.B: Computation of weighted Gramians 106

5 A new transfer function sensitivity measure   109
  5.1 Introduction 109
  5.2 Minimization of an L_2 sensitivity measure 110
    5.2.1   The L_2 sensitivity measure of a realization . . . 110
    5.2.2   Optimal L_2 sensitivity realizations 111
    5.2.3   Solution of the optimal realization problem . . . 113
  5.3 Relationship between L_1/L_2 and L_2 sensitivity measures 118
  5.4 An example 122
  5.5 Conclusions 125

6 Pole and zero sensitivity minimization   127
  6.1 Introduction 127
  6.2 A pole-zero sensitivity measure 129
  6.3 The eigenvalue sensitivity problem 131
  6.4 Pole sensitivity minimization and normal matrices . . . 139
  6.5 Zero sensitivity measure 142
  6.6 Pole-zero sensitivity coordinate dependence 144
  6.7 Optimal realizations for pole-zero sensitivity minimization 146
  6.8 Numerical example 151
  6.9 Conclusions 156

7 A synthetic sensitivity - roundoff design   159
  7.1 Introduction 159
  7.2 A synthetic FWL noise gain 160
  7.3 Optimizing the Total Noise Gain 168
  7.4 A numerical example 175
  7.5 Conclusions 179
  Appendix 7: Existence of a constrained minimum 180

8 Sparse optimal and suboptimal realizations   183
  8.1 Introduction 183
  8.2 Sparse optimal realizations 186

	8.2.1	Hessenberg optimal realizations 187
	8.2.2	Schur optimal realizations 190
	8.2.3	Sparse block-balanced realizations 192
8.3	Theoretical versus actual sensitivity measure 193
	8.3.1	Pole sensitivity of Schur realizations 194
	8.3.2	A numerical example 197
8.4	Sparse quasi-optimal realizations 201
	8.4.1	Constrained similarity transformations 202
	8.4.2	Extensions of the Bomar and Hung algorithm . . 209
8.5	Sparse suboptimal realizations 211
8.6	Conclusion 214

9 Parametrizations in control problems 217

9.1 Introduction 217
9.2 Implementation of a pole placement controller 223
    9.2.1 The ideal pole placement controller 223
    9.2.2 Finite precision aspects in a closed loop compensator: problem formulation 225
    9.2.3 Sensitivity analysis and optimal structures 227
    9.2.4 Roundoff noise study and optimal realizations . . 231
    9.2.5 Design example 235
9.3 FWL LQG controller design 244
    9.3.1 The 'ideal' LQG controller 245
    9.3.2 Roundoff noise study of an LQG controller . . . 248
    9.3.3 Optimal implementations of FWL LQG controllers 250
    9.3.4 Optimal FWL LQG controllers 253
9.4 Conclusions 255
Appendix 9: Sensitivity functions of closed loop system . . . 257

10 Synthetic FWL compensator design 261

10.1 Introduction 261
10.2 State space description of a compensator 263
10.3 Analysis of FWL effects of a compensator 268
10.4 Optimal FWL compensator realizations 276
10.5 A design example 279
10.6 Conclusion 285

11 Parametrizations in the Delta operator 289

11.1 Introduction and motivation 289
11.2 Delta operator parametrizations 293

11.3	Sensitivity of delta parametrizations	296
	11.3.1 Sensitivity measure	297
	11.3.2 Optimal realization set	299
	11.3.3 Numerical example	302
11.4	Roundoff noise analysis	305
	11.4.1 Introduction	305
	11.4.2 The roundoff noise gain of shift and delta operator realizations	305
	11.4.3 Minimization of the roundoff noise gain	308
	11.4.4 New conditions for superiority of δ-realizations	310
	11.4.5 Numerical examples	314
11.5	Conclusions	315
	Appendix 11: Proof of Theorem 11.5	317

12 Generalized transfer function parametrizations for adaptive estimation   319

12.1	Introduction	319
12.2	Parameter estimation and the information matrix	321
12.3	Connection between parametrization and data filtering	327
12.4	Optimal and suboptimal design choices	332
	12.4.1 Manipulating the information matrix with data information	333
	12.4.2 A robustness property	334
12.5	γ-operator parametrizations	335
	12.5.1 A transfer function in γ-operator	336
	12.5.2 Computing the information matrix	337
	12.5.3 Analysis of the information matrix	341
	12.5.4 Comparison with δ-operator parametrizations	346
12.6	Applications in estimation and adaptive filtering	348
12.7	Conclusions	356

References   359

Index   369

Preface

This book is all about finite wordlength errors in digital filters, controllers and estimators, and how to minimize the deleterious effects of these errors on the performance of these devices. This does by no means imply that all about finite wordlength errors in filters, controllers and estimators is to be found in this book.

We first ventured into the world of finite wordlength effects in 1987 when Gang Li began his PhD thesis in this area. Our more experienced readers might well say '*This shows*', but we believe that the extent of our new contributions largely offsets our relative inexperience about the subject that might surface here and there in the book. Our naive view on the subject of finite wordlength errors in 1987 could probably be summarized as follows:

- numerical errors due to finite wordlength encoding and roundoff are something that one has to live with, and there is probably not much that can be done about them except to increase the wordlength by improvements on the hardware;

- these errors are as old as finite arithmetic and numerical analysis and they must therefore be well understood by now;

- thus, if something can be done to minimize their effects, it must have been analysed and put into practice a long time ago.

It is almost fair to say that we were wrong on all counts. Indeed, whereas numerical analysis has been in a fairly advanced state of development for some years, its practical implications in terms of filter design has only begun to be understood in the late seventies. The first results came out in 1976-77 and gave rise to the now celebrated MRH structures (for Mullis, Roberts and Hwang, the inventors of digital filter realizations that minimize the roundoff noise gain). More surprisingly, the first hard results on the design of digital filters minimizing a certain sensitivity measure of the transfer function with respect to coefficient errors date back only to 1986. Our own first effort in this area was to address the corresponding feedback controller design problem, with a first solution published in 1990. At the same time, other authors were producing alternative results addressing different versions of the controller design problem. The late eighties saw a suddenly increased interest in finite wordlength design problems, with solutions being proposed by several authors for both filter design and controller design using different sensitivity and roundoff noise measures.

Our contribution in this book covers a very wide range of new optimal filter, controller and estimator realization problems in the face of numerical errors. The problems differ in terms of either the application (filtering, control, identification), the performance objective (minimizing the output variance, the error on the transfer function, or the movement of the poles and/or zeros, optimizing the convergence speed of parameter estimators) and the measures in which these performance objectives are expressed (variances, L_1 sensitivity measures, L_2 sensitivity measures, combinations of roundoff noise measures and sensitivity measures, etc...).

Our methodology throughout the book is based on the following idea. A digital filter can be implemented in many different ways, each way corresponding to a particular combination of memory registers and shifts, multiplications by parameters and additions of signals in a computer program or in a hardware circuit. Each such particular implementation is called a realization or parametrization of the filter and, most often, these can be represented by a set of state space equations.

Theoretically, all these realizations are equivalent in that they produce an identical input-output relation. However, different realizations of, say, the same input-output transfer function have different properties in terms of their sensitivity with respect to parameter errors, or in terms of the propagation of errors in signals.

This leads one to formulate optimal realization (or parametrization) problems in which one seeks a realization that minimizes some measure of the performance degradation due to errors. Depending on whether the filter to be implemented is used as a signal processor, or as a regulator in a control loop, or as a parameter estimator in an adaptive loop, a wide range of optimal parametrization problems can be formulated. *The main theme of this book is therefore how to minimize the performance degradation of filters, controllers, identifiers due to numerical errors, by optimally choosing the realization of these filters, controllers, identifiers.* An important practical case, widely discussed in this book, is where these errors are due to finite wordlength implementation of coefficients and/or signals. A range of optimal realization problems are solved using classical shift-operator realizations, the more recent δ-operator realizations, as well as new generalized transfer function representations.

A typical chapter therefore consists of a problem formulation including the choice of a performance measure, a discussion of its relevance, the expression of the performance measure in terms of the set of possible realizations, and the minimization of this measure. These last two aspects often constitute the key novel contributions.

With the exception of Chapters 11 and 12, we have restricted our candidate realizations to the class of all *equivalent state space realizations*.[1] By doing so, we are fully aware that we are leaving aside some important classes of filter realizations, such as lattice filters. Our reasons for limiting our discussions mainly to equivalent state space realizations stems as much from a desire to keep a unified and easily understandable framework all through the book, as from a requirement to keep the page numbers reasonable. Perhaps, the authors' laziness is another less avowable reason.

We believe that the subject of optimal finite wordlength state space realizations is thoroughly covered in this book. Because of the relative freshness of the area - as mentioned above - many of the results in this book are new, but we have tried to cover the available results as extensively as could possibly be.

This book will be of interest to graduate students, scientists and engineers working in signal processing and telecommunications, systems theory, adaptive and robust control.

Acknowledgements

The project of writing this book emerged at the end of Gang Li's thesis, and about half of the material originates from that thesis, which was produced under Michel Gevers' supervision. Thus, the first author would like to take the unusual step of thanking the second author for making the material of his thesis available for the book. In addition, he takes pleasure in acknowledging that most of the technical work is Gang Li's contribution. As for the second author, he takes pleasure in thanking Michel Gevers for his invaluable guidance and encouragement,

---

[1]Or, more precisely, *similar realizations* in system theory jargon.

and for the tutorial writing of this book. In the end, the book does not bear much resemblance with the thesis that served as its starting point, due to the combined work of both authors. Finally, the order of the authors was decided at the request of the publishers.

These unusual niceties having been exchanged, we would both like to thank a number of people who contributed to the successful realization of this book. First and foremost, we would like to thank Yves Genin who volunteered to proof-read the whole book and to give us feedback both on the technical and the tutorial quality of the book. Most books are proof-read at some stage or another, but we do not believe that many authors have benefited from a proof-reading of the quality and thoroughness that Yves has given us. Greg Allen had started proof-reading the book at an earlier stage during his sabbatical visit at Louvain la Neuve and gave us invaluable and expert feedback and criticism. Brian Anderson has been a co-author of the material of Chapter 4 and some of Chapter 8, and has given us much encouragement and valuable feedback on other parts of our work. Stephen Boyd, Graham Goodwin, Jean Meinguet, Jane Perkins, Bob Skelton, Paul Van Dooren, Vincent Wertz, Darrell Williamson and Karl-Johan Åström gave us positive feedback about our work at various stages of its development. Finally, the first author would like to thank his great friend Bob Bitmead: even though Bob did not contribute to this particular piece of work, it just feels good to blame him for everything that goes right in my scientific endeavours.

Besides the people who contributed to the technical part of our book in various ways, we would also like to thank the Editors Brad Dickinson and Eduardo Sontag for their confidence and enthusiastic support for our project, the Springer Verlag staff in the New York and London offices, the secretaries Lydia De Boeck, Isabelle Hisette and Michèle Termolle for the many tedious hours they spent struggling with LaTeX and our colleagues of the Laboratoire d'Automatique at the Université Catholique de Louvain for coping with the frequent lapses of our minds during our sojourn in the world of finite wordlength errors. In particular, Michel Gevers would like to thank his friends Georges Bastin and Guy Campion, and Gang Li his friends Libei Chen, Marc Haest and Rodolphe Sepulchre for their various forms of support.

During the writing of this book, Gang Li was supported by the Belgian Programme on Inter-University Poles of Attraction initiated by the Belgian State, Prime Minister's Office. The financial support of this research programme is gratefully acknowledged.

Last but not least, Michel Gevers would like to thank Muriel for enduring his endless weekends and evenings of typing on the home computer, and our future twins for patiently waiting to see the light of day until the book was finished.

Michel Gevers and Gang Li
Louvain la Neuve
September 1992.

1

Introduction

1.1 Motivation and general statement of objectives

Digital signal processing has been considered as one of the most important and attractive fields in engineering for many years, especially in electrical engineering, and this will continue due to the rapid developments and improvements in digital computers. The recent advances in digital hardware capacities have opened up many new possibilities for filter and controller implementations. One of the basic problems in digital signal processing is the precision of finite arithmetic computations.

A rational filter is usually defined by its transfer function, which uniquely describes its input-output relation. This transfer function is classically described by a ratio of polynomials in the shift operator, but there exists a whole class of alternative parametrizations (also called realizations or representations)[1] which, in infinite precision, are all equivalent: a ratio of polynomials in the delta operator, canonical or non-canonical state space realizations, and many others. When the parameters that describe the filter are implemented with *finite wordlength* (FWL), these different representations are no longer equivalent. Thus, the transfer function, say, obtained from one parametrization will be different from that obtained from another, and both differ from the ideal one. Similarly, when the signals processed by the digital filter are encoded in finite wordlength, then roundoff must occur at every arithmetic operation and the input-output behaviour of the filter is again different from what it would be in infinite precision. The error caused by the roundoff operation is called roundoff noise. In practice, these two FWL effects occur jointly, and both result in a performance degradation of the filter compared with its ideal (infinite wordlength) behaviour. It turns out that different representations that are equivalent in infinite precision are no longer so in a FWL implementation, and they thus lead to different performance degradations. Given this very

---

[1] Throughout this book we shall indiscriminately use the terms *parametrization*, *realization* and *representation* to mean the same thing.

fundamental observation, it is then natural to formulate a range of optimal parametrization problems, where optimality refers to objectives like minimizing the sensitivity of the transfer function with respect to parameter errors, minimizing the roundoff noise gain, etc. A lot of work has been accomplished in this direction in the signal processing area, where the optimal FWL state space design of digital filters is considered as one of the most successful approaches. *A first objective of this book will be to collect the existing results on optimal state space design and present them in a coherent framework.*

The 'classical' sensitivity measure that is commonly minimized with respect to all possible realizations in FWL design is based on averaging the sensitivity of the transfer function with respect to the parameters of the state space realization over all frequencies. In this averaging operation all frequencies are evenly weighted. In practice, it is often the case that a very accurate implementation is required in some frequency band while more looseness is allowed elsewhere. In addition, this 'classical' sensitivity measure for which optimal FWL parametrizations have been derived in the mid-eighties has no physical meaning and is rather illogical in that it mixes different norms. New and more logical measures are therefore desirable, as well as extensions to sensitivities of quantities other than the transfer function. For example, some measure of the sensitivity of the poles and zeros of a filter to coefficient errors would be a useful quantity to optimize in many applications. *A second objective of the book is to present a number of new optimal filter realization results for a collection of interesting measures and situations.*

Given a particular objective, the methodology of optimal FWL design is to construct a measure that best reflects the performance degradation of this objective due to FWL errors, to compute the dependence of this measure on the set of theoretically equivalent parametrizations, and to then minimize this measure with respect to all candidate realizations. The result of this optimization step is typically a *set* of optimal realizations. Given a choice between two optimal (from a FWL error point of view) realizations, one will clearly prefer one that uses fewer nontrivial parameters if that is possible. Parametrizations that use few nontrivial parameters[2] are called *sparse*. This entails fewer multipliers (and hence a greater computational speed) and more error-free computations. *A third objective of the book is to produce FWL optimal or*

---

[2]By trivial parameters we mean coefficients such as 0 or 1 that lead to trivial multiplications; the other parameters are called 'nontrivial' or 'free'.

suboptimal realizations that are as sparse as possible.

While the effects of FWL errors have been much studied by the digital signal processing community, by comparison these FWL effects have received much less attention in control problems despite the widespread use of microprocessors in control applications. The numerical questions in digital control are just as important as those in digital signal processing, particularly with the recent trend towards the use of fast sampling control. Fast sampling keeps enough information about the continuous time signals in their discretized version, but it also brings some serious numerical problems. One particular property of a control system with fast sampling is that the corresponding controller has its poles and zeros around $z = +1$ in the complex plane and hence the transfer function of the controller is quite sensitive to perturbations of its coefficients. *A fourth objective of this book is to investigate the optimal parametrizations of controllers in terms of reducing the finite precision effects on the closed loop system.*

The search for optimal state space realizations of a given shift operator transfer function is one way of minimizing the effects of FWL errors on some performance measure (a sensitivity measure, or a weighted sensitivity measure, or a roundoff noise gain, or a cost function index, or some combination thereof). But an alternative approach is to operate a transformation on the shift operator transfer function itself, which effectively amounts to operating a transformation on the data. In the last few years, both Peterka [Pet86] and Middleton and Goodwin [MG90] have promoted the use of the delta operator as opposed to the shift operator in estimation and control applications. The δ-operator used in [Pet86] and [MG90] is of the form $\delta = (z - 1)/T_s$, where T_s is the sampling period. With this definition the δ-operator tends towards the time derivative operator, and hence the discrete and continuous time estimation and control theories can be unified. In addition, Middleton and Goodwin have shown that δ-operator implementations have a number of numerical advantages over shift operator implementations. Once a transfer function has been expressed in δ-operator form, one can similarly search for optimal δ-operator state space realizations that minimize some performance measure with respect to FWL errors. *A fifth objective of this book will be to compare the robustness to FWL errors of optimal shift operator state space realizations vis-à-vis optimal δ-operator realizations.*

Forgetting about the physical meaning of T_s, one recognizes that the

δ-operator is nothing more than a linear transformation of the usual shift operator z. This motivates us to generalize this algebraic transformation idea. A shift operator polynomial representation can be seen as a representation of a polynomial in a particular set of basis functions, the powers of z. The same polynomial can be represented in δ-form using the powers of $\delta = (z-1)/\Delta$ as the basis functions. This idea can be further extended by introducing generalized polynomial basis functions leading to new classes of parametrizations. It can be shown that transforming the basis functions, and hence the parametrization, amounts to a particular prefiltering of the data. *A final objective of this book is to investigate the role played by such generalized basis functions (or, equivalently, by a prefiltering of the data) in estimation, identification and adaptive filtering problems, and to develop new and more robust operators for these applications.*

In order to give the reader a foretaste of the problems that will be dealt with in this book, we present some examples and simulations in the next section.

1.2 A historical view and some motivating examples

Digital signal processing could be traced back to the numerical procedure invented by Newton and Gauss in the 17th and 19th centuries, but it was during the decade of 1960 - 1970 that it became of importance. New technologies in the past 20 years have reduced the cost of digital hardware, and its speed has increased to such an extent that digital signal processing has replaced a great deal of analog signal processing. One of the basic problems in replacing analog signals and processors is the finite precision problem. This problem was first considered in digital filter design [RG75] and [RR72]. In implementing a system with digital processors, the finite precision aspects are twofold: the finite wordlength representation of system parameters and the finite precision arithmetic operations in which roundoff occurs. Traditionally, finite precision analysis is separated into sensitivity and roundoff studies. In the early days, the digital filter was implemented with direct forms[3] and the parametrization problem was not seriously considered. Later, the poor performance of a filter implemented in these forms made one realize the importance of the structure of a digital filter. The key point is that for a fixed input/output relation, there exists an infinite class of parametrizations (or realizations). Different parametrizations of the same input-output relation yield different internal structures which have different numerical properties. For example, let the input-output behaviour of a dynamical system be described by the following equation,

$$f(y, u, \rho, \theta_0, t) = 0, \qquad (1.1)$$

where ρ is a generalized operator: for a continuous time system, $\rho = d(.)/dt$ and for a discrete time system, $\rho = z$; θ_0 is the nominal parameter vector that corresponds to a specific parametrization. For a fixed input $u(t)$, the output $y(t, \theta_0)$, that is the solution of (1.1), is a function of this parameter vector. When θ_0 is perturbed by $\Delta\theta$, the perturbed output will be $y(t, \theta_0 + \Delta\theta)$ which can be evaluated via a Taylor series expression and approximated as follows if the perturbation $\Delta\theta$ is much smaller than the absolute value of the nominal θ_0:

$$y(t, \theta_0 + \Delta\theta) \approx y(t, \theta_0) + \frac{\partial y}{\partial \theta}|_{\theta=\theta_0} \Delta\theta. \qquad (1.2)$$

Clearly, the deviation of the output from its true value $y(t, \theta_0)$ can be estimated by the last term of the right side of (1.2), which is the product of the sensitivity of the output with respect to the parameter vector at its nominal value θ_0 and the parameter perturbation $\Delta\theta$. In many practical cases, the latter is bounded by some specific value (for example, the number of bits used in the digital implementation of a filter), so that the degradation of the output is essentially determined by the sensitivity function $\partial y/\partial \theta$ evaluated at the nominal parameter θ_0.

If another parametrization is used, the dynamical system equation is given, say, by

$$F(y, u, \rho, \gamma_0, t) = 0. \qquad (1.3)$$

At their nominal values, the two parametrizations (1.1) and (1.3) yield the same output, but with perturbations of the same magnitude,

---

[3] A direct form is one in which the coefficients of the numerator and denominator of the transfer function appear directly as parameters in the realization; it is often called a companion form in control theory because it corresponds to a state space realization in which the state transition matrix is in companion form.

they could give different output deviations since the sensitivity behaviour of the output with respect to the two parametrizations might be quite different. A reasonable question is to search for those parametrizations that lead to a minimal sensitivity.

State space realization theory has been well established for decades and it is well known that the infinite set of all similar[4] realizations have the same dynamical input-output behaviour (see e.g. [Kai80]), but it was not until 1968 that the problem of how to exploit this class of parametrizations for the purpose of optimal FWL implementation of a digital system came under scrutiny. In [Man68], Mantey examined how to use a similarity transformation to obtain realizations that minimize some measure of the pole sensitivity.

That different parametrizations of an input-output transfer function, say, yield different sensitivities of this transfer function to parameter errors is reasonably easy to understand. What is perhaps more remarkable - and not quite so intuitive - is that different state space realizations also affect differently the way in which arithmetic roundoff errors are reflected in the output of the filter. The ratio of the variance of this output error to the variance of the arithmetic roundoff error is called the roundoff noise gain of the realization. It turns out that the first significant results on the design of state space realizations minimizing FWL error effects are results on the minimization of the roundoff noise gain rather than on the minimization of transfer function sensitivity measures. Indeed, the pioneering work in exploring the use of different realizations can be found in Mullis and Roberts's seminal 1976 paper on the synthesis of digital filters with minimum roundoff noise due to fixed point multiplication [MR76a]. In their work, similarity transformations were explored to determine the state space realizations that minimize the output error variance due to arithmetic roundoff errors. Independently, Hwang [Hwa77] considered the same problem and gave a constructive procedure to determine the optimal transformations or realizations. The realizations obtained by this procedure are now called MRH structures, for Mullis, Roberts and Hwang.

An alternative objective to the minimization of the roundoff noise gain over all possible realizations is to identify those state space realizations that minimize some measure of the sensitivity of the transfer function with respect to the coefficients in the matrices (A, B, C)

---

[4]Similar realizations are state space realizations whose states are related by a real nonsingular transformation matrix.

of the realization. It is well known [Doo81] and [DD81] that the numerical accuracy of a problem depends crucially on which particular state space representation is used. Therefore, it is desirable to define appropriate measures which represent the sensitivity of the realized transfer function with respect to parameter perturbations. The first analytical attempt to synthesize minimum sensitivity state space realizations for linear systems was due to Tavsanoglu and Thiele [TT84] and [Thi84]. They defined a rather sophistical frequency independent sensitivity measure of a transfer function with respect to the parameters of a corresponding realization (A, B, C), which we shall call L_1/L_2 sensitivity measure and denote by $M_{L_{12}}$ from now on (the name will be explained in Chapters 3 and 4). The success of their method stems from the fact that an upper bound of this sensitivity measure can be obtained whose minimization with respect to all similar realizations is particularly simple. In [Thi86] it is shown that not only the upper bound but also the sensitivity measure itself is minimized by all realizations whose controllability and observability Gramian matrices are identical[5]. In [Gra89], the same conclusion was achieved using a geometric approach. These results have since been extended to Multiple Input and Multiple Output (MIMO, in short) systems [LH88].

To demonstrate the fact that different realizations have different numerical properties we consider a sixth order low pass narrow band filter, which will be fully described in Chapter 3 (see Example 3.2). In this example, we compute and compare the frequency responses for several realizations whose coefficients are implemented with B_c bits for the fractional parts. One realization is the controllable canonical form, R_c, corresponding to the transfer function parametrization, another one is the so-called balanced realization, R_b, which is one of the realizations that minimize the Tavsanoglu and Thiele sensitivity measure $M_{L_{12}}$. Figure 1.1 represents the magnitude frequency responses of the ideal filter, of the controllable canonical form, R_c, with coefficients truncated to 14 bits, and of the balanced form, R_b, with coefficients truncated to 8 bits. Further details about this example will be given in Chapter 3.

The figure clearly shows that the optimal realization R_b is signif-

---

[5]In this introductory chapter, we use a number of concepts from system theory such as Gramians, controllable and observable canonical forms, etc. Those readers who might not be familiar with these concepts should not worry: these will be explained in Chapter 3. The idea in the present chapter is to give the reader a flavour of what can be achieved with FWL optimal realizations.

8   1. Introduction

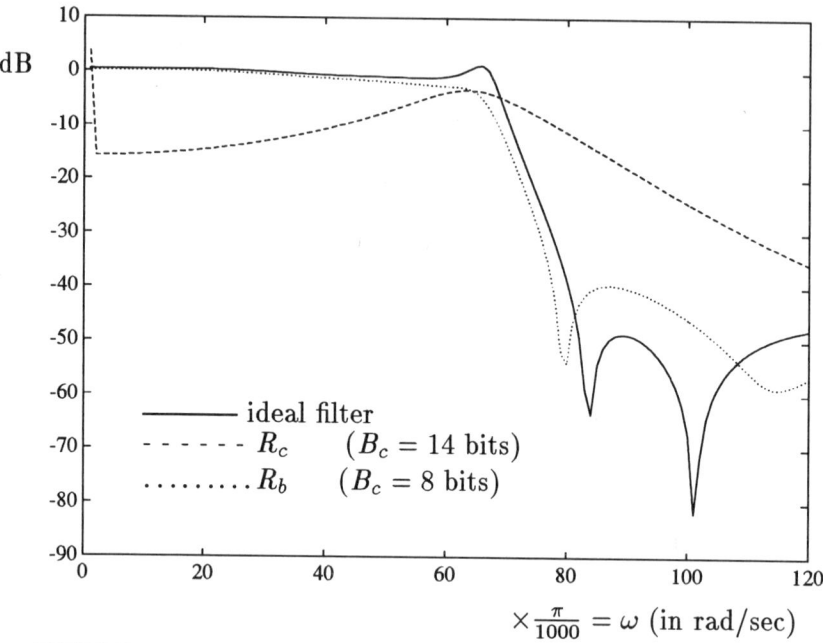

FIGURE 1.1. Magnitude frequency responses of ideal filter, of R_c and of R_b

icantly less sensitive to coefficient perturbations than the canonical realization R_c, particularly given that the fractional parts of the coefficients of R_b are encoded with 6 fewer bits than those of R_c. Even though the canonical (or direct) forms are particularly simple to implement, similar examples throughout this book will illustrate how poorly they resist to coefficient errors. Conversely, these examples illustrate the vastly superior FWL performance of the various FWL-optimal realizations.

One observes from Figure 1.1 that, even for the 'optimal' realization R_b, the performance of the filter in the face of FWL errors is not very satisfactory for frequencies around $\frac{70}{1000}\pi$, which belong to the bandwidth of the filter. In many applications, an accurate implementation is required in some frequency band (for example within the bandwidth of the filter) while more slackness is allowed elsewhere. To obtain a better performance in a particular frequency interval, a frequency weighted sensitivity measure can be used. In Figure 1.2, we present a new realization, $R^{opt}(M^*_{L_{12}})$, which minimizes a frequency weighted sensitivity measure $M^*_{L_{12}}$, where the weighting function is a filter of order 10 whose magnitude frequency response will be given in Chapter 4. The perfor-

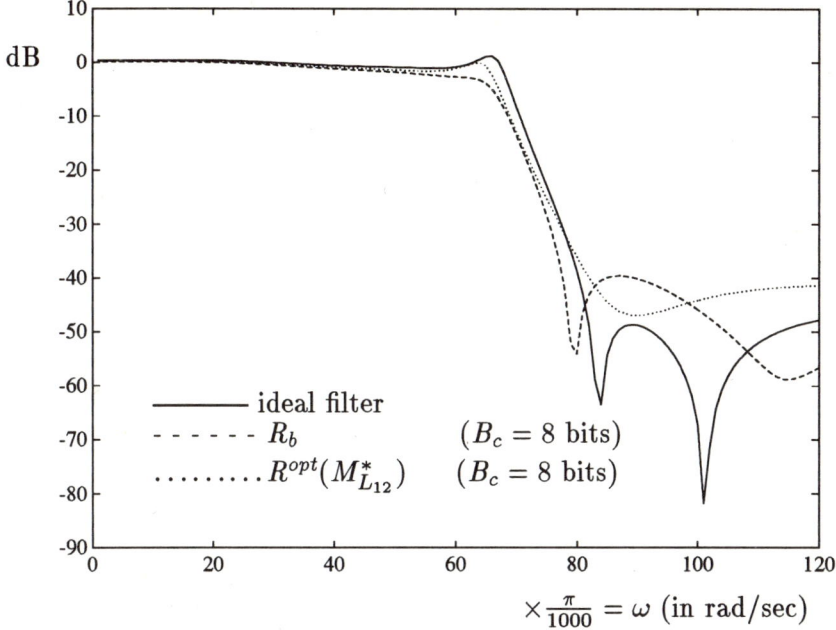

FIGURE 1.2. Magnitude frequency responses of ideal filter, of R_b and of $R^{opt}(M^*_{L_{12}})$

mance improvement within the filter bandwidth is evident compared with that of R_b. This new frequency weighted optimal design technique is the object of Chapter 4 where the example is fully described.

With the same example, we now examine the sensitivity of the poles and zeros of the realizations R_c and R_b to coefficient errors. We have seen how much the frequency response of R_c is degraded by coefficient errors. Figure 1.3 shows the unit step responses of the ideal filter (solid line), the 8-bit realization R_b (dotted line) and the 14-bit realization R_c (dashed line): it shows that the direct form realization R_c is in fact unstable, even though the ideal filter, of which R_c is a truncated realization, is stable.

The vectors of poles and zeros of the ideal filter, denoted by V_p and V_z, respectively, are as follows:

$$V_p = \begin{pmatrix} 0.9723 \pm j0.1989 \\ 0.9389 \pm j0.1623 \\ 0.9152 \pm j0.0646 \end{pmatrix}, \quad V_z = \begin{pmatrix} 0.7447 \pm j0.6674 \\ 0.9531 \pm j0.3027 \\ 0.9681 \pm j0.2507 \end{pmatrix}.$$

To illustrate the effect of coefficient truncation errors on the pole and zero locations, we present in Figure 1.4 (resp. Figure 1.5) the pole

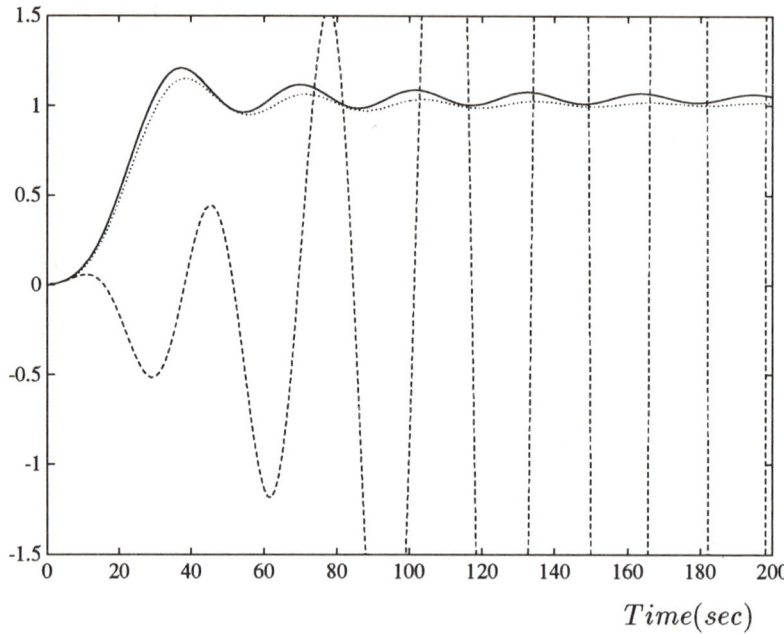

FIGURE 1.3. Unit step responses of ideal filter (–), of R_b (...) and of R_c (- -)

locations (resp. the zero locations) of the ideal filter (+), the 14-bit implemented R_c (x) and the 8-bit implemented R_b (*).

We observe that R_b has a better pole and zero stability than R_c. However, if pole and/or zero locations are important for the application at hand, then a much better performance can be achieved by designing a realization that minimizes some sensitivity measure of the poles and/or zeros with respect to coefficient errors. It has been known for some time that a minimum pole sensitivity is achieved if the state transition matrix A of the state space realization (A, B, C) of the filter is in *normal form*[6] (see, e.g. [SW84]), but no explicit result seemed to be available on how to transform an arbitrary realization into a normal realization. In Chapter 6, we provide a solution to this problem. To illustrate the pole behaviour of a normal realization, we have computed a normal realization, denoted R_n, for the example above; that is the matrix A in this realization is normal. We now compare the FWL degradation of its pole locations compared to that of the balanced realization R_b by truncating the fractional parts of the coefficients of the A matrices

---

[6]A real matrix A is normal (or in normal form) if $AA^T = A^T A$ (see Chapter 6).

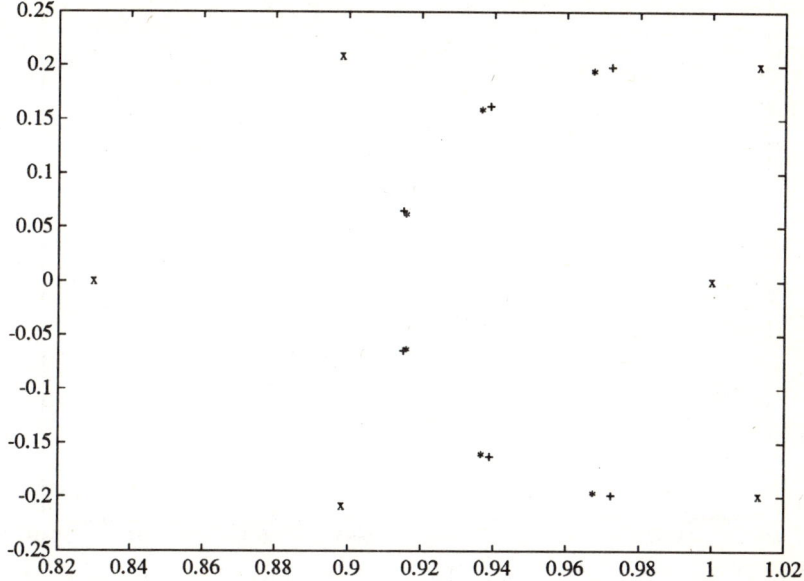
FIGURE 1.4. Pole distribution of the ideal filter (+), of R_c (x) and of R_b (*)

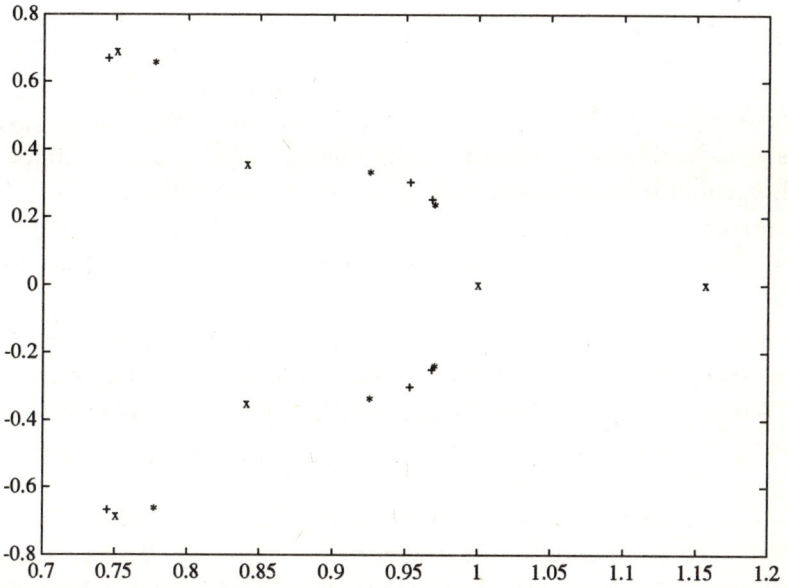
FIGURE 1.5. Zero distribution of the ideal filter (+), of R_c (x) and of R_b (*)

12   1. Introduction

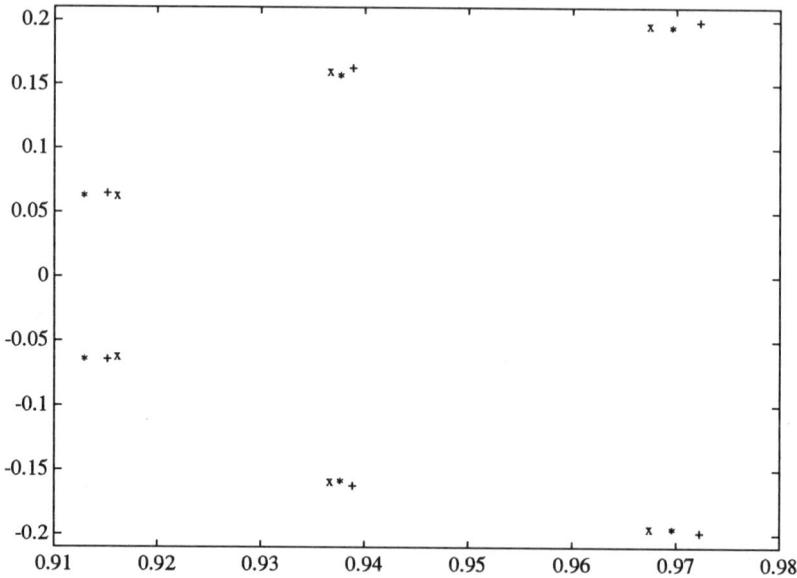

FIGURE 1.6. Pole distribution of the ideal filter (+), of R_b (x) and of R_n (*)

of both realizations to 8 bits. Figure 1.6 shows the pole distributions, respectively, of the ideal filter (denoted +), of the truncated realization of R_b (denoted x) and of the truncated realization of R_n (denoted *). We observe that the FWL pole stability of R_n is slightly better than that of R_b, this last realization being optimal in terms of the preservation of the transfer function in the face of FWL errors. This is a first illustration of the fact that different objectives lead to different FWL-optimal realizations, a central theme of this book. We note also, by comparing with Figure 1.4, that both R_b and R_n have a significantly better pole stability performance than the commonly used direct form R_c.

Now, suppose that in some application the preservation of the zero locations, rather than the pole locations, is an important objective. We first observe that the realization R_n, which minimizes a pole sensitivity measure, is far from optimal in terms of the preservation of the zero locations. Figure 1.7 shows the zero distributions, respectively, for the ideal filter (denoted +), for the 8-bit truncation of R_b (denoted x) and for the 8-bit truncation of R_n (denoted *). Whereas R_n has a much better pole sensitivity performance than R_b, its zero stability behaviour is worse than that of R_b, especially for the smallest (in magnitude) zero that is far from its exact value.

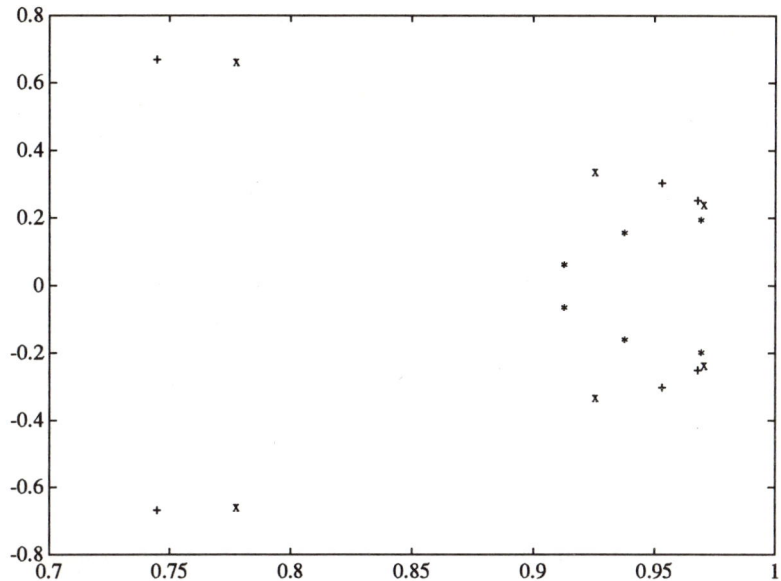

FIGURE 1.7. Zero distribution of the ideal filter (+), of R_b (x) and of R_n (*)

To construct realizations that minimize the errors on the zero locations, a zero sensitivity measure can be defined and minimized over the set of all similar realizations. The results on pole sensitivity minimization were directly extended to such a zero sensitivity minimization problem by Williamson [Wil86]. Classically, the pole and the zero sensitivity measures were studied separately. Since the values taken by these two measures are determined by the same realization (A, B, C) and since their minimization may constitute two conflicting objectives, it would be better to define some unified measure that integrates these two objectives in some weighted fashion, and hence to balance the pole and zero sensitivities according to the application at hand. This is the object of Chapter 6 where a solution to this problem can be found.

The results discussed so far have been on the minimization of the errors due to FWL effects in the realization of digital filters. By comparison, the FWL effects have received much less attention in the control literature. Knowles and Edwards [KE65] and Curry [Cur67] have analysed roundoff noise in sampled data systems, while Sripad [Sri78] has looked in some depth at the roundoff noise and finite precision coefficient performance of the discrete time Kalman filter and the Linear Quadratic Gaussian controller. Rink and Chong [RC79] have derived bounds on the effects of quantization errors in floating point regulators.

Moroney, Willsky and Houpt [MWH80] have investigated the FWL effects of coefficients on LQG compensators using a stochastic estimation approach. It is not until 1990 that the first results appeared on optimal compensator design problems. Williamson and Kadiman [WK89] and Liu and Skelton [LS90] have studied the optimal FWL design for the LQG problem in terms of minimizing the roundoff noise effect on the total LQG cost. Our own investigations have first focused on optimal FWL realization problems for a state estimate pole placement controller, with the separate minimizations of a closed loop sensitivity measure and of a closed loop roundoff noise gain measure [LG90b]. These results were later extended to a general two-degree-of-freedom controller with a synthetic measure that incorporates both coefficient error effects and a roundoff noise gain measure [LG91].

Clearly, there exist some parallels between filter and controller implementation [Wil79]. However, because of the presence of a feedback loop around the digital compensator, many of the existing concepts on filter design do not directly apply to control problems, and adaptations are necessary. A controller, just as an open loop system, can be defined by its input-output relation, but the optimal realization of the controller, looked upon as a filter realization problem, would not make the closed loop system have the best possible FWL performance. The key point is that in the design of a FWL controller, the behaviour of the over-all closed loop system should be considered, and not only that of the controller itself. To show this, let us consider an example that will be used in Chapter 9. This is a state estimate feedback controller obtained using a pole placement design. From the theoretical (infinite wordlength) controller expressed in transfer function form, one can first compute a balanced realization. This particular realization minimizes the L_1/L_2 sensitivity measure of the controller, considered as a filter. We shall see in Chapter 9 that an alternative - and better - strategy is to minimize a sensitivity measure of the *closed loop system*. To illustrate the difference between these two strategies, we present in Figure 1.8 the magnitude frequency responses of the closed loop system for three different implementations of the compensator: the ideal (infinite wordlength) compensator, the balanced realization of the compensator truncated to 12 bits, denoted R_b, and the closed loop optimal realization also truncated to 12 bits, denoted $R^{opt}(\bar{M}_{cl})$. Clearly, the FWL performance of the compensator realization obtained via a closed loop strategy is vastly superior to that of the compensator R_b, even though

R_b would be the best FWL realization of the compensator considered as an independent filter.

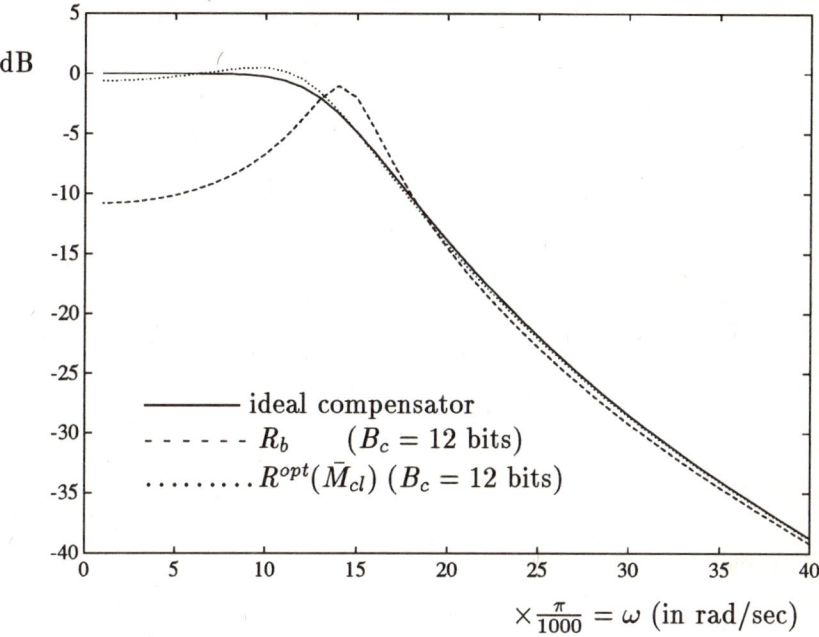

FIGURE 1.8. Magnitude frequency responses of closed loop system with ideal and with 2 FWL compensators

In modern control, fast sampling is often used. In such case, and with the usual shift operator parametrizations, the poles and zeros of the controller cluster around the point $+1$ in the complex plane [ÅSH84], [MG90]. A simple analysis of the root sensitivity (see Chapter 6) shows that these poles and zeros are very sensitive to parameter perturbations, which may lead to FWL implementations that are unstable. Better numerical properties can be achieved by simply pulling the poles and zeros back to the origin 0 from the point $+1$ in the complex plane. This is one of the motivations for the use of δ-operator parametrizations.

Although the idea of this shift to the origin for numerical accuracy reasons can be traced back to Agarwal and Burrus [AB75] in the digital filtering literature, the promotion of what is now called the *delta operator* in estimation and control applications is due to Peterka [Pet86] and to Middleton and Goodwin [MG90]. In [MG90] two major advantages are claimed for the delta operator formulation: a theoretically interesting unified formulation of continuous time and discrete time control theory which entails a better understanding of discrete time control under fast sampling, and a range of practically interesting numerical advantages connected with finite wordlength effects. Two problems not studied in [MG90] are those of comparing the sensitivity and round-off noise performances of shift operator versus δ-operator implementations of digital filters. These questions constitute the major objective of Chapter 11 of this book.

Numerical problems in estimation and identification have been studied by several authors. In 1985, Ljung and Ljung examined the numerical error propagation properties of different recursive algorithms [LL85]. In 1986, Verhaegen and Van Dooren [VD86] studied the same error propagation problems but in different Kalman filter implementation schemes. Later, Verhaegen investigated these questions for more general algorithms [Ver89]. These investigations have only studied the effects of different numerical algorithms. As far as we know, not much work has been accomplished in estimation, identification and optimal filtering to investigate the role played by different parametrizations of a system in terms of the reduction of FWL error effects. One notable exception is Goodwin's plenary address [Goo88][7], in which he studied the issue of numerical accuracy in these applications and showed how it can be improved if δ-operator parametrizations are used instead of the usual shift operator parametrizations.

An important issue in an adaptive estimation and/or filtering problem is the convergence speed. This issue is usually studied by comparing different algorithms. In Chapter 12 we will show that it can alternatively be investigated by comparing different parametrizations. To illustrate this, let us consider the following identification example taken from Chapter 12. The system to be identified from input and output data sequences is a Finite Impulse Response (FIR) model given in the usual shift operator representation:

$$H(z) = 1 - 2.1000z^{-1} + 1.4600z^{-2} - 0.3360z^{-3}.$$

Without giving too much detail here, we mention that this system can also be parametrized by what we shall call a γ-operator parametrization in Chapter 12. For both parametrizations, the four unknown parameters of the FIR model are estimated by an adaptive Least Mean

---

[7]Graham Goodwin, always a step ahead of all the rest of us.

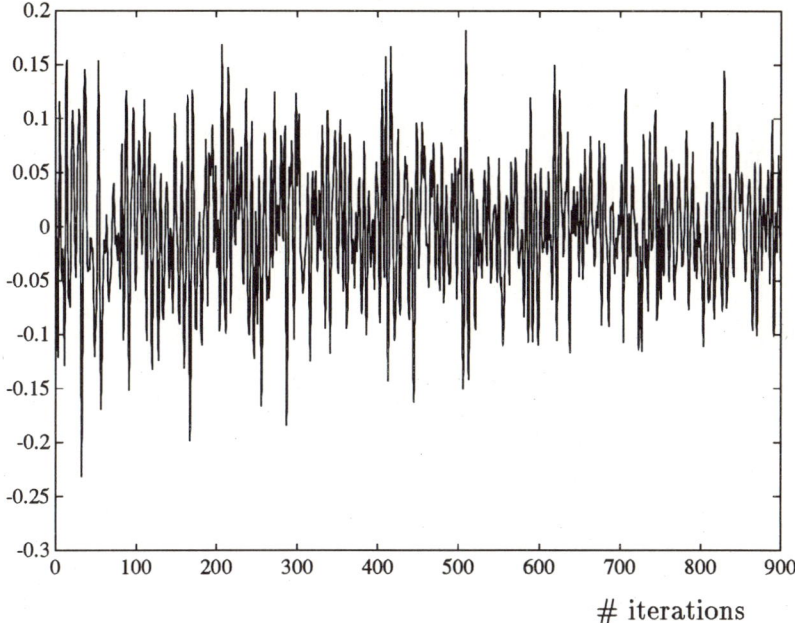

FIGURE 1.9. Prediction error sequence $e_z(t)$

Square (LMS) algorithm; the same step size is used in both simulations (again, we refer the reader to Chapter 12 for details). With the estimated parameters, the prediction errors are computed on line; they are denoted $e_z(t)$ for the shift operator parametrization and $e_\gamma(t)$ for the γ-operator parametrization. Figures 1.9 and 1.10 show the prediction error sequences obtained with this adaptive identification algorithm using shift- and γ-operator parametrizations, respectively.

The figures show the tremendous improvement in convergence speed of the parameter estimation algorithm when the system is parametrized by the gamma operator rather than by the usual shift operator. A detailed discussion of the parametrization issue in adaptive estimation problems is given in Chapter 12.

1.3 Outline of the book.

A brief outline of the contents of this book is as follows. We start in Chapter 2 with an introduction to finite precision problems and finite wordlength computations.

The FWL optimal state space realization design of a digital filter

18   1. Introduction

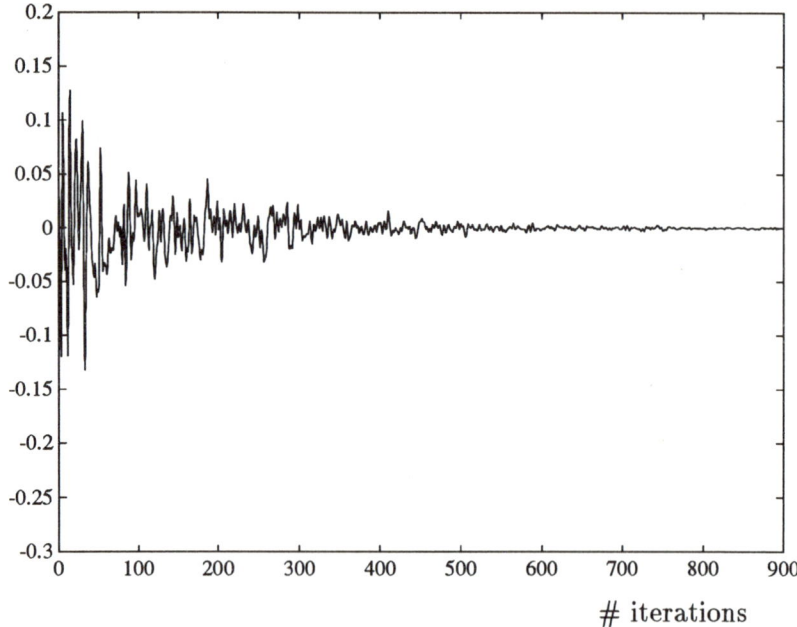

FIGURE 1.10. Prediction error sequence $e_\gamma(t)$

is addressed in Chapter 3, where the classical results are given. The commonly used criteria for optimal FWL design are presented and the corresponding classes of optimal realizations are derived. These classical criteria are a sensitivity measure that mixes an L_1 and an L_2 norm [TT84], [Thi86], and a roundoff noise gain between the roundoff noise variance at the states of the realization and that at the output [MR76a], [Hwa77], [Wil86]. The roles of different parametrizations in reducing the FWL effects are exhibited by a few numerical examples.

The classical transfer function sensitivity measure is based on averaging the sensitivity of the transfer function with respect to all parameters of a realization over all frequencies. The sensitivity behaviour of the transfer function at one frequency point is thus considered to be as important as at another frequency point. In practice, one is usually interested in the performance of the transfer function within a specified frequency range, for example the bandwidth of the transfer function. To achieve this, we introduce a frequency weighted sensitivity function, and hence a corresponding measure, in Chapter 4. The optimal FWL realization problem for such a frequency weighted sensitivity measure is formulated and then solved using an iterative algorithm.

The selection, by Tavsanoglu and Thiele, of a sensitivity measure that mixes an L_1 norm and an L_2 norm is essentially motivated by the simple mathematical treatment it leads to, but this choice is not very logical. In Chapter 5, we introduce a more logical sensitivity measure that contains L_2 norms only. The corresponding optimal realization problem is formulated and solved. The relationship between this newly defined sensitivity measure and the classical one (that is the Tavsanoglu and Thiele definition) is discussed. In Chapter 7 this newly defined L_2 sensitivity measure will be shown to have some interesting physical interpretation.

In the previous section we have discussed and illustrated the problem of designing realizations that maintain the pole and/or zero locations as much as possible in the face of numerical errors on the parameters of these realizations. This issue is fully discussed in Chapter 6. Classically, the pole and zero sensitivities are studied separately. The pole and zero sensitivity measures can then be studied as a general eigenvalue sensitivity problem for a matrix M, and this sensitivity is minimized when M is normal. One of our new results in this chapter is to give an analytic expression for the optimal transformation from an initial realization to such a normal realization. Since FWL errors on the parameters affect both the poles and the zeros, it makes sense to attempt minimizing some measure that contains a weighted combination of pole and zero sensitivities. This is the other main contribution of Chapter 6.

Classically, sensitivity and roundoff noise problems are investigated separately. Since the quantization errors on the coefficients and the signal roundoff errors appear simultaneously in an actually implemented digital system, it is desirable to have a measure that unifies these two FWL effects. In Chapter 7 such a combined sensitivity plus roundoff noise measure, called *Total Noise Gain* (G_T in short), is introduced. It is shown that this measure is nothing but a linear combination of the classical roundoff noise gain and the new L_2 sensitivity measure introduced in Chapter 5. The physical meaning of this L_2 sensitivity measure now becomes apparent. The minimization problem for this synthetic measure G_T is formulated under the usual l_2-scaling on the states of the realization, and a solution is provided. An example is given to illustrate the theoretical results.

The optimal realizations that minimize any one of the measures presented in the preceding chapters are generically fully parametrized. As a consequence, the implementation of a filter with these realizations contains a maximal number of multipliers and hence the processing is slowed down. Another important point is that trivial parameters such as 0 and ±1 produce no FWL errors, and this is not taken into account in the theoretical measures used in the previous chapters. It turns out that the sets of realizations that minimize many of the FWL measures studied in this book also contain thin subsets of sparser realizations. In Chapter 8, we examine how to obtain optimal sparse realizations, and also suboptimal realizations that contain even fewer free parameters while still preserving nice numerical properties.

In Chapter 9, we deal with the optimal FWL design of a controller. The main differences between the FWL design of a digital system (or filter) and that of a controller are discussed. We present some existing results on the FWL Linear Quadratic Gaussian (LQG) regulator implementation problem and on a FWL pole placement controller design problem. The discussion is based on the work of Moroney et al [MWH80], Williamson et al [WK90], and Liu et al [LS90], [LSG92] for the LQG controller design and on that of the authors [LG90a,b] for the pole placement controller design problem. An example is given to show the FWL design procedure of a compensator by using the optimal parametrization methods. In this chapter the classical FWL analysis is adopted, that is the two FWL effects of coefficient errors and signal roundoff are studied separately.

To overcome the disadvantage of these separate FWL analyses, we propose a synthetic FWL compensator design procedure in Chapter 10. We analyse the FWL effects in a compensator for a general two-degree-of-freedom controller using the synthetic sensitivity plus roundoff noise measure introduced in Chapter 7 in the filter design context. A solution to this optimal synthetic compensator design problem is obtained.

In Chapter 11 we compare the effects on sensitivity and roundoff noise gain of the usual shift operator parametrization and of a δ-operator parametrization. This comparison shows the clear numerical superiority of δ-operator parametrizations in the case of fast sampling. One conclusion here is that by using algebraic transformations on the representations of the transfer functions, one can explore new equivalence classes of parametrizations that have better numerical properties than the usual shift operator parametrizations.

Motivated by that last conclusion, more general equivalence classes of parametrizations, called generalized polynomial parametrizations, are explored in Chapter 12. They correspond to expressing the numerator and denominator of the transfer function in polynomial basis functions that are not limited to the powers of z. The use of these polynomial parametrizations in estimation and identification are examined by exhibiting the effect of the choice of parametrization on the information matrix, and therefore on the convergence speed of parameter estimation algorithms. It is shown that some existing results on the numerical accuracy of parameter estimators can be recast within this polynomial parametrization concept. A new operator based on the *bilinear transformation*, which we call γ-operator, is shown to have an interesting numerical robustness property for estimation and identification applications, in that it tends to decouple the components of the transformed regression vector, and hence to improve the properties of the information matrix. Applications of this new class of parametrizations are illustrated by several examples.

And now, enough for introductory 'balivernes'. Let's get on with the real thing.

2

Finite Word Length errors and computations

2.1 Introduction

The technological developments of the 1960's have made it easier, more convenient and more practical to represent a signal in digital instead of analog form. These technological developments have given rise to major progresses in digital signal processing theory. In particular, digital system (or filter) design has become an important field in electrical engineering and in systems and control theory. One of the important problems in digital filter design is associated with the minimization of the performance degradation due to the finite precision of the filter implementation and of the computations. This finite precision is due to digital quantization. Among the effects resulting from finite precision are:

1. the quantization of the signals into a set of discrete levels and the ensuing quantization errors;

2. the overflow of the computations, which occurs when the computation results are out of the range of the hardware capacity and can lead to strange effects, including instability of an otherwise stable filter;

3. the accumulation of roundoff errors that occur at arithmetic operations;

4. the quantization of the coefficients into a finite number of bits and the ensuing quantization errors.

The first effect, the input signal quantization error, depends on the analog-to-digital converter (ADC) that is used. It is usually modelled as an additive uniformly distributed white noise process (see e.g. [SS77]) and has nothing to do with the filter structure or parametrization. The other three effects are influenced by the particular implementation chosen for the filter.

The overflow effect is related to the wordlength capacity of the machine on which the computations are performed. Overflow results when the dynamic range of the signals exceeds the wordlength capacity. It can cause serious errors and must be avoided by a proper scaling of the signals, itself a function of the dynamic range of the signals, of the available wordlength and of the filter realization. The dynamic range and the wordlength have a direct bearing on the roundoff noise. We will briefly explain these issues, and in particular the overflow - roundoff noise trade-off, later in this chapter.

Assuming that the overflow has been taken care of by proper scaling, then the accumulation of roundoff errors at arithmetic operations and the errors induced by the finite wordlength encoding of the filter coefficients become the main objects of concern. They are usually called Finite Word Length (FWL) effects. The errors induced by these two effects are to a large extent influenced by the particular way in which the filter is implemented, i.e. by what is called the particular *parametrization* or *realization* of the filter. That is to say that, as soon as an ideal filter, estimator or controller is implemented in a FWL machine (a computer or a hardware device) some degradation of performance due to FWL errors becomes inevitable. However, by cleverly designing this filter, estimator or controller, one can often reduce this performance degradation. The study of the effects of different implementations on FWL errors, induced by these two mechanisms, for a variety of filtering, estimation and control problems is a central theme of this book.

Since much of the book is concerned with the development of filter designs and controller designs that aim at reducing or minimizing the effects of finite precision errors on the performance of these filters or controllers, it seems reasonable, at this starting point, to provide the reader with at least a basic understanding of how such errors arise, of their effects, and of their relative importance. Without attempting to turn our readers into digital signal processing experts - something the authors do not claim to be - we present in this chapter a brief introduction to the world of Finite Word Length errors and computations: it is indeed appropriate to briefly describe the FWL effects before considering how they can be defeated. Much of the material in this section is derived from [OS89], [RM87] and [Wil91]. The last two books, in particular, contain elaborate and clearly written presentations of all issues dealing with finite word length representations and computations, and we refer the interested reader to these references for more detail.

2.2 Representations of binary numbers

The practical problems of numerical computation begin with the finite representation of real numbers. In most computers, microprocessors and other specific or general processors, a real number is represented using binary codes. The representation of an arbitrary real number in binary code will not be exact unless the number of bits tends towards infinity. Since the memory capacity in the computing machine is limited, a real number with an infinite number of bits has to be rounded off or truncated to an acceptable finite number of binary digits, called bits. The error due to this is the so-called Finite Word Length (FWL) effect. Note that two types of numbers are present in the (software or hardware) operation of a digital filter: the coefficients, which are usually fixed (i.e. constant) numbers[1], and the signals, which change at every sampling time. These two types of numbers are often implemented with two different wordlengths chosen so that the larger wordlength is used for the type having a greater FWL contribution to the total error, in an attempt to make errors due to each type of the same order of magnitude.

There are two basic ways of representing real numbers. One is fixed point representation; the other is floating point representation. We give a very brief introduction to these representations here.

In *fixed point binary representation*, several formats exist that differ by the way in which the algebraic sign of the number is encoded. The most common representation is called 'two's complement'. In two's complement format, a real number x is represented as

$$x = K(-x_0 + \sum_{i=1}^{\infty} x_i 2^{-i}) \quad (2.1)$$

where K is an arbitrary scale factor and where the x_i are binary digits which are either 0 or 1. The first bit, x_0, is called the *sign bit*. If $x_0 = 0$, then x is a positive number in the range $0 \leq x \leq K$, and if $x_0 = 1$, then x is a negative number in the range $-K \leq x < 0$. The number inside the parenthesis in (2.1) is between -1 and $+1$. Hence, any number whose magnitude is less than K can be represented exactly by (2.1). The range of numbers that can be represented, $[-K, +K]$, is called the *dynamic range* of the representation. We note that this range is entirely

---

[1] Adaptive filters or controllers are a notable exception, of course.

2. Finite Word Length errors and computations

determined by the scaling constant K.

If only a finite length register of $B+1$ bits is available, then the above representation must be replaced by a finite wordlength estimate. For example, by deleting all the terms beyond 2^{-B} one gets the following truncated version of the infinite wordlength representation (2.1):

$$Q[x] = K(-x_0 + \sum_{i=1}^{B} x_i 2^{-i}) \qquad (2.2)$$

We shall adopt the convention throughout this book to call such a number a B-bit number or the B-bit representation of x, meaning that the fractional part of the number contains B bits, even though encoding such number with its sign requires $B + 1$ bits. The representation of the number (2.2), without its scaling constant K, is often graphically illustrated as in Figure 2.1.

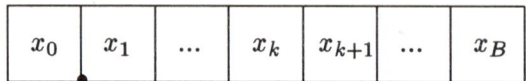

FIGURE 2.1. Fixed point representation with fractional part only

The quantized numbers that can be represented by (2.2) are in the range $-K \leq Q[x] \leq K - q$, but each one of them is a multiple of the smallest quantum $q = K2^{-B}$. This number q is called the *quantization step size* and is the smallest difference between any two of the 2^{B+1} numbers that can be represented with registers of length $B + 1$. It is common to assume that all signals and coefficients are binary fractions. The number K is then a scaling factor that allows one to represent numbers whose magnitude is larger than one. The scaling factors are then usually taken as integer powers of 2: $K = 2^c$, with c an integer.

Alternative unscaled fixed point representations exist in which the binary point is not at 'the extreme-left-but-one' position as in (2.2), i.e. in which a first fixed number k of bits is used for the representation of the *integer part* of x, and the last $B - k$ bits are used for the fractional part. The position of the binary point is then predetermined and the dynamic range is then essentially determined by the range of the integer part. This is illustrated in Figure 2.2.

More details on the different fixed point representations and implementation formats are available in several textbooks (e.g. [RM87] and [Wil91]) and are not required for an understanding of this book.

2. Finite Word Length errors and computations   27

FIGURE 2.2. Fixed point representation with integer part

The number $Q[x]$ represented in (2.2) is the *truncated* version of the exact number x - each number x is truncated to the largest quantized number that does not exceed x. This produces an error $e_T \triangleq x - Q[x]$, called *truncation error*, that is in the range $0 \leq e_T < q$. Hence truncated numbers have an average bias of $-q/2$. More often, quantization is performed by a *roundoff* operation - each number x is replaced by the nearest quantized number. With roundoff, the *roundoff error* $e_R \triangleq x - Q[x]$ is in the range $-q/2 \leq e_R < q/2$ and is unbiased. *Unless otherwise specified we shall throughout this book consider that a roundoff operation, rather than truncation, is used to quantize the numbers.* The roundoff operation is graphically illustrated in Figure 2.3.

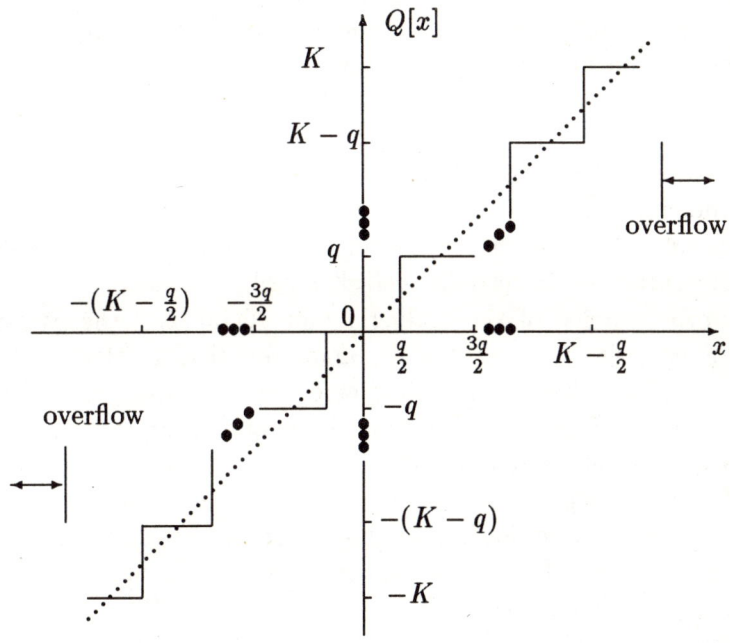

FIGURE 2.3. ADC characteristic for rounding of two's complement numbers

The major drawback of unscaled fixed point arithmetic is its limited dynamic range which forces the designer to implement a careful scaling of the signals so that the input signal and all intermediate signals (e.g. realization states) remain within the dynamic range. One way of significantly increasing the dynamic range of the signals and coefficients is to use *floating point binary arithmetic*. The floating point binary representation of a number is

$$x = x_m 2^{x_e} \qquad (2.3)$$

where x_m is a fractional part called the *mantissa*, and x_e is called the *exponent*. Floating point numbers are typically normalized with the mantissa being represented as a fixed point fraction with the first bit used for its sign and such that $0.5 \leq |x_m| < 1$, and with the exponent represented as a fixed point signed integer. For example, with four bits available for the mantissa and three for the exponent, the number 2.5 could be represented as 0.625×2^2. In binary code, this number can be represented as $x_m 2^{x_e}$ with x_m encoded as 0101 and x_e encoded as 010. The dynamic range of floating point numbers is primarily determined by the number of bits used for the exponent. If $e+1$ bits are used for the representation of the integer x_e (including one for its sign), then the dynamic range is approximately 2^e. We mention that the number e is an integer number here - it has no connection with the base e of natural logarithms. The resolution between two successive floating point numbers depends on the magnitude of these numbers, with the quantization error being proportional to this magnitude. Thus, for floating point representations we have $|x - Q[x]| = \epsilon |x|$.

With proper choice of the scaling factor K in (2.2) the dynamic range of fixed point numbers can be made identical to that of floating point numbers when both are implemented using the same number of bits. An essential difference is that the quantization error on fixed point representations is identical all through the dynamic range, while for floating point representations the quantization error is small for small numbers and large for large numbers. In other words, with fixed point representations the total dynamic range $2K$ is divided into a grid made up of 2^{B+1} elements, each of equal size $q = K2^{-B}$. The precision with which the numbers are represented is therefore identical throughout the range: it is equal to the quantization step size q. For a fixed number of bits B, the larger the dynamic range the smaller the precision. This range - precision tradeoff is also present with floating point representations since a larger dynamic range requires more bits for the exponent x_e at the expense of fewer bits for the mantissa x_m. In addition, with floating point numbers the precision is not uniform through the range: smaller numbers are represented with greater precision than larger ones. Since the relative error is often more relevant than the absolute one, floating point representations typically deliver a better range - precision tradeoff than fixed point ones. However, the error analysis for floating point arithmetic is much more complicated than for fixed point arithmetic: see e.g. [Liu71], [Hwa77] and [RC79]. A recent attempt to analyse the effects of floating point errors in digital filters by embedding the roundoff errors in coefficient sensitivities can be found in [ZN91]. From a practical point of view, arithmetic operations with floating point numbers are also more complicated and hence slower than with fixed point numbers if the hardware available has only fixed point arithmetic operations. In order to speed up the operations, floating point coprocessors are often added in modern day computers when speed of computation is an important factor.

Good discussions on properties of fixed point and floating point arithmetic can be found in [Wil63], [Liu71], [Hen82] and [Wil91]. *In this book we consider mainly fixed point arithmetic, and in the remainder of this section on FWL effects and computations we consider fixed point representations and fixed point arithmetic unless otherwise specified.*

2.3 Overflow and quantization errors

An overflow occurs if a number x is outside the range $[-K, K)$. This typically occurs when the number x is the result of an addition or multiplication of two numbers that are in the range but whose result is outside the range. To understand this mechanism, consider the addition of two numbers in two's complement representation, as illustrated by Figure 2.4.

For $K = 1$ Figure 2.4 represents the addition of the two positive numbers $\frac{3}{8}$ and $\frac{1}{4}$ to produce the result $\frac{5}{8}$, which is inside the range $[-1, +\frac{7}{8}]$ of numbers that can be represented by this $(B+1)$-bit register. No overflow has occurred. However, overflow occurs if the addition of two positive numbers, say, exceeds the available range of positive numbers, thereby producing a carry into the sign bit (the extreme left bit), in which case the sign changes. The finite wordlength representation of the result may then be very different from x, and hence large errors typically occur by overflow. This is illustrated in Figure 2.5.

30   2. Finite Word Length errors and computations

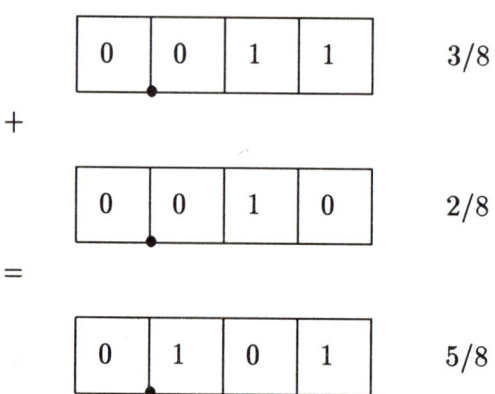

FIGURE 2.4. Illustration of addition

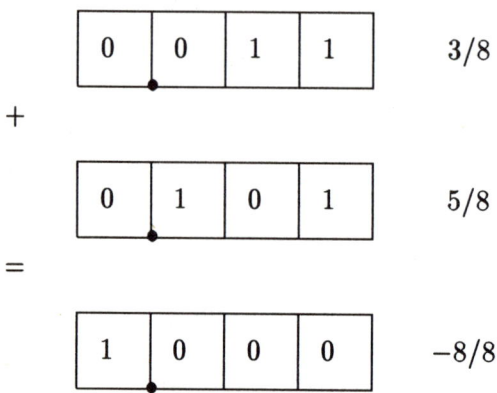

FIGURE 2.5. Addition causing overflow

With $K = 1$ Figure 2.5 illustrates that the addition of $\frac{3}{8}$ and $\frac{5}{8}$ produces the 4-bit number 1000 which, in two's complement representation, is -1 instead of $+1$. This represents an error of magnitude $2K$, a very large error indeed.

Overflow errors must therefore be avoided at all cost, and this can only be done by increasing the dynamic range. In fixed point representation and for a fixed wordlength, this amounts to increasing the value of K. In such a case the quantization step q will also increase since $q = K2^{-B}$. Thus, if the wordlength is fixed, a decrease in the probability of overflow can only be obtained at the expense of an increase in the quantization error.

There are several methods for dealing with overflow numbers, the most common ones being to either do nothing at all (two's complement overflow characteristic) or to replace the overflown number by the nearest representable number (saturation overflow characteristic). The characteristics illustrating these two strategies are represented graphically in Figures 2.6 and 2.7.

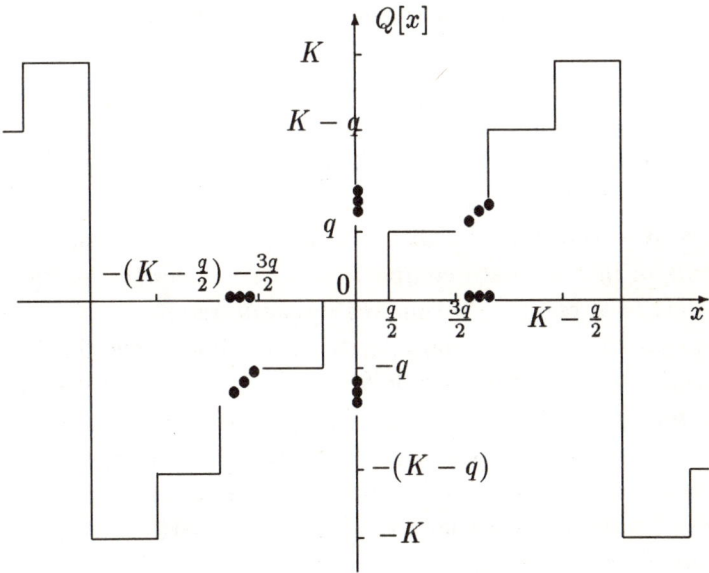

FIGURE 2.6. Two's complement overflow characteristic

With the two's complement overflow characteristic, nothing is done to modify the sum of two numbers when that sum overflows. This can

32   2. Finite Word Length errors and computations

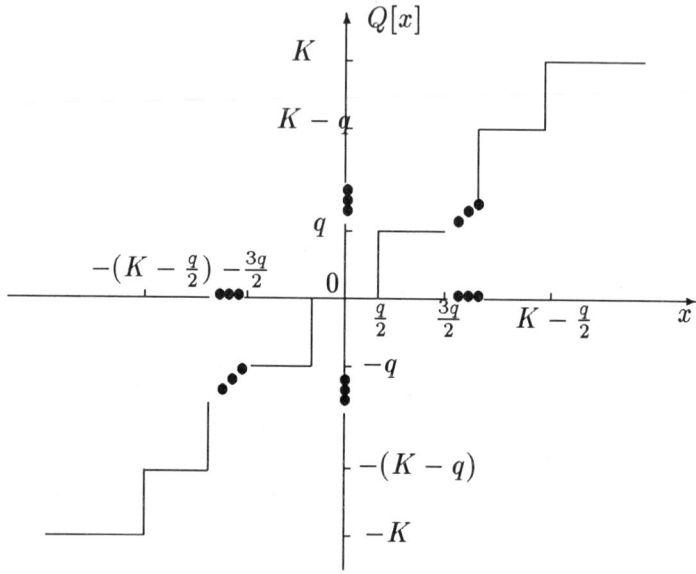

FIGURE 2.7. Saturation overflow characteristic

cause large errors as explained above, but a major advantage is that if the sum of several numbers does not overflow, the result of two's complement addition of these numbers is correct even if some intermediate sums do overflow. With a saturation overflow characteristic (most often used in ADC converters) small overflows cause small errors, but overflow errors in intermediate computations carry over to the final result even if that final result is within the dynamic range.

Given that the overflow characteristic is nonlinear, overflow may produce oscillations even for a stable filter: the filter output does not converge to zero even for a zero input. This occurs when some of the state components increase in value and overflow after multiplication by the state transition matrix, while others do not. This can be prevented by a proper choice of state space realization: in particular, if the state transition matrix A is made *normal* by a similarity transformation, such overflow oscillations will not occur.[2] We do not pursue this study of overflow oscillations here, and we refer the interested reader to the

---

[2] A matrix A is normal if $A^T A = AA^T$. Normal matrices have a number of interesting properties that make them attractive candidates for realizations minimizing FWL error effects. We shall examine these properties in detail in Chapter 6.

excellent analysis of overflow and overflow oscillations in [RM87].

We have seen in the previous subsection that the finite wordlength representation of a number produces a *quantization error*:

$$e = x - Q[x]. \qquad (2.4)$$

When roundoff is used (as opposed to truncation), this error lies between $-q/2$ and $q/2$. It is then common practice, supported by simulations and theoretical studies, to consider e as white noise, uncorrelated with x, and with a uniform distribution on $[-q/2, q/2]$: see [SS77]. The variance of the noise e associated with this distribution is then

$$\sigma^2 = E\{e^2\} = \frac{q^2}{12}. \qquad (2.5)$$

The important message from our discussion is that overflow produces such unacceptable errors, including possibly oscillations, that sufficient care must be taken to prevent it. This is usually done by taking the scaling factor K large enough so that the signals in the digital filter realization stay within the dynamic range. Of course, this increases the quantization step size q and hence the quantization noise and the roundoff noise produced by arithmetic computations. This observation provides a first glimpse at the 'overflow - roundoff error tradeoff'.

2.4 Arithmetic computations and roundoff errors

In realizations of linear digital filters, three basic operations are performed: multiplications of signals by filter parameters, additions of these products, and storage into registers. Unless otherwise specified, we shall consider throughout this book that coefficients are implemented in fixed point as B_c bit numbers, while signals are implemented in fixed point as B_s bit numbers.[3] In fixed point arithmetic, the three basic operations have the following properties:

1. The addition of two numbers is exact provided there is no overflow - this has been illustrated in Figure 2.4.

2. The product of two numbers having, respectively, a B_s-bit and a B_c-bit fractional part produces a number having a $B_s + B_c$-bit

---

[3] Following our earlier stated convention this implies that the encoding of a coefficient uses registers of length $B_c + 1$, while the encoding of signals requires registers of length $B_s + 1$.

fractional part, which must thus be rounded off or truncated to a B_s-bit number in order to be stored in the finite wordlength register. Assuming again that there is no overflow, this may or may not produce an error. This is illustrated in Figures 2.8 and 2.9. This last figure also shows that truncation and rounding may lead to different results. In practice, rounding is used and the resulting rounding error is bounded by $q/2$ if two's complement format is used.

3. All numbers in these arithmetic operations must be limited in magnitude by K.

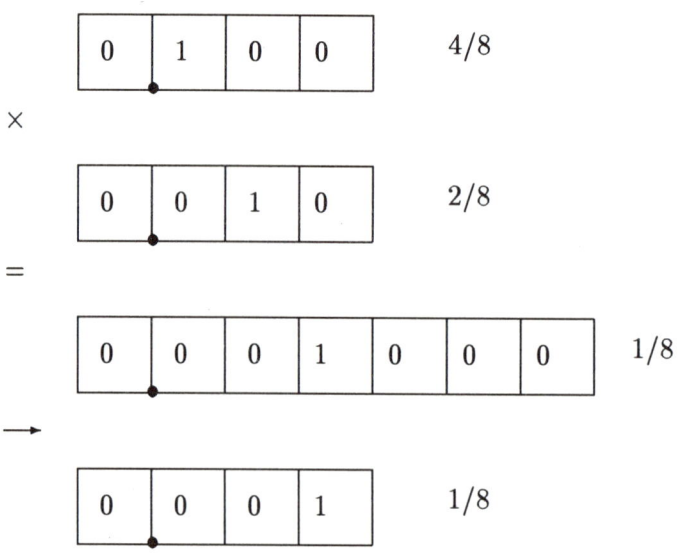

FIGURE 2.8. Illustration of multiplication without error

The error that occurs at the rounding of products to B_s bits is called *roundoff noise*. The same model is used for roundoff noise as for the quantization error produced when a signal is quantized: the roundoff noise is assumed to be zero mean white noise, uniformly distributed over $[-q/2, q/2]$. More details on the characteristics of fixed point multiplicative noise can be found in [Won90].

Depending on the hardware, two different roundoff versions are being used in arithmetic operations, which we successively examine: roundoff before multiplication and roundoff after multiplication [Hwa77], [Wil86]. To keep things simple, we explain these two different implementations on a scalar equation.

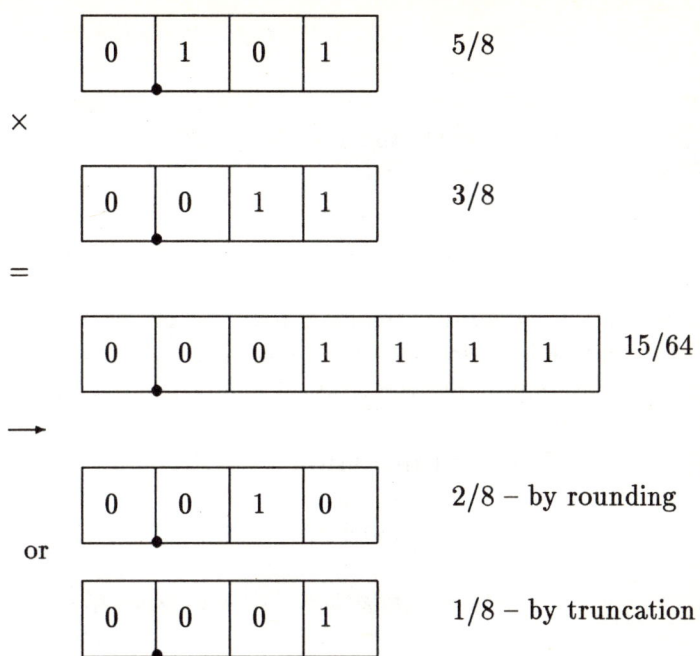

FIGURE 2.9. Illustration of multiplication with error

Consider the following scalar recursive equation

$$x(t+1) = \beta x(t) + \mu u(t) \qquad (2.6)$$

where β and μ are two real constants. A schematic representation of this recursion is given in Figure 2.10.

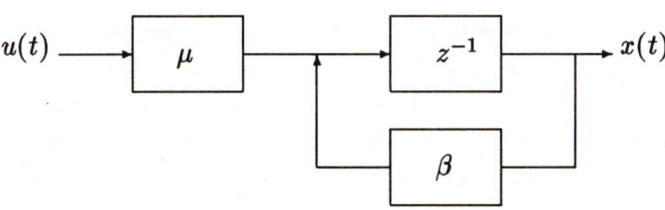

FIGURE 2.10. A schema of a recursive equation

Assuming that the scale factor K is 1 in our example, then the implemented versions of β and μ are

$$\beta^* \quad \text{with} \quad |\beta - \beta^*| < 2^{-(B_c+1)}$$

36   2. Finite Word Length errors and computations

$$\mu^* \text{ with } |\mu - \mu^*| < 2^{-(B_c+1)} \tag{2.7}$$

assuming that they are rounded off to the nearest multiple of q.

Without any rounding, the products $\beta x(t)$ and $\mu u(t)$ in equation (2.6) as well as the result $x(t+1)$ would be $(B_c + B_s)$-bit numbers, and the next iteration would produce $(2B_c + B_s)$-bit numbers. The computation of difference equations such as (2.6) is typically performed in *accumulators* of length $(B_c + B_s)$ bits or more, but to maintain a B_s-bit wordlength for the signals, the numbers must be rounded off every time a multiplication is performed. Two implementations exist depending on whether the numbers are rounded off before or after multiplication.

Roundoff Before Multiplication – RBM

In this case, the actual model of equation (2.6) is (see e.g. [Wil86]):

$$x^*(t+1) = \beta^* Q[x^*(t)] + \mu^* Q[u(t)]. \tag{2.8}$$

This is illustrated diagrammatically in Figure 2.11

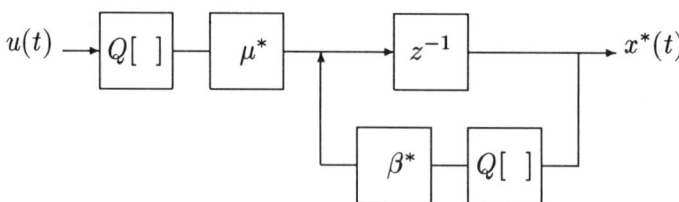

FIGURE 2.11. A schema of a recursive equation implemented with RBM

The $(B_c + B_s)$-bit signal $x^*(t)$ obtained from the previous iteration has been rounded to B_s bits by the quantizer and stored for the next iteration. The products $\beta^* Q[x^*(t)]$ and $\mu^* Q[u(t)]$ are accumulated in the $(B_c + B_s)$-bit accumulator to form the result $x(t+1)$, which is again rounded to B_s bits. Equation (2.8) can therefore be rewritten as

$$x^*(t+1) = \beta^* x^*(t) + \mu^* u(t) - \beta^* e^x(t) - \mu^* e^u(t), \tag{2.9}$$

where

$$e^x(t) \triangleq x^*(t) - Q[x^*(t)], \quad e^u(t) \triangleq u(t) - Q[u(t)] \tag{2.10}$$

2. Finite Word Length errors and computations

are the roundoff errors, which are both bounded by 2^{-B_s}. The roundoff process is usually modelled as zero mean uniformly distributed white noise with variance σ^2 [Liu71], [Hwa77], [Won90]:

$$\sigma^2 = (1/12)2^{-2B_s}. \tag{2.11}$$

Denote

$$\Delta\beta \triangleq \beta - \beta^*, \quad \Delta\mu \triangleq \mu - \mu^*. \tag{2.12}$$

The actual implementation of equation (2.6) can then be written as (see (2.9), (2.12)):

$$x^*(t+1) = \beta x^*(t) + \mu u(t) - \Delta\beta x^*(t) - \Delta\mu u(t) - \beta^* e^x(t) - \mu^* e^u(t). \tag{2.13}$$

Roundoff After Multiplication – RAM

In this case, one has

$$x^*(t+1) = Q[\beta^* x^*(t) + \mu^* Q[u(t)]]. \tag{2.14}$$

This implementation is diagrammatically represented in Figure 2.12.

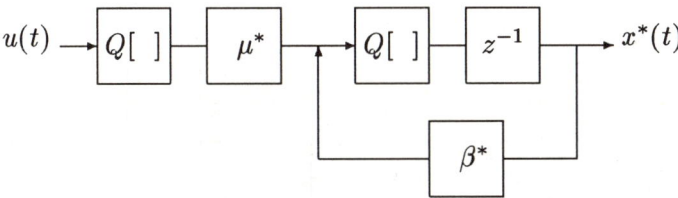

FIGURE 2.12. A schema of a recursive equation implemented with RAM

Unlike for the RBM case, the state $x^*(t)$ obtained from the previous iteration has length B_s and is directly stored in a storage unit. The products, $\beta^* x^*(t)$ and $\mu^* Q[u(t)]$, are available as $(B_c + B_s)$-bit numbers in the accumulator where they are added. They are then rounded to form the B_s-bit number $x^*(t+1)$. Similarly to (2.13), equation (2.14) can be rewritten as

$$x^*(t+1) = \beta x^*(t) + \mu u(t) - \Delta\beta x^*(t) - \Delta\mu u(t) - \eta^s(t), \tag{2.15}$$

where
$$\eta^s(t) = \beta^* x^*(t) + \mu^* Q[u(t)] - Q[\beta^* x^*(t) + \mu^* Q[u(t)]] + \mu^* e^u(t) \quad (2.16)$$
is a roundoff error.

Since in equation (2.14), $x^*(t)$ and $x^*(t+1)$ are both B_s-bit numbers, the roundoff error will be zero if the coefficients β and μ are ± 1 or 0. Clearly, whatever roundoff version is used, the actually implemented scalar recursive equation (2.6) has the following form:
$$x^*(t+1) = \beta x^*(t) + \mu u(t) - \eta^c(t) - \eta^s(t), \quad (2.17)$$
where
$$\eta^c(t) = \Delta\beta x^*(t) + \Delta\mu u(t) \quad (2.18)$$
$$\eta^s(t) = \beta^* e^x(t) + \mu^* e^u(t) \text{ for RBM}$$
$$= \beta^* x^*(t) + \mu^* Q[u(t)]$$
$$\quad - Q[\beta^* x^*(t) + \mu^* Q[u(t)]] + \mu^* e^u(t) \text{ for RAM.}$$
$$(2.19)$$

This can be schematically represented by the diagram of Figure 2.13, where $e(t)$ represents the total FWL error: $e(t) = -\eta^c(t) - \eta^s(t)$.

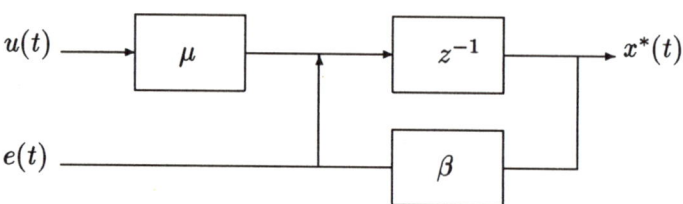

FIGURE 2.13. A general schema of a recursive equation implemented with FWL

Denote
$$E(t) = x(t) - x^*(t). \quad (2.20)$$
From equations (2.6) and (2.17)-(2.19), one has
$$E(t+1) = \beta E(t) + \eta^c(t) + \eta^s(t). \quad (2.21)$$

It is thus clear that the actual process, $x^*(t)$, computed in FWL will deviate from the desired one, $x(t)$.

The mechanism for the generation of FWL errors illustrated by our scalar example can be generalized without much difficulty to the case of a system of recursive linear equations implemented in finite wordlength. From the above discussion, we see that the deterioration of the performance of a realization (or parametrization) of a digital system due to the finite wordlength can be separated into two effects as shown in (2.21): one is due to the FWL implementation of the coefficients of the realization, characterized by the error $\eta^c(t)$. The other is due to roundoff of the signals during arithmetic operations, characterized by the error $\eta^s(t)$. The first effect is related to the *sensitivity* of the system with respect to errors in its parameters, while the other is usually measured by the *roundoff noise gain*, which is the ratio between the output error variance and the variance σ^2 of the unit roundoff noise generated at every multiplication. Much of this book is concerned with defining appropriate measures for the sensitivity and the roundoff noise, in a variety of different applications, and with constructing realizations that minimize these measures. Before we address these issues, we end this chapter on finite wordlength effects by a brief discussion of the important subject of scaling.

2.5 Dynamic range and scaling

We have already stressed the importance of avoiding overflow in arithmetic computations. The proper way to achieve this is by scaling the variables so that the dynamic range of internal variables, obtained after multiplications, never exceeds the available wordlength. There are several ways of performing a scaling operation, going from very simple input scaling to the more sophisticated dynamical scaling of the internal variables as the computations proceed.

It is important that the external variables (i.e. the 'physical' inputs and outputs) be scaled so that their expected ranges are roughly the same. This preliminary operation amounts to replacing physical variables (such as *volts*, kg/m^2, m/sec) by dimensionless variables and is known as *normalization*. This normalization must be performed before any digital filter design. We illustrate this normalization procedure with the following simple example.

Consider a simple *R-L* electrical circuit. The mathematical model is

$$L\frac{di(t)}{dt} + Ri(t) = e(t),$$

where L is the inductance in Henry (H), $i(t)$ is the current in Ampere (A), R is the resistance in Ohm (Ω) and $e(t)$ is the control voltage in Volt (V).

Suppose that we know the current is bounded by $2.5A$ for a bounded control voltage with a bound of, say, $10V$, that is $|i(t)| \leq 2.5, |e(t)| \leq 10$. In order to make the variables in the equivalent model dimensionless and bounded by one, we define the following variables:

$$x(t) \triangleq \frac{i(t)}{2.5}, \quad u(t) \triangleq \frac{e(t)}{10}.$$

The above equation becomes

$$(2.5L)\frac{dx(t)}{dt} + (2.5R)x(t) = 10u(t),$$

or

$$\frac{dx(t)}{dt} + ax(t) = bu(t).$$

This is now a standard first order differential equation in which all the variables are dimensionless and bounded by 1.

We shall henceforth look at the filter scaling design problem assuming that this prior normalization has been performed, and that the inputs and outputs to the filter have roughly the same expected dynamic range.

The only foolproof way to avoid overflow would be to know exactly what input signal sequence will be fed to the digital filter. This is a very unrealistic situation, and so one has to make do with prior assumptions on or prior models of the likely input signals. We have already shown that scaling affects not only overflow, but also quantization noise. Assuming that the available word length is fixed, one therefore has to strike a compromise between an excessively cautious scaling that results in an available dynamic range that far exceeds the likely range of the signals (this would totally prevent overflow but at the expense of high quantization noise), and an insufficient scaling that would result in likely overflow.

Consider first a scalar relation between the input signal $u(t)$ and some internal variable $x(t)$ at an internal node of a filter, and let $f(k), k = 0, 1, 2, \ldots$ be the impulse response of that filter from u to x:

$$x(t) = \sum_{k=0}^{\infty} f(k)u(t-k). \tag{2.22}$$

By requiring that the input signal u and the state x of this filter remain roughly within the same dynamic range $[-K, K]$, we are in fact imposing that the 'gain' of the impulse response be smaller than or equal to 1. To make this statement more precise, we have to decide on a particular definition of impulse response gain, and this choice depends in turn on what type of input signal is expected to occur.

For example, if the input is bounded, $|u(t)| \leq K$ $\forall t$, then $|x(t)| \leq K$ $\forall t$ if

$$\sum_{k=0}^{\infty} |f(k)| \triangleq \|f\|_1 = 1. \tag{2.23}$$

Alternatively, if the input has finite energy, $\sum_{l \leq t} u^2(l) \leq K$, then by the Cauchy Schwartz inequality $|x(t)| \leq K$ if

$$\left[\sum_{k=0}^{\infty} f^2(k)\right]^{\frac{1}{2}} \triangleq \|f\|_2 = 1. \tag{2.24}$$

The more common form of scaling constraint, which we shall adopt in this book unless otherwise specified, is the latter constraint, called l_2 scaling. In order to handle white noise input signals, a more cautious version of l_2 scaling is often adopted. Indeed, with a white noise input signal $u(t)$ of unit variance, we have

$$\{E[x^2(t)]\}^{\frac{1}{2}} = \|f\|_2. \tag{2.25}$$

In order to guarantee that several standard deviations (rather than just one) of the state remain within the allocated dynamic range, one therefore performs an l_2 scaling on $\{f\}$ such that

$$\alpha \|f\|_2 = 1 \tag{2.26}$$

where the parameter α is typically between 1 and 5.

Scaling the relationship between the input $u(t)$ and the internal variable $x(t)$ of a filter must of course not alter the input/output relationship. Assume that the transfer function $H(z)$ from the input u to the output y through the internal state x can be factored as

$$H(z) = G(z)F(z). \tag{2.27}$$

Then any scaling of the-input-to-state transfer function $F(z)$ by a scaling factor α must be compensated for by an inverse scaling of the transfer function $G(z)$:

$$X(z) = \alpha F(z) U(z), \quad Y(z) = \frac{1}{\alpha} G(z) X(z). \tag{2.28}$$

2. Finite Word Length errors and computations

Now consider the case where the input-output filter is realized via a vector state-variable description:

$$x(t+1) = Ax(t) + Bu(t), \qquad (2.29)$$

with $A \in \mathbb{R}^{n \times n}$ and $B \in \mathbb{R}^n$.[4] We assume in the sequel that A is a stable matrix, that is $|\lambda_i(A)| < 1$ for $i = 1, 2, \ldots, n$. Each component $x_i(t)$ of the state vector is typically an accumulator node of the realization, and the input-to-state-component transfer functions must therefore all be scaled as described above. We now show that, if l_2 constraints are imposed, this can easily be achieved by diagonal similarity transformations performed on the system realization.

First notice that with $x(0) = 0$, we have

$$x(t) = \sum_{k=0}^{t-1} A^k B u(t - k - 1), \qquad (2.30)$$

so that the impulse response elements of the system from $u(.)$ to $x(.)$ are given by

$$f(k) \triangleq A^k B = [f_1(k) \ldots f_n(k)]^T \quad k = 0, 1, \ldots \qquad (2.31)$$

To ensure equal probability of overflow between all components of $x(.)$, the following l_2 norm scaling is introduced on the scalar impulse response sequences:[5]

$$\|f_i\|_2^2 \triangleq \sum_{k=0}^{\infty} f_i^2(k) = 1 \quad i = 1, \ldots, n. \qquad (2.32)$$

Now consider

$$W_c \triangleq \sum_{k=0}^{\infty} f(k) f^T(k) = \sum_{k=0}^{\infty} A^k BB^T (A^T)^k. \qquad (2.33)$$

This matrix is called the controllability Gramian of the realization (2.29): it is positive definite if the pair $[A, B]$ is completely reachable

---

[4] $\mathbb{R}$ is used to denote the real numbers, $\mathbb{R}^n$ an n-vector of real numbers, and $\mathbb{R}^{n \times n}$ an $n \times n$ matrix of real numbers.

[5] Note that when we write the l_2 norm of the impulse response sequence, we implicitly assume that the system is stable, i.e. all eigenvalues of the matrix A are absolutely smaller than 1.

(see e.g. [Kai80]). It can be computed by solving the linear matrix equation
$$W_c = AW_cA^T + BB^T. \tag{2.34}$$
Such equation is called a Lyapunov equation. It clearly follows from its definition that
$$(W_c)_{ii} = \|f_i\|_2^2. \tag{2.35}$$
The l_2 scaling on the impulse responses f_i, $i = 1, \ldots, n$ can then be achieved by a similarity transformation $x = T\xi$. Indeed, this transforms the triple (A, B, W_c) in the original coordinate system into another triple $(T^{-1}AT, T^{-1}B, T^{-1}W_cT^{-T})$ in the new coordinate system. If
$$T \triangleq diag\{t_1, t_2, \ldots, t_n\},$$
then clearly
$$(T^{-1}W_cT^{-T})_{ii} = \frac{(W_c)_{ii}}{(t_i)^2}. \tag{2.36}$$
By choosing $t_i = \sqrt{(W_c)_{ii}}$, $i = 1, \ldots, n$, we obtain
$$(\tilde{W}_c)_{ii} = (T^{-1}W_cT^{-T})_{ii} = 1, \quad i = 1, \ldots, n. \tag{2.37}$$
That is the controllability Gramian has its diagonal elements all equal to unity. If a more conservative scaling is desired such that $\alpha(W_c)_{ii} = 1, i = 1, \ldots, n$ with α larger than one, as explained above, then the diagonal elements of T are defined as $t_i = \sqrt{\alpha(W_c)_{ii}}$, $i = 1, \ldots, n$.

A more sophisticated but elegant procedure is obtained if the diagonal scaling matrix T is replaced at each iteration by a state dependent diagonal similarity transformation $T(t)$. This adaptive scaling procedure leads to *dynamically scaled* fixed point arithmetic. It is described in detail in [Wil91].

2.6 Conclusions

We have now fought our way through the basic concepts and methods of finite wordlength representation and arithmetic. It is a fascinating and highly specialized subject to which hundreds of pages could have been devoted. However, an excessive indulgence into the vagaries (and the beauties) of FWL arithmetic and implementation issues would have distracted the reader from what is the essential contribution of this book: the study of the relationship between the choice of a particular realization for a digital filter or a digital controller and the numerical accuracy of this filter or this controller. We therefore move on to what is the essence of this book and we begin our journey with the study of state space realizations of filters and of a survey of classical results on the realization of minimal FWL error filters.

3

Parametrizations in digital system design

3.1 Introduction

In the early days, a digital filter characterized by its transfer function was usually implemented directly in one of several possible 'canonical forms'- more often called 'direct forms' in the signal processing community - because these forms minimize the number of coefficients and arithmetic operations that are required. A minimal number of coefficients corresponds to minimal complexity, an attribute that may be important if the operation count or memory space plays a significant role. It was found, however, that the performance of a digital filter implemented in these forms can be seriously degraded by Finite Word Length (FWL) effects. In fact, these structures have undesirable numerical properties as will be made clear throughout this book. This problem can be partially alleviated by the use of other parametrizations of the filter such as the parallel and cascade forms of a series of first or second order sub-filters.

Consider a finite dimensional digital filter described by its rational transfer function, $H(z)$ say. This transfer function is usually parametrized as a ratio of two polynomials, although other parametrizations are possible. For example, the numerator and denominator can both be factored as ratios of first and second order terms. Whatever the parametrization of such a transfer function, its use as a data filter requires an implementation using memory registers, multipliers and adders. This implementation can take the form of a software program in a computer, or of an application oriented hardware circuit. By defining as states the internal variables of the software program or the contents of the register memories in a hardware circuit, a particular implementation of a transfer function is then most often equivalent to a particular state space realization of this transfer function[1]. The effects of finite word lengths on the performance degradation of various implementations of a digital filter can then best be studied using the state space realizations of these implementations, within the framework of linear system theory. This approach to digital filter design has been adopted by various authors [MR76a], [MR76b], [Hwa77], [Thi86], [Gra89], and will be exploited throughout this book. For the totally ignorant among our readers[2], a state space realization of a transfer function $H(z)$ is a linear system of equations,

$$\begin{aligned} x(t+1) &= Ax(t) + Bu(t) \\ y(t) &= Cx(t) + Du(t) \end{aligned} \quad (3.1)$$

with $A \in \mathbb{R}^{n \times n}, B \in \mathbb{R}^n, C^T \in \mathbb{R}^n$ and $d \in \mathbb{R}$, such that $H(z) = C(zI - A)^{-1}B + D$.

It is well known that the number of state space realizations of a rational transfer function is infinite. These realizations are said to be equivalent - they are all related by what is called a 'similarity transformation'. This means that if (A_1, B_1, C_1, D_1) and (A_2, B_2, C_2, D_2) are two equivalent realizations of a same transfer function $H(z)$, then there exists a nonsingular matrix T such that

$$A_2 = T^{-1}A_1T, B_2 = T^{-1}B_1, C_2 = C_1T, D_2 = D_1. \quad (3.2)$$

This equivalence between realizations breaks down when the elements of these matrices $\{A_i, B_i, C_i, D_i\}$ are stored in finite word length, or when the computations are performed using finite word length, i.e. when the signals on which the filter operates and the internal states of the filter are rounded off or truncated. It is illustrated extensively in this book that different FWL state space realizations of the same transfer function may produce significantly different responses to the same input signal. The basic idea is then to identify which realizations have superior numerical properties in terms of some optimal design criterion so that in practice their performance degradation due to FWL effects

---

[1]There are some implementations that cannot be described as state space realizations, a notable special case being the lattice filters: see [RM87] for a discussion of this issue. In this book we shall focus almost exclusively on state space realizations.

[2]The authors have this mental picture that their thousands of readers include electrical engineers, but also professional football players, philosophers, artists and others. Some of these might be excused for not having been exposed to state space realizations.

is minimized. In optimal state space digital filter design, two tracks have been pursued: one is to minimize, over the set of equivalent state space realizations, some appropriately defined measure of the sensitivity of the transfer function with respect to coefficient errors; the other is to minimize, over the set of equivalent state space realizations, some appropriately defined roundoff noise gain of the filter.

In this chapter we review some key classical[3] results on optimal filter design, where the optimality is with respect to a minimization of the effects of finite wordlength errors. The objective is as much to familiarize the unsuspecting reader with this rather esoteric material as it is to introduce notations and concepts that will be slowly and progressively spoon-fed to him or her.

In 1976, Mullis and Roberts published a seminal paper on the synthesis of digital filters with minimum roundoff noise due to fixed point multiplication [MR76a]. In their approach, similarity transformations were explored to determine the state space realizations that minimize the output error variance. Independently, Hwang [Hwa77] considered the same problem and gave a minimization procedure to determine the optimal transformations or realizations. The fundamental result of this early work, which will be presented in Section 3.6, was to show that realizations achieve minimal roundoff noise gain if and only if their controllability and observability Gramians are proportional to one another. More shall be said about this in the fullness of time.[4]

A parallel approach to roundoff noise minimization is to identify the realizations that minimize some sensitivity measure with respect to errors in the coefficients. Several sensitivity measures can be and have been adopted: sensitivity of the transfer function with respect to coefficient errors, or of the pole and/or zero locations with respect to such errors, and others. We shall examine a number of these measures and their interconnections later in this book.

The first purely analytical attempt to synthesize minimum sensitivity state space realizations for linear systems was given in 1984 by Tavsanoglu and Thiele (see [TT84] and [Thi84]). They defined a sensitivity measure of a transfer function with respect to the parameters

---

[3]The word 'classical' is always a dangerous one to use, because what is classical to one person is novel to another. We shall here adopt our own definition and call 'classical' whatever material was available to the authors of this book at the time (1987) they embarked on their study of optimal FWL implementations.

[4]In this particular instance the fullness of time is to be found in Section 3.6.

of a corresponding realization (A, B, C) in the form of an integral over all frequencies of a frequency dependent sensitivity function. For this sensitivity measure, they obtained a particularly simple upper bound, for which they posed and solved an optimal realization problem. Their main result was to show that their upper bound was minimized by realizations whose controllability and observability Gramian matrices are identical. In 1986, Thiele [Thi86] showed that these realizations minimize not only the upper bound, but also the sensitivity measure itself. In [Gra89], the same conclusion was obtained using a geometric approach.

The main aim of this chapter is to clearly establish that the FWL error properties of a digital filter depend critically on the state space realization chosen for its implementation, and to review the existing results on optimal digital filter design. Our development is based on [MR76a], [TT84], [Hwa77] and [Thi86]. An outline of the chapter is as follows. Section 3.2 defines the class of equivalent realizations and Section 3.3 introduces the classical definitions of sensitivity and sensitivity measure of a transfer function with respect to the parameters of a realization. We then review the main classical results on sensitivity-optimal state space digital system design in Section 3.4. In Section 3.5 we present the most commonly adopted roundoff noise gain measure for the realization of a digital filter. The set of optimal realizations that minimize this measure is described in Section 3.6. The relationship between the sensitivity measure and the roundoff noise gain measure and between their corresponding optimal realization sets are established in Section 3.7. Some examples and simulations illustrate the importance of these, by now, classical results; these are the subject of Section 3.8. Besides optimal state space design, a few alternative approaches have been proposed for the minimization of FWL errors in digital filters. These are briefly presented in Section 3.9. Finally Section 3.10 contains some concluding remarks which summarize the essential results of this chapter.

3.2 State space realization set

Consider a proper discrete scalar rational stable transfer function described in the usual shift operator notation,

$$H(z) = \frac{\sum_{i=0}^{n} b_i z^{-i}}{1 + \sum_{i=1}^{n} a_i z^{-i}} \qquad (3.3)$$

and a minimal[5] state space realization of $H(z)$:

$$\begin{aligned} x(t+1) &= Ax(t) + Bu(t) \\ y(t) &= Cx(t) + du(t) \end{aligned} \qquad (3.4)$$

with $A \in \mathbb{R}^{n \times n}, B \in \mathbb{R}^n, C^T \in \mathbb{R}^n$ and $d \in \mathbb{R}$. The transfer function can be expressed in terms of the state matrices as

$$H(z) = C(zI - A)^{-1}B + d. \qquad (3.5)$$

So one sees that in the state space model the system is parametrized by the system matrices (A, B, C, d) which, unlike the $(\{a_i\}, \{b_i\})$ for the transfer function parametrization (3.3), are not unique. Unless certain entries are fixed as is the case in canonical forms, the parametrization (A, B, C, d) contains $n^2 + 2n + 1$ parameters whereas the parametrization of $H(z)$ in the form (3.3) contains only $2n + 1$ parameters, indicating a redundancy in the representation of the state variable realization. The set composed of all realizations (A, B, C, d) of a given $H(z)$ is called *realization set*, and is denoted by S_H:

$$S_H \triangleq \{(A, B, C, d): H(z) = C(zI - A)^{-1}B + d\}. \qquad (3.6)$$

If (A, B, C, d) belongs to S_H, then so does $(T^{-1}AT, T^{-1}B, CT, d)$ for any real nonsingular matrix T. Therefore, with any given initial realization $(A_0, B_0, C_0, d) \in S_H$, S_H can be characterized by the set $(T^{-1}A_0T, T^{-1}B_0, C_0T, d)$ for all nonsingular matrices T:

$$S_H = \{(T^{-1}A_0T, T^{-1}B_0, C_0T, d), \; det T \neq 0\}. \qquad (3.7)$$

---

[5]The term *minimal* refers to the fact that the transfer function $H(z)$ cannot be represented by a realization whose state vector x has a smaller dimension.

50   3. Parametrizations in digital system design

The transformation from (A, B, C, d) to $(T^{-1}AT, T^{-1}B, CT, d)$ is called a similarity transformation. Since d is left unchanged by a similarity transformation T, we shall in the sequel refer to a realization (A, B, C) instead of (A, B, C, d).

In the ideal case of infinite precision implementation, any realization in S_H is equivalent to any other one, since they yield exactly the same transfer function, $H(z)$. But they have different numerical properties, which leads to the fact that *different realizations of the same system yield different performance when implemented with finite wordlength. The optimal FWL state space design is to find, in the realization set S_H, those realizations that optimize some specified design performance criterion.* These performance criteria are in terms of the robustness of the realization with respect to errors due to the FWL implementation. As explained in the introduction, the most commonly used performance criteria are a sensitivity measure and a roundoff noise gain.

The first part of this book is concerned with optimal filter design. A number of different sensitivity measures will be proposed, analysed and compared. A new synthetic design method will also be presented in which a new measure synthesizing the effects of both the coefficient errors and the roundoff noise will be formulated and minimized. In this chapter we introduce the whole philosophy of optimal filter design by presenting earlier 'classical' sensitivity measures and roundoff noise gains, as well as the available results concerning their minimization. We start by a transfer function sensitivity measure.

3.3 Sensitivity measure of a state space realization.

If the coefficients in $(A, B, C, d) \in S_H$ are implemented in a computer program or in a hardware circuit with finite wordlength (FWL), the actual state space model is the model (3.4) but with (A, B, C, d) replaced by their FWL versions (A^*, B^*, C^*, d^*):

$$\begin{aligned} x^*(t+1) &= A^*x^*(t) + B^*u(t) \\ y^*(t) &= C^*x^*(t) + d^*u(t). \end{aligned} \quad (3.8)$$

This means that the actual output y^* will deviate from the desired output $y^*(t)$ because the actually implemented transfer function $H^*(z)$ deviates from the desired one, $H(z)$. The amount of this deviation can be measured by the sensitivity of the system transfer function $H(z)$ with respect to changes in the coefficients of the matrices A, B and C.

There are of course several ways of defining an overall transfer function sensitivity measure: obviously, it will be some function of the partial derivatives of the transfer function $H(z)$ with respect to the coefficients in A, B, C. Probably the first measure that was analysed and that led to computable optimal realizations was one proposed by Tavsanoglu and Thiele [TT84], which we present as our Definition 3.2 below. We first need to define the sensitivity function of a complex function with respect to a matrix.

Definition 3.1 : *Let $M \in \mathbb{R}^{n \times m}$ be a matrix and let $f(M) \in \mathbb{C}$ be a scalar complex function of M, differentiable w.r.t. all the elements of M. We then define the sensitivity function of f w.r.t. M as*

$$S_M \triangleq \frac{\partial f}{\partial M} \text{ with } (S_M)_{ij} \triangleq \frac{\partial f}{\partial m_{ij}} \qquad (3.9)$$

where m_{ij} denotes the (i,j)-th element of the matrix M.
∎

With this notation it is easy to show (see [TT84]) that the sensitivities $S_A(z)$, $S_B(z)$ and $S_C(z)$ of $H(z)$ to the coefficients of A, B and C, are given, respectively, by

$$\begin{aligned} S_A(z) &\triangleq \frac{\partial H(z)}{\partial A} = G(z)F^T(z) \\ S_B(z) &\triangleq \frac{\partial H(z)}{\partial B} = G(z) \\ S_C(z) &\triangleq \frac{\partial H(z)}{\partial C^T} = F(z), \end{aligned} \qquad (3.10)$$

where

$$\begin{aligned} F(z) &\triangleq (zI - A)^{-1}B = [f_1(z) \ldots f_n(z)]^T \\ G^T(z) &\triangleq C(zI - A)^{-1} = [g_1(z) \ldots g_n(z)]. \end{aligned} \qquad (3.11)$$

We note that the sensitivity functions $S_A(z), S_B(z), S_C(z)$ are matrix or vector functions of the complex variable z. A further definition is required to be able to measure the collective effects, averaged over all frequencies, of the elements of $S_A(z), S_B(z)$ and $S_C(z)$, respectively.

52   3. Parametrizations in digital system design

Definition 3.2 : *Let $f(z) \in \mathbb{C}^{n \times m}$ be any complex matrix valued function of the complex variable z. We then define the L_p norm of $f(z)$ as*

$$\| f \|_p \triangleq \left(\frac{1}{2\pi} \int_0^{2\pi} \| f(e^{j\omega}) \|_F^p \, d\omega \right)^{1/p} \qquad (3.12)$$

where $\| f(e^{j\omega}) \|_F$ is the Frobenius norm of the matrix $f(e^{j\omega})$ given by:

$$\begin{aligned} \| f(e^{j\omega}) \|_F &\triangleq \left(\sum_{i=1}^n \sum_{k=1}^m |f_{ik}(e^{j\omega})|^2 \right)^{1/2} \\ &= \{tr[f^T(e^{-j\omega})f(e^{j\omega})]\}^{1/2} \end{aligned} \qquad (3.13)$$

■

Taking the L_p norm of S_A, S_B or S_C now yields nonnegative real numbers. The overall sensitivity measure of the transfer function $H(z)$ with respect to the parameters in the realization A, B and C was then defined in [TT84] as follows:

$$M_{L_{12}} \triangleq \| \frac{\partial H}{\partial A} \|_1^2 + \| \frac{\partial H}{\partial B} \|_2^2 + \| \frac{\partial H}{\partial C^T} \|_2^2 \qquad (3.14)$$

One notes that in (3.14) an L_1 norm is used for the sensitivity function of $H(z)$ with respect to A, and an L_2 norm for the other two sensitivity functions: this explains our use of the notation $M_{L_{12}}$ for the sensitivity measure. We shall therefore refer to this measure as an L_1/L_2 sensitivity measure. This mixed L_1/L_2 measure has become widely accepted for the simple reason that it lends itself to state realization optimization problems with computable solutions, but the mixing of two different norms is rather illogical, and we shall replace it in Chapter 5 by a more appropriate global L_2 measure.

We now examine the effect of a coordinate transformation by a nonsingular matrix T on the sensitivity measure $M_{L_{12}}$. Consider an arbitrary realization $(A, B, C, d) \in S_H$ and the corresponding vector functions $F(z)$ and $G(z)$ defined in (3.11). A similarity transformation transforms $(A, B, C, d, F(z), G(z))$ into $(T^{-1}AT, T^{-1}B, CT, d, T^{-1}F(z), T^TG(z))$. The sensitivity measure $M_{L_{12}}$ therefore depends on the particular coordinate space chosen for the state space realization (via the transformation matrix T) in the following way:

$$M_{L_{12}} = \|T^T G(z) F^T(z) T^{-T}\|_1^2 + \|T^T G(z)\|_2^2 + \|T^{-1} F(z)\|_2^2. \qquad (3.15)$$

3. Parametrizations in digital system design 53

As it turns out, the minimization of $M_{L_{12}}$ over the equivalence class $S_H = \{T^{-1}AT, T^{-1}B, CT\}$ of realizations is very difficult, but the problem becomes feasible if $M_{L_{12}}$ is replaced by an upper bound containing only L_2 norms. In addition, it was shown by Thiele [Thi86] that the solution that minimizes the upper bound also happens to minimize the measure $M_{L_{12}}$ itself.[6] This optimization problem and its solution will be detailed in the following section. Here we show that the adoption of the mixed L_1/L_2 sensitivity measure (3.14) allows one to bound $M_{L_{12}}$ by a function containing only L_2 norms.

It is not difficult to show that

$$\| \frac{\partial H}{\partial A} \|_1^2 \leq \| \frac{\partial H}{\partial B} \|_2^2 \| \frac{\partial H}{\partial C^T} \|_2^2. \tag{3.16}$$

Indeed, we note that

$$\| \frac{\partial H}{\partial A} \|_1^2 = \left(\frac{1}{2\pi} \int_0^{2\pi} \| \frac{\partial H}{\partial A} \|_F d\omega \right)^2$$

where

$$\begin{aligned}
\| \frac{\partial H}{\partial A} \|_F &= \| G(e^{j\omega}) F^T(e^{j\omega}) \|_F \\
&= [G^T(e^{-j\omega}) G(e^{j\omega}) F^T(e^{j\omega}) F(e^{-j\omega})]^{1/2} \\
&= [G^T(e^{-j\omega}) G(e^{j\omega})]^{1/2} [F^T(e^{-j\omega}) F(e^{j\omega})]^{1/2}
\end{aligned}$$

So, by the Cauchy-Schwartz inequality,

$$\begin{aligned}
\| \frac{\partial H}{\partial A} \|_1^2 &= \left(\frac{1}{2\pi} \int_0^{2\pi} [G^T(e^{-j\omega})G(e^{j\omega})]^{1/2}[F^T(e^{-j\omega})F(e^{j\omega})]^{1/2} d\omega \right)^2 \\
&\leq \left(\frac{1}{2\pi} \int_0^{2\pi} [G^T(e^{-j\omega})G(e^{j\omega})] d\omega \right) \\
&\quad \left(\frac{1}{2\pi} \int_0^{2\pi} [F^T(e^{-j\omega})F(e^{j\omega})] d\omega \right) \\
&= \| \frac{\partial H}{\partial B} \|_2^2 \| \frac{\partial H}{\partial C^T} \|_2^2. \tag{3.17}
\end{aligned}$$

It is a property of the Cauchy-Schwartz inequality that equality holds in (3.17) if and only if

$$F^T(e^{-j\omega}) F(e^{j\omega}) = \rho^2 G^T(e^{-j\omega}) G(e^{j\omega})$$

---

[6]This situation where an apparently impossible problem is replaced by an easier one, and where the solution of the easier one fortuitously solves the hard one as well is called a 'rare event' in statistics.

or, equivalently,
$$F^H(z)F(z) = \rho^2 G^H(z)G(z) \tag{3.18}$$
for all z in the unit disc and for some non-zero constant $\rho \in \mathbb{R}$, where H denotes conjugate transpose. From equations (3.14)-(3.17) it follows that an upper bound for $M_{L_{12}}$ is given by

$$\begin{aligned}\bar{M}_{L_{12}} &= \|\frac{\partial H}{\partial B}\|_2^2 \|\frac{\partial H}{\partial C^T}\|_2^2 + \|\frac{\partial H}{\partial B}\|_2^2 + \|\frac{\partial H}{\partial C^T}\|_2^2 \\ &= \|G(z)\|_2^2 \|F(z)\|_2^2 + \|G(z)\|_2^2 + \|F(z)\|_2^2. \end{aligned} \tag{3.19}$$

We note that this upper bound now contains only L_2 norms. This allows one to rewrite $\bar{M}_{L_{12}}$ in terms of functions of the controllability and observability Gramians of the state space model. The controllability Gramian, W_c, has been defined in (2.33). The observability Gramian, W_o, is defined as follows:

$$W_o = \sum_{k=0}^{\infty} (A^T)^k C^T C A^k. \tag{3.20}$$

It is positive definite if the pair $[C, A]$ is completely observable, and it is the solution of the Lyapunov equation:

$$W_o = A^T W_o A + C^T C. \tag{3.21}$$

With these definitions we can now write

$$\bar{M}_{L_{12}} = tr(W_o)tr(W_c) + tr(W_o) + tr(W_c) \tag{3.22}$$

where $tr(W)$ denotes the trace of W and where

$$\begin{aligned}\|\frac{\partial H}{\partial B}\|_2^2 &= \|G\|_2^2 \triangleq \frac{1}{2\pi} \int_0^{2\pi} G^T(e^{-j\omega})G(e^{j\omega})d\omega \\ &= tr\left(\frac{1}{2\pi j} \oint_{|z|=1} G(z)G^H(z)z^{-1}dz\right) \\ &= tr(\sum_{k=0}^{\infty}(A^T)^k C^T C A^k) = tr(W_o) \end{aligned} \tag{3.23}$$

$$\begin{aligned}\|\frac{\partial H}{\partial C}\|_2^2 &= \|F\|_2^2 \triangleq \frac{1}{2\pi} \int_0^{2\pi} F^T(e^{-j\omega})F(e^{j\omega})d\omega \\ &= tr\left(\frac{1}{2\pi j} \oint_{|z|=1} F(z)F^T(z)z^{-1}dz\right) \\ &= tr(\sum_{k=0}^{\infty} A^k BB^T (A^T)^k) = tr(W_c). \end{aligned} \tag{3.24}$$

The controllability and observability Gramians have important system theoretical significance: see e.g. [Kai80]. A similarity transformation $x = Tz$ transforms $\{A, B, C, W_c, W_o\}$ into $\{T^{-1}AT, T^{-1}B, CT, T^{-1}W_cT^{-T}, T^TW_oT\}$. These relationships can now be used to establish the relationships between the upper bounds $\bar{M}_{L_{12}}$ of two similar realizations[7]:

$$\bar{M}_{L_{12}} = tr(T^TW_oT)tr(T^{-1}W_cT^{-T}) + tr(T^TW_oT) + tr(T^{-1}W_cT^{-T}). \quad (3.25)$$

Using the expressions above, we shall in the next section solve the problem of minimizing $M_{L_{12}}$ and $\bar{M}_{L_{12}}$ with respect to all similar realizations in S_H. Before moving on to this optimization problem, we note the following relation between the measure $M_{L_{12}}$ and its upper bound $\bar{M}_{L_{12}}$.

Theorem 3.1 : $\bar{M}_{L_{12}} = M_{L_{12}}$ *if and only if the realization satisfies the following condition* :

$$F^H(z)F(z) = \rho^2 G^H(z)G(z) \quad (3.26)$$

for all z in the unit disc and for some non-zero constant $\rho \in \mathbb{R}$.

Proof: The theorem follows directly from (3.14)-(3.19). ∎

We have now defined an L_1/L_2 sensitivity measure $M_{L_{12}}$ of a transfer function $H(z)$ with respect to the parameters of a state space realization, as well as an upper bound $\bar{M}_{L_{12}}$. We have shown how these measures depend on the particular realization chosen within the set S_H via the expressions (3.15) and (3.25). In the next section we show how to compute realizations that minimize both $\bar{M}_{L_{12}}$ and $M_{L_{12}}$.

3.4 Optimal realizations with respect to a sensitivity measure

Having defined a global (i.e. over the whole frequency range) measure of the effect of FWL coefficient errors on the transfer function, we can now pose the corresponding optimal sensitivity problem as follows: find

---

[7]Two state space realizations A, B, C and A^*, B^*, C^* are called *similar* if they are related by a similarity transformation, i.e. if $A^* = T^{-1}AT, B^* = T^{-1}B, C^* = CT$ for some nonsingular matrix T.

the realization subset that minimizes $M_{L_{12}}$ over all similar realizations in S_H. In equation form,

$$\min_{(A,B,C)\in S_H} M_{L_{12}} \qquad (3.27)$$

A direct solution to this optimization is rather intractable. Instead, one solves the easier problem of finding the realizations that minimize the upper bound $\bar{M}_{L_{12}}$, that is

$$\min_{(A,B,C)\in S_H} \bar{M}_{L_{12}} \qquad (3.28)$$

The following two inequalities, due to Thiele [Thi84], are very important. They enable us to find representations (A, B, C) that minimize $\bar{M}_{L_{12}}$.

Theorem 3.2 : *Let (A, B, C) be a completely reachable and completely observable system, and let W_c and W_o be the Gramians defined in (2.33) and (3.20). Then the following relations hold:*

$$tr(W_c)tr(W_o) \geq (\sum_{i=1}^{n} \sigma_i)^2 \qquad (3.29)$$

$$tr(W_c) + tr(W_o) \geq 2\sum_{i=1}^{n} \sigma_i \qquad (3.30)$$

where $\{\sigma_i^2 = \lambda_i(W_c W_o), i = 1, \ldots, n\}$ is the eigenvalue set of the product $W_c W_o$. The equality sign holds

* *for (3.29) if and only if* $W_o = \dfrac{tr(W_o)}{tr(W_c)} W_c$ \qquad (3.31)

* *for (3.30) if and only if* $W_c = W_o$. \qquad (3.32)

Proof: the proof was first given in [Thi84]. An alternative proof can be found in Appendix 3. ∎

The σ_i defined above are called the *Hankel singular values* or *second-order modes* of the transfer function. They are invariant under similarity transformations, and they play an important role in many problems of system theory, including in FWL error minimization problems. From Theorem 3.2, we get the following result for the upper bound $\bar{M}_{L_{12}}$ defined in (3.19).

Corollary 3.1 : *The minimum sensitivity upper bound is*

$$\bar{M}_{L_{12}}^{min} \triangleq \min_{(A,B,C) \in S_H} \bar{M}_{L_{12}} = (\sum_{i=1}^{n} \sigma_i)^2 + 2 \sum_{i=1}^{n} \sigma_i \qquad (3.33)$$

and the optimal realization set, i.e. the solutions (A,B,C) of the minimization problem (3.28), are those that satisfy (3.32).

Proof: The proof follows directly from Theorem 3.2. ∎

The set of realizations that satisfy (3.32) is a nonempty subset of S_H. A particular subset of realizations that satisfy (3.32) are the *internally balanced realizations*. A realization is called internally balanced if its controllability and observability Gramians are *identical and diagonal*. We show in the next Lemma that an arbitrary realization (A,B,C) can always be transformed to internally balanced from by a similarity transformation matrix.

Lemma 3.1 : *Given an initial stable minimal realization (A^0, B^0, C^0) with Gramians W_c^0 and W_o^0, there exists a similarity transformation matrix that transforms (A^0, B^0, C^0) to internally balanced form.*

Proof: Since W_o is symmetric and nonnegative definite, we can perform a Cholesky factorization [Ste73]:

$$W_o = R^T R. \qquad (3.34)$$

Now consider the matrix $RW_c R^T$. Since it is symmetric and nonnegative definite, it has a Singular Value Decomposition[8] (see e.g. [Ste73]) that takes the form

$$RW_c R^T = U\Sigma^2 U^T, \qquad (3.35)$$

where U is an orthogonal matrix and

$$\Sigma = diag\{\sigma_1, \ldots, \sigma_n\}; \quad \sigma_1 \geq \ldots \geq \sigma_n. \qquad (3.36)$$

We note that $\{\sigma_i, \; i = 1, \ldots, n\}$ are the singular values of $W_c W_o$. Now define the transformation matrix

$$T = R^{-1} U \Sigma^{\frac{1}{2}}. \qquad (3.37)$$

Then

$$T^{-1} W_c T^{-T} = T^T W_o T = \Sigma. \qquad (3.38)$$

∎

---

[8] For a symmetric nonnegative definite matrix, this Singular Value Decomposition is also an eigenvalue decomposition.

58   3. Parametrizations in digital system design

We have thus established the existence of realizations that minimize the upper bound $\bar{M}_{L_{12}}$. In fact, internally balanced realizations are not unique, and in addition they are not the only realizations that have the property (3.32). In the next Theorem, we characterize the set of all realizations that satisfy (3.32).

Theorem 3.3 : *Let (A_b, B_b, C_b) be an internally balanced realization of $H(z)$, i.e. a realization for which $W_c = W_o = \Sigma > 0$, where Σ is diagonal. Then all realizations satisfying $W_c = W_o$ are characterized by*

$$A = U^T A_b U, \quad B = U^T B_b, \quad C = C_b U. \tag{3.39}$$

Proof: First we note that, if (A, B, C) are related to (A_b, B_b, C_b) as in (3.39), then the controllability and observability Gramians of (A, B, C) are, respectively,

$$U^T W_c U = U^T \Sigma U \quad \text{and} \quad U^T W_o U = U^T \Sigma U,$$

and hence $W_c = W_o$. Conversely, any $(A, B, C) \in S_H$ is related to (A_b, B_b, C_b) by some similarity transformation T. We show that, if for some (A, B, C), $W_c = W_o$, then T must be orthogonal. Indeed, $W_c = W_o$ implies that $T^{-1}\Sigma T^{-T} = T^T \Sigma T$. This implies $T T^T \Sigma T T^T = \Sigma$, which in turn implies $T T^T = I$ by the uniqueness of the Singular Value Decomposition (SVD) of Σ. ∎

The following theorem yields some important properties of the realizations satisfying (3.31).

Theorem 3.4 : *If a realization (A, B, C) satisfies (3.31), then*

$$G(z) = |\rho| U^* F(z) \text{ for some orthogonal matrix } U^*$$

$$\text{and for } \rho = \sqrt{\frac{tr(W_o)}{tr(W_c)}}; \tag{3.40}$$

$$M_{L_{12}} = \bar{M}_{L_{12}} \quad \text{for this realization}. \tag{3.41}$$

Proof: Let (A, B, C) be an initial realization whose Gramians satisfy (3.31) and consider first the similarity transformation $T_1 = |\rho|^{-1/2} I$. Then

$$W_o^{(1)} = T_1^T W_o T_1 = |\rho| W_c, \quad W_c^{(1)} = T_1^{-1} W_c T_1^{-T} = |\rho| W_c.$$

By SVD decomposition, $|\rho|W_c = U\Sigma U^T$ with U orthogonal and Σ diagonal. Let $T_2 = U$, then one has

$$W_o^{(2)} = T_2^T W_o^{(1)} T_2 = \Sigma, \quad W_c^{(2)} = T_2^{-1} W_c^{(1)} T_2^{-T} = \Sigma,$$

which means that the corresponding realization (A_2, B_2, C_2) is internally balanced. The balanced forms are not unique and there exists one such balanced realization satisfying (see [Kun78], [WK83])

$$A_2 = QA_2^T Q, B_2 = QC_2^T \text{ for some signature matrix } Q.$$

It then follows that $F_2(z) = QG_2(z)$ for this realization. Denote $T = T_1 T_2 = |\rho|^{-1/2} U$, then $A = TA_2^T T^{-1}, B = TB_2, C = C_2 T^{-1}$ and hence

$$\begin{aligned} G(z) &= T^{-T} G_2(z) = T^{-T} Q F_2(z) \\ &= T^{-T} Q T^{-1} F(z) = |\rho|(UQU^T) F(z) \triangleq |\rho| U^* F(z). \end{aligned}$$

The second part follows from Theorem 3.1 on using the first part. ∎

So it is clear that the realizations defined by (3.31) satisfy

$$M_{L_{12}} = \bar{M}_{L_{12}} \geq (\sum_{i=1}^{n} \sigma_i)^2 + 2\sum_{i=1}^{n} \sigma_i, \qquad (3.42)$$

with equality if, in addition, they satisfy (3.32). In [Thi86] Thiele showed that the first term in the sensitivity measure $M_{L_{12}}$ of (3.14) is also lower bounded by $(\sum_{i=1}^{n} \sigma_i)^2$. That is,

$$\| \frac{\partial H}{\partial A} \|_1^2 \geq (\sum_{i=1}^{n} \sigma_i)^2. \qquad (3.43)$$

It follows that the lower bound on $M_{L_{12}}$ in (3.42) is valid for all realizations, not just those that satisfy the relationship (3.31). Since the lower bound is achieved (for both $\bar{M}_{L_{12}}$ and $M_{L_{12}}$) by those realizations that satisfy (3.32), we have the following remarkable result first obtained by Thiele.

Theorem 3.5 : *The realizations satisfying (3.32) minimize the sensitivity measure $M_{L_{12}}$.*

Proof: see [Thi86]. ∎

The outcome of this 'classical' sensitivity study is that we have a characterization of the state space realizations that minimize the L_1/L_2 sensitivity measure defined in (3.14). This characterization, as Theorem 3.5 shows, is via the relationship (3.32) on their Gramians. We have shown that such realizations always exist: we have shown in Lemma 3.1 how to construct a particular one, and in Theorem 3.3 how to obtain all of them. We shall return to sensitivity minimization problems in subsequent chapters using 'non classical' but probably more realistic sensitivity measures. We now continue our voyage through classical FWL filter design techniques by studying the roundoff noise gain of a digital filter and procedures to minimize it.

3.5 Roundoff noise gain of state space realizations

Limited wordlength effects on the signals cause another source of error on the output $y(t)$ of the realization (3.4), as we have seen in Chapter 2. This error, known as roundoff noise, is due to the fact that the signals are rounded off before or after each arithmetic operation. We have seen in Section 2.4 that the actual model of the realization (3.4), when implemented with FWL coefficients and with rounding of the signals before or after multiplication, can be written as

$$\begin{aligned} x^*(t+1) &= Ax^*(t) + Bu(t) - \eta^c(t) - \eta^s(t) \\ y^*(t) &= Cx^*(t) + du(t) - \zeta^c(t) - \zeta^s(t), \end{aligned} \qquad (3.44)$$

where $\{\eta^c(t), \zeta^c(t)\}$ are the noise (error) processes produced by the FWL effects on the coefficients while $\{\eta^s(t), \zeta^s(t)\}$ are white noise errors due to state roundoff (see expression (2.17)). In the representation (3.44), (A, B, C, d) are the exact (infinite wordlength) coefficient matrices.

Since the effects of FWL coefficient errors can be measured by the sensitivity measure studied in the previous subsection, we will assume in the sequel that the noise processes $\eta^c(t)$ and $\zeta^c(t)$ are zero and concentrate here on the roundoff effects. This means that the coefficients of the system are now assumed to be implemented exactly. In addition we assume that the input signal $u(t)$ can be described exactly within the available B_s-bit wordlength, i.e. $Q[u(t)] = u(t)$. This is often the case in practice when the input signal itself has been generated by a FWL device.

Denote
$$E(t) \triangleq x(t) - x^*(t) \text{ and } \Delta y(t) \triangleq y(t) - y^*(t). \tag{3.45}$$

It follows from (3.4) and (3.44) that
$$\begin{aligned} E(t+1) &= AE(t) + \eta^s(t) \\ \Delta y(t) &= CE(t) + \zeta^s(t) \end{aligned} \tag{3.46}$$

The steady state output error variance $\sigma^2_{\Delta y}$ can be easily shown to be
$$\sigma^2_{\Delta y} \triangleq \lim_{t \to \infty} E[\{\Delta y(t)\}^2] = \sum_{i=0}^{\infty} CA^i R^s_\eta (A^T)^i C^T + R^s_\zeta, \tag{3.47}$$

where
$$R^s_\eta \triangleq E[\eta^s(t)\eta^s(t)^T], \quad R^s_\zeta \triangleq E[[\zeta^s(t)]^2]. \tag{3.48}$$

For **RBM**[9] implementation, and given our assumption on the input signal wordlength, we have $\eta^s(t) = A^* e^x(t) \simeq A e^x(t)$ and $\zeta^s(t) = C^* e^x(t) \simeq C e^x(t)$ (see (2.19)). Hence
$$R^s_\zeta = CC^T \sigma^2, \quad R^s_\eta = AA^T \sigma^2 \tag{3.49}$$

where σ^2 is the variance of the unit roundoff error on every state: see (2.11).

For **RAM** implementation, $\eta^s(t)$ and $\zeta^s(t)$ depend on the number of elements that are neither 0 nor ± 1 in A, B, C and d. In this case,
$$R^s_\zeta = m_{n+1}\sigma^2, R^s_\eta = Q\sigma^2 \tag{3.50}$$

with σ^2 as above, where the state $x^*(t)$ is assumed to have a B_s-bit representation, m_{n+1} is the number of non trivial (i.e. not zero or one) elements in $(C \ d)$ and $Q = diag(m_1 \ m_2 \ldots m_n)$ with m_i the number of non trivial elements in the i-th row of the matrix $(A \ B)$.

So one has
$$\sigma^2_{\Delta y} = \begin{cases} \sum_{i=0}^{\infty} CA^i AA^T (A^i)^T C^T \sigma^2 + CC^T \sigma^2 \\ \quad = tr(W_o)\sigma^2 \quad \text{for } RBM \\ \\ \sum_{i=0}^{\infty} CA^i Q (A^i)^T C^T \sigma^2 + m_{n+1}\sigma^2 \\ \quad = [tr(W_o Q) + m_{n+1}]\sigma^2 \quad \text{for } RAM \end{cases} \tag{3.51}$$

---

[9] Remember that **RBM** denotes roundoff before multiplication while **RAM** denotes roundoff after multiplication.

where W_o is the observability Gramian defined in (3.21).

The roundoff noise gain is commonly defined as (see e.g. [Hwa77] and [Wil86])

$$G \triangleq tr(W_o). \qquad (3.52)$$

For a RBM implementation, the physical interpretation of G is quite clear. It is the ratio between the output error variance and the variance of the roundoff noise at each accumulation point, that is at each state component. For RAM implementation, the interpretation is not so obvious. When a realization is full of non trivial elements, then $Q = (n+1)I$. In this case

$$\frac{\sigma_{\Delta y}^2}{(n+1)\sigma^2} = tr(W_o) + 1,$$

and the interpretation of G is again clear. It is difficult to minimize the roundoff noise gain $tr(W_o Q) + m_{n+1}$ for RAM since Q depends on the number of non trivial elements in the realization, but experiments have shown [AS88] that the realization that minimizes the roundoff noise gain defined in (3.52) also gives a very good performance for RAM. Furthermore, starting from a fully parametrized realization that minimizes (3.52), one can further improve the performance by a succession of constrained similarity transformations that move a number of coefficients to 0 or ± 1. This procedure, due to Bomar and Hung [BH84] and further refined by Amit and Shaked [AS88], will be discussed in Chapter 8.

Whether roundoff is performed before or after multiplication, we shall in the sequel take (3.52) as our definition of the roundoff noise gain of a state space realization, thereby following common practice.

The problem of minimizing the roundoff noise gain G over all equivalent minimal state space realizations of $H(z)$ can be formulated as follows: minimize G over the set $(A, B, C) \in S_H$ of all equivalent realizations. That is,

$$\min_{(A,B,C) \in S_H} G \qquad (3.53)$$

Assuming that one starts from an arbitrary initial realization (A^0, B^0, C^0) and its corresponding Gramian W_o^0, and recalling that a similarity transformation T transforms W_o^0 into $T^T W_o^0 T$, then this optimization problem can be reformulated as

$$\min_{T:\det T\neq 0} tr(T^T W_o^o T). \qquad (3.54)$$

As such, the solution only depends on the choice of C and A. However, the problem (3.53), or alternatively (3.54), is ill-posed since G can be made arbitrarily small by choosing $T = \epsilon I$ with ϵ arbitrarily small.

To render the problem meaningful, an l_2-norm scaling is introduced on the input-to-state transfer functions to maintain the amplitudes of the different state components within the same dynamic range and thereby reduce the probability of overflow. As we have explained in Section 2.5, such constraint imposes conditions on the controllability Gramian (and hence on the input vector B) of the realization, namely

$$(W_c)_{ii} = 1, i = 1, \ldots, n.$$

This turns the minimization of the roundoff noise gain into a well posed constrained minimization problem, as we now explain.

3.6 Minimal roundoff noise gain realizations

We can now formulate the l_2-constrained minimal roundoff noise realization problem as follows :

$$\min_{(A,B,C)\in S_H} G \text{ or, equivalently, } \min_{(A,B,C)\in S_H} tr(W_o) \qquad (3.55)$$

subject to

$$(W_c)_{ii} = 1 \quad i = 1, \ldots, n. \qquad (3.56)$$

As a consequence of Theorem 3.2, we have the following result.

Theorem 3.6 : *A realization (A, B, C) minimizes G subject to the l_2-scaling constraints (3.56) if and only if its two Gramians are related by*

$$W_o = (\frac{1}{n}\sum_{i=1}^n \sigma_i)^2 W_c, \qquad (3.57)$$

where $\sigma_i^2 = \lambda_i(W_c W_o)$, $i = 1, \ldots, n$.

3. Parametrizations in digital system design

Proof: First we note that with $(W_c)_{ii} = 1$, $i = 1, \ldots, n$, (3.29) gives

$$G \triangleq tr(W_o) \geq \frac{1}{n}(\sum_{i=1}^{n} \sigma_i)^2. \tag{3.58}$$

According to Theorem 3.2, equality holds if and only if (3.31) holds. Since the subset of S_H that satisfies (3.31) is non-empty (see Lemma 3.1 for example), it follows that

$$G_{\min} = \frac{1}{n}(\sum_{i=1}^{n} \sigma_i)^2. \tag{3.59}$$

Therefore

$$\rho^2 = \frac{tr(W_o)}{tr(W_c)} = \frac{G_{\min}}{n} = (\frac{1}{n}\sum_{i=1}^{n} \sigma_i)^2, \tag{3.60}$$

from which the theorem follows. ∎

This result was first obtained by Mullis and Roberts in [MR76a], but it was Hwang who showed how to construct optimal realizations [Hwa77]. These roundoff noise optimal realizations are nowadays often called MRH structures. The minimum is achieved by a set of optimal realizations that are related by orthogonal similarity transformations, all of which satisfy the dynamic range constraint. Hwang's procedure shows how to construct this optimal realization set by starting from an initial realization (A^0, B^0, C^0). His procedure is described in the following result.

Theorem 3.7 *[Hwa77]: Given an arbitrary realization (A, B, C) of an n-th order digital transfer function, with Gramians W_c and W_o, the set of optimal realizations that minimize the roundoff noise gain G subject to the l_2 constraint (3.56) on W_c are obtained from this realization by the set of similarity transformations T^{opt} constructed as follows:*

$$T^{opt} = T_0 U X V^T \tag{3.61}$$

where

1. T_0 is a square root of W_c: $T_0 T_0^T = W_c$,

2. $X = diag(x_1, \ldots, x_n)$, where $x_i = (\frac{\sum_{k=1}^{n} \sigma_k}{n\sigma_i})^{1/2}$ for all i with $\{\sigma_i\}$ as defined before,

3. U is an orthogonal matrix such that $U^T(T_0^T W_o T_0)U$ is diagonal,

4. V is an orthogonal matrix such that all the diagonal elements of $VX^{-2}V^T$ are equal to one.

Proof: see [Hwa77]. ∎

The orthogonal matrix U can be obtained by the SVD decomposition of $T_0^T W_o T_0$. The existence of the orthogonal matrix V and an algorithm for computing it by a succession of $(n-1)$ Givens rotations are obtained from the following lemma.

Lemma 3.2 *[Hwa77]: Let M be an n-th order diagonal nonnegative definite matrix. Then there exists an orthogonal matrix V such that the diagonal elements of VMV^T are all equal to one if and only if $tr(M) = n$.*

Proof: Let $\{m_i\}$ be the diagonal elements of M with $m_i \geq 0$. Without loss of generality, we can assume that $m_1 < 1$ and $m_2 > 1$ since $tr(M) = n$. Consider the following Givens rotation [GL83], which is an orthogonal transformation:

$$V_1 = \begin{pmatrix} \cos\theta_1 & \sin\theta_1 \\ -\sin\theta_1 & \cos\theta_1 \end{pmatrix}. \tag{3.62}$$

With $\cos\theta_1 = (\frac{m_2-1}{m_2-m_1})^{1/2}$ and $\sin\theta_1 = (\frac{1-m_1}{m_2-m_1})^{1/2}$, one can see that

$$\begin{pmatrix} V_1 & 0 \\ 0 & I_{n-2} \end{pmatrix} M \begin{pmatrix} V_1 & 0 \\ 0 & I_{n-2} \end{pmatrix}^T = \begin{pmatrix} 1 & x^T \\ x & M_1 \end{pmatrix}, \tag{3.63}$$

where I_{n-2} is the unit matrix of order $n-2$, $x \in \mathbb{R}^{n-1}$, and M_1 is a $(n-1)$-th order diagonal nonnegative definite matrix with $tr(M_1) = n-1$. A similar transformation can be applied to this reduced order matrix M_1. Thus, the entire process ends in, at most, $n-1$ steps. This completes the proof. ∎

Comment: One of the remarkable properties of MRH structures is that their roundoff noise gain depends only on the Hankel singular values $\{\sigma_i\}$. Now it turns out that these singular values are not only invariants of the transfer function, but they are also invariant under so-called *frequency transformations*: see [RM87] for detail. These are transformations of $H(z)$ into $H(H_0(z))$, where $H_0(z)$ has the form:

$$H_0(z) = \pm \prod_{i=1}^{k} \left(\frac{z - \alpha_i^*}{1 - \alpha_i z} \right) \quad \text{for } |\alpha_i| < 1, \tag{3.64}$$

where α_i^* is the conjugate of α_i. By replacing z by $H_0(z)$, these frequency transformations allow one to change the passband of the original $H(z)$, and thus these transformations are very useful in filter design. The invariance mentioned above thus means that this roundoff noise gain is effectively independent of the filter's bandwidth. This is in contrast with the direct form realizations (such as the control canonical form, say) for which Mullis and Roberts have shown that the roundoff noise gain tends to infinity as the filter bandwidth goes to zero [MR76c]. This result suggests that the computation of minimal roundoff noise structures is most relevant in the case of narrow band filters, whether these are low pass or high pass. The roundoff noise gain goes to infinity as the poles of the filter cluster together. The same observation holds for the transfer function sensitivity, and - as we shall see in Chapter 6 - for the pole sensitivity measure.

We have seen in Section 3.4 that realizations that minimize the L_1/L_2 sensitivity, without any constraint on the dynamic range of their state components, must have identical Gramians. In this section we have shown that realizations that minimize the roundoff noise gain (and this problem is only well posed if state constraints are imposed) must have Gramians that are proportional, with a proportionality constant that is determined by the filter transfer function.[10] We shall now observe that, when an l_2-constraint is imposed on the states, the realizations that minimize the roundoff noise gain also minimize the upper bound of the L_1/L_2 sensitivity measure.

3.7 Relationship between sensitivity measure and roundoff noise gain.

It follows from (3.22) and (3.52) that, *under the dynamic range constraint* $(W_c)_{ii} = 1, i = 1, \ldots, n$, the following relationship holds:

$$\bar{M}_{L_{12}} = (1+n)tr(W_o) + n = (1+n)G + n. \tag{3.65}$$

Since $\bar{M}_{L_{12}}$ and G are both positive quantities, it follows from this relationship that, under the dynamic range constraint, the sensitivity upper bound $\bar{M}_{L_{12}}$ and the roundoff noise gain G are simultaneously minimized, i.e. they are minimized by the subset of S_H characterized by the relation (3.57) between their two Gramians.

---

[10] We recall that the Hankel singular values σ_i are invariants of $H(z)$.

We can thus summarize the main optimal realization results of this chapter as follows.

Result 1 : The sensitivity measure $M_{L_{12}}$ and its upper bound $\bar{M}_{L_{12}}$ are both minimized by the same subset of realizations $(A, B, C) \in S_H$. These realizations are completely characterized by the identity between their controllability and observability Gramians, that is by $W_c = W_o$. This set is non-empty; in particular, it includes the internally balanced forms. For those realizations that satisfy $W_c = W_o$, the L_1/L_2 coefficient sensitivity measure, $M_{L_{12}}$, is equal to its upper bound, $\bar{M}_{L_{12}}$. Such realizations thus have the property that errors in the implementation of their coefficients produce minimal errors in the resulting transfer function as measured by the L_1/L_2 measure.

Result 2 : For those realizations $(A, B, C) \in S_H$ whose state components have been l_2-scaled such that $(W_c)_{ii} = 1, i = 1, \ldots, n$, the relation (3.65) holds between the roundoff noise gain, G, and the upper bound of the sensitivity, $\bar{M}_{L_{12}}$. The roundoff noise gain, G, is a measure of the gain between the variance of the roundoff error at the accumulator nodes and the variance of the output error due to this roundoff noise. Within this subset of l_2-constrained realizations, this roundoff noise gain G (and thus also $\bar{M}_{L_{12}}$) is minimized for those realizations that satisfy $W_o = (\frac{1}{n}\sum_{i=1}^{n} \sigma_i)^2 W_c$. Again, this set is non-empty.

The roundoff noise error is typically the more serious Finite Word Length error, assuming that sufficient precautions have been taken to avoid overflow errors. It is then a pleasing outcome of the analysis of this chapter to observe that, when l_2 constraints are imposed on the state of a realization - and the purpose of this l_2 normalization is to maintain the different state components within the same dynamic range - the minimization of the roundoff noise gain simultaneously leads to realizations that minimize a reasonable upper bound of the sensitivity. This pleasing result is obtained provided the L_1/L_2 coefficient sensitivity measure is adopted. We note that, because this connection between the roundoff noise gain and the sensitivity upper bound is established for the case of l_2-constrained realizations only, the realizations that minimize the roundoff noise gain do not minimize the sensitivity measure itself, even among the set of these constrained realizations. However, since $\bar{M}_{L_{12}}$ upper bounds the actual sensitivity measure $M_{L_{12}}$ only through the Cauchy Schwartz inequality, it is reasonable to assume that, among the l_2-constrained realizations, those that minimize the roundoff noise gain also achieve a good sensitivity behaviour.

3.8 Examples and simulations.

In this section, we present several examples and simulations to illustrate the theoretical results of this chapter. All the computations and simulations have been done in MATLAB.[11]

Example 3.1: This example has been used in [Hwa77]. It is a low pass filter whose transfer function is given by

$$H(z) = \frac{0.01594(z+1)^3}{z^3 - 1.9749z^2 + 1.5562z - 0.4538}.$$

It is chosen because it has a triple zero at $z = -1$ and so the zero positions are very sensitive to the coefficients when realized directly. The state space realization in control canonical form, denoted R_c, is given as follows:

$\boxed{R_c :}$

$$A_c = \begin{pmatrix} 1.9749 & -1.5562 & 0.4538 \\ 1 & 0 & 0 \\ 0 & 1 & 0 \end{pmatrix} \quad B_c = \begin{pmatrix} 1 \\ 0 \\ 0 \end{pmatrix}$$

$$C_c = (0.0793 \ 0.0230 \ 0.0232), \quad d = 0.0159.$$

Taking this form as an initial realization, and using (2.33) and (3.20), we can compute the corresponding Gramians, denoted W_c^c and W_o^c, and hence the sensitivity upper bound and the roundoff noise gain.

We also compute the l_2-scaled control canonical realization, denoted R_c^{sc}. This realization can be obtained from R_c by simply applying a diagonal similarity transformation $D_{sc} = \text{diag}\{W_c^c(i,i)^{1/2}\}I$ to the realization R_c.

---

[11]MATLAB is a product of The Mathworks, Inc. It uses double precision - a 60 bit floating point processor for PC's - but the results are quoted only to 4 significant digits for the fractional part.

$\boxed{R_c^{sc}:}$

$$A_c^{sc} = \begin{pmatrix} 1.9749 & -1.5562 & 0.4538 \\ 1 & 0 & 0 \\ 0 & 1 & 0 \end{pmatrix} \quad B_c^{sc} = \begin{pmatrix} 0.2421 \\ 0 \\ 0 \end{pmatrix}$$

$C_c^{sc} = (0.3276\ 0.0951\ 0.0957).$

The optimal designs can be done with the procedures given in the previous sections. We present two optimal realizations: the first one minimizes the L_1/L_2 sensitivity among unscaled realizations and is denoted by $R^{opt}(M_{L_{12}})$; the second one, denoted $R^{opt}(G)$, minimizes the roundoff noise gain among all l_2-scaled realizations.

$\boxed{R^{opt}(M_{L_{12}}):}$

$$A^{opt}(M_{L_{12}}) = \begin{pmatrix} 0.8236 & 0.3999 & -0.0165 \\ -0.3999 & 0.5935 & 0.3425 \\ -0.0165 & -0.3425 & 0.5577 \end{pmatrix}$$

$B^{opt}(M_{L_{12}}) = (0.4424\ 0.3799\ 0.1671)^T$

$C^{opt}(M_{L_{12}}) = (0.4424\ -0.3799\ 0.1671)$

$\boxed{R^{opt}(G):}$

$$A^{opt}(G) = \begin{pmatrix} 0.6282 & -0.0345 & 0.4719 \\ -0.1061 & 0.6711 & 0.1050 \\ -0.5786 & 0.1192 & 0.6757 \end{pmatrix}$$

$B^{opt}(G) = (0.6400\ -0.6141\ 0.0479)^T$

$C^{opt}(G) = (-0.1444\ -0.2989\ -0.2478).$

We now compare the sensitivity upper bounds and the roundoff noise gains of these four realizations.

A. For the unscaled realizations:

$$R_c: \qquad \bar{M}_{L_{12}} = 81.9891 \qquad G = 0.5903$$

$$R^{opt}(M_{L_{12}}): \quad \bar{M}_{L_{12}} = 4.7560 \qquad G = 1.3992$$

B. For the l_2-scaled realizations:

$$R_c^{sc}: \qquad \bar{M}_{L_{12}} = (n+1)G + n = 43.2840 \qquad G = 10.0710$$

$$R^{opt}(G): \quad \bar{M}_{L_{12}} = (n+1)G + n = 5.6104 \qquad G = 0.6526$$

These computations confirm the theoretical results. The sensitivity upper bound $\bar{M}_{L_{12}}$ of the optimal realization $R^{opt}(M_{L_{12}})$ (which, for this optimal realization, is identical to the sensitivity $M_{L_{12}}$ itself), is drastically smaller than that of the canonical realization, and the same holds true for the roundoff noise gains of the optimal and canonical l_2-scaled realizations. In addition, we observe that the realization that minimizes the roundoff noise gain, $R^{opt}(G)$, has a sensitivity upper bound that is close to the sensitivity of the optimal sensitivity minimizing realization $R^{opt}(M_{L_{12}})$. This confirms our concluding remarks of the previous section. It is worth noting, however, that the converse is not necessarily true: in this example the unscaled realization that minimizes the sensitivity has a larger roundoff noise gain than the unscaled control canonical form.

Since the comparison between optimal realization and canonical form is based not on the sensitivity measure $M_{L_{12}}$ but on an upper bound $\bar{M}_{L_{12}}$, it is important to evaluate whether the superiority of the realization $R^{opt}(M_{L_{12}})$ over the realization R_c carries through when the actual sensitivity measure is evaluated rather than its upper bound. To test this, simulations have been performed in which the fractional part of every coefficient in the realizations R_c and $R^{opt}(M_{L_{12}})$ have been truncated to B_c bits. The actual frequency responses of each of these B_c-bit realizations have been computed and compared with the frequency response of the ideal (infinite wordlength) filter realization. The simulation results are given in Figures 3.1 and 3.2 for $B_c = 6$ and

$B_c = 8$ bit truncations, respectively.

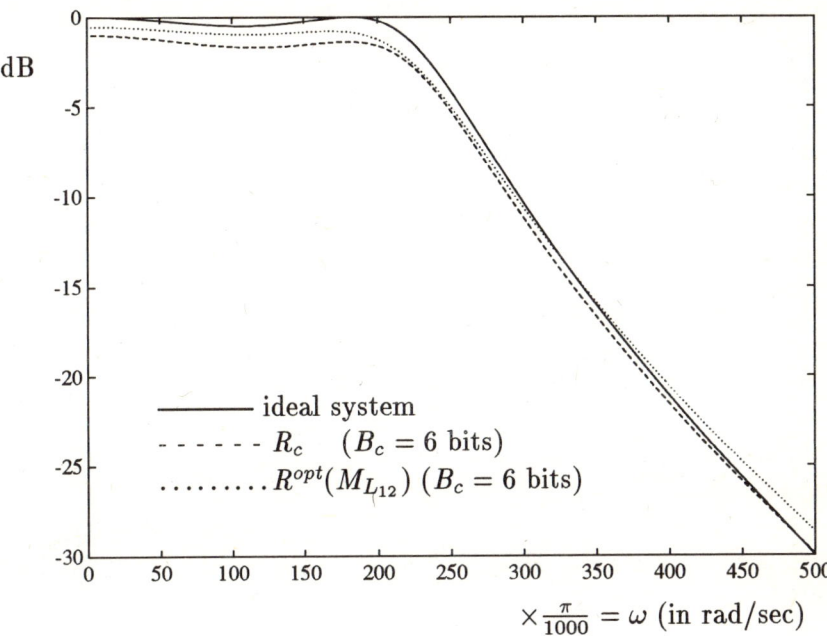

FIGURE 3.1. Magnitude frequency responses of ideal system, of R_c and of $R^{opt}(M_{L_{12}})$

Comments:

1. Figures 3.1 and 3.2 show the superiority of the optimal sensitivity realization $R^{opt}(M_{L_{12}})$ over the controllable realization R_c (also called direct form). With $B_c = 8$ bits the optimal realization $R^{opt}(M_{L_{12}})$ yields a response very close to that of the exact transfer function: see Figure 3.2.

2. The difference between the performance of the truncated optimal and controllable realizations is not very important for this example. The poles of this filter are at 0.6579, $0.6585 \pm j0.5061$; they are well separated from one another. This means that the pole sensitivity to the perturbations of the coefficients is low, while in addition the pole behaviour dominates the frequency response characteristics of this filter except around $z = -1$. We will discuss the concept of pole sensitivity thoroughly in Chapter 6.

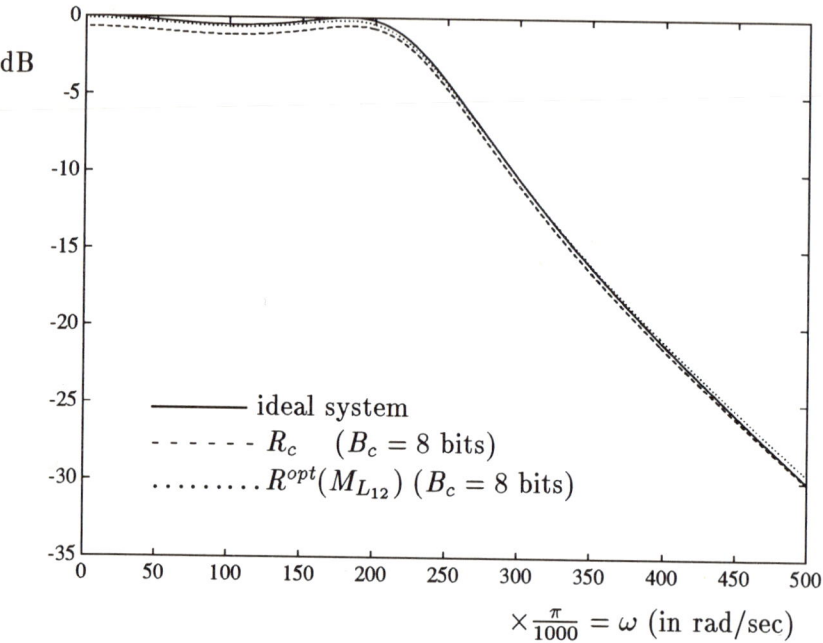

FIGURE 3.2. Magnitude frequency responses of ideal system, of R_c and of $R^{opt}(M_{L_{12}})$

Example 3.2: This is a sixth-order narrow band low pass filter with a normalized sampling frequency $f_s = 1$ and filter design parameters $f_p = 0.03125$ (passband frequency), $F_s = 0.0390625$ (stopband frequency), and $\epsilon_p = 1\ dB$ (passband ripple). The upper limit of the frequency response is $0\ dB$, while the stopband attenuation is $46.68 dB$. This example has been used in [Wil86]. The poles are very close together and so - anticipating on the analysis of Chapter 6 - we can expect sensitivity to coefficients.

As in Example 3.1, we consider the four realizations R_c, R_c^{sc}, $R^{opt}(M_{L_{12}})$ and $R^{opt}(G)$. They are given as follows, respectively:

$\boxed{R_c :}$

$$A_c = \begin{pmatrix} 5.6526 & -13.3818 & 16.9792 & -12.1765 & 4.6789 & -0.7526 \\ 1 & 0 & 0 & 0 & 0 & 0 \\ 0 & 1 & 0 & 0 & 0 & 0 \\ 0 & 0 & 1 & 0 & 0 & 0 \\ 0 & 0 & 0 & 1 & 0 & 0 \\ 0 & 0 & 0 & 0 & 1 & 0 \end{pmatrix}$$

$$B_c = (1\ 0\ 0\ 0\ 0\ 0)^T \qquad d = 0.4708 \times 10^{-2}$$

$$C_c = (0.1511\ -0.4558\ 0.3855\ 0.116\ -0.3074\ 0.1165) \times 10^{-2}.$$

$\boxed{R_c^{sc}:}$

$$A_c^{sc} = A_c \qquad B_c^{sc} = (0.2455 \times 10^{-4}\ 0\ 0\ 0\ 0\ 0)^T$$

$$C_c^{sc} =$$
$$(61.5364\ -185.6992\ 157.0320\ 45.4684\ -125.2118\ 47.4582).$$

$\boxed{R^{opt}(M_{L_{12}}):}$

$$A^{opt}(M_{L_{12}}) =$$
$$\begin{pmatrix}
0.9878 & 0.1123 & -0.0063 & 0.0275 & 0.0088 & -0.0057 \\
-0.1123 & 0.9709 & 0.1354 & -0.0195 & -0.0221 & 0.0090 \\
-0.0063 & -0.1354 & 0.9530 & 0.1163 & 0.0282 & -0.0191 \\
-0.0275 & -0.0195 & -0.1163 & 0.9264 & -0.1142 & 0.0336 \\
0.0088 & 0.0221 & -0.0282 & 0.1142 & 0.9095 & 0.1212 \\
0.0057 & 0.0090 & 0.0191 & 0.0336 & -0.1212 & 0.9041
\end{pmatrix}$$

$$B^{opt}(M_{L_{12}}) =$$
$$(0.1121\ 0.1543\ 0.1901\ 0.1783\ -0.1072\ -0.0553)^T$$

$$C^{opt}(M_{L_{12}}) =$$
$$(0.1121\ -0.1543\ 0.1901\ -0.1783\ -0.1072\ -0.0553)$$

$\boxed{R^{opt}(G):}$

$$A^{opt}(G) =$$
$$\begin{pmatrix}
0.9476 & -0.0749 & -0.0972 & 0.0850 & 0.0826 & -0.0798 \\
0.0796 & 0.9434 & 0.0857 & -0.0049 & -0.0175 & -0.1443 \\
0.0765 & -0.0161 & 0.9352 & 0.0397 & -0.1127 & -0.0246 \\
-0.0362 & 0.0185 & -0.0029 & 0.9219 & 0.0320 & -0.0444 \\
-0.1011 & -0.0154 & 0.1194 & 0.0506 & 0.9476 & -0.0013 \\
0.0529 & 0.1630 & 0.0062 & 0.0660 & 0.0291 & 0.9569
\end{pmatrix}$$

$$B^{opt}(G) =$$
$$(0.0388\ 0.1570\ -0.2601\ -0.3596\ 0.0774\ 0.1520)^T$$
$$C^{opt}(G) =$$
$$(-0.1760\ 0.1047\ -0.0596\ 0.0181\ -0.0347\ -0.0947)$$

The same computations as before produce the following results.

A. For the unscaled realizations:

$R_c:$ $\qquad \bar{M}_{L_{12}} = 1.2836 \times 10^{11}$ $\qquad G = 11.8902$

$R^{opt}(M_{L_{12}}):$ $\bar{M}_{L_{12}} = 13.6537$ $\qquad G = 2.8280$

B. For the l_2-scaled realizations:

$R_c^{sc}:$ $\qquad \bar{M}_{L_{12}} = 1.3814 \times 10^{11}$ $\qquad G = 1.9734 \times 10^{10}$

$R^{opt}(G):$ $\bar{M}_{L_{12}} = 15.3306$ $\qquad G = 1.3329$

Note that, for this example, the theoretical results show an enormous difference between the sensitivity upper bounds of the optimal realizations and those of the controllable realizations. The same holds true for the comparison of the roundoff noise gains. Once again, the realization that minimizes the roundoff noise gain, $R^{opt}(G)$, has a sensitivity upper bound that is very close to the sensitivity of the optimal sensitivity minimizing realization, $R^{sc}(M_{L_{12}})$. This will be confirmed by the corresponding simulation results.

Simulations have again been performed to confirm that the upper bounds $\bar{M}_{L_{12}}$ used for comparing these different realizations are a good reflection of the actual sensitivities of the transfer functions with respect to parameter errors. These simulations have been performed identically to those of Example 3.1. The results are presented in Figures 3.3 and 3.4.

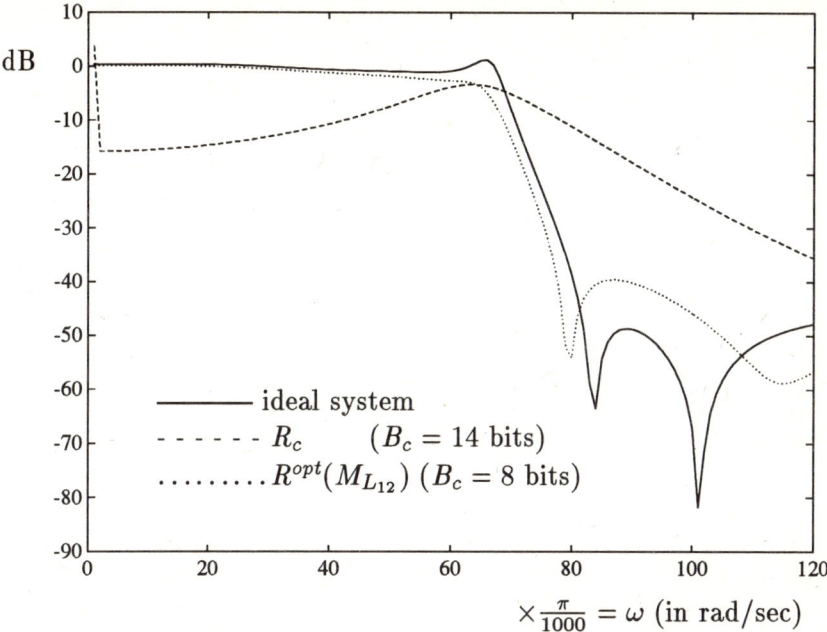

FIGURE 3.3. Magnitude frequency responses of ideal system, of R_c and of $R^{opt}(M_{L_{12}})$

Comments:

1. In Figure 3.3 one observes that even with $B_c = 16$ bits the FWL controllable realization R_c yields a frequency response that is far from the desired one, while with only $B_c = 8$ bits the optimal FWL realization $R^{opt}(M_{L_{12}})$ gives a satisfactory consistency with the ideal frequency response of the filter.

2. With $B_c = 12$ bits the optimal realization yields almost the exact frequency response. In fact, it gives a performance that is even better than that of the FWL controllable realization with 24 bits. This means that, for comparable performances, the optimal realization could save 12 bits compared with the canonical (or direct form) realization.

3. Finally note that the controllable realizations R_c and R_c^{sc} have a very large coefficient dynamic range. The ratios between the maximum and minimum absolute values of the coefficients in R_c and R_c^{sc} are larger than 10^4, while for the optimal realizations $R^{opt}(M_{L_{12}})$ and $R^{opt}(G)$, these ratios are of the order of 10^2. This

76   3. Parametrizations in digital system design

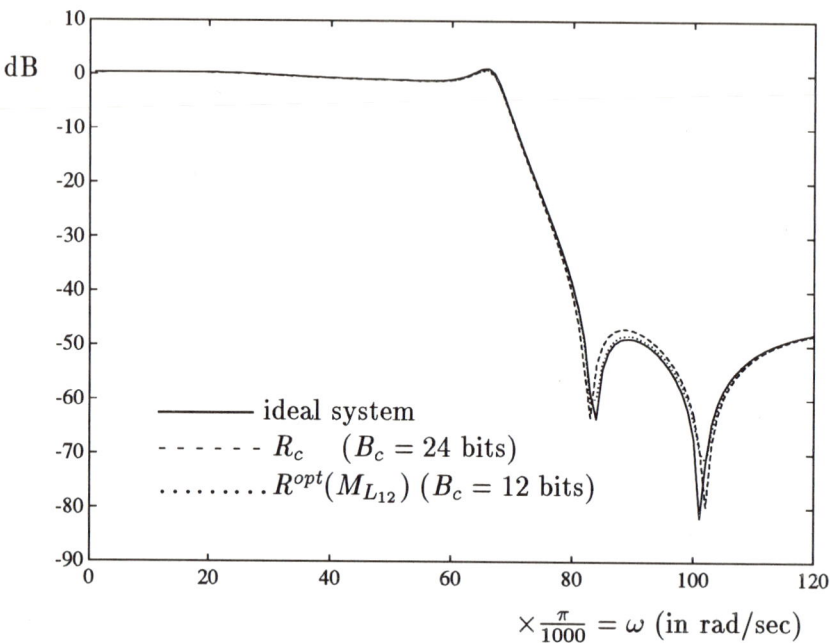

FIGURE 3.4. Magnitude frequency responses of ideal system, of R_c and of $R^{opt}(M_{L_{12}})$

is a typical feature of direct form realizations and is an additional reason for their poor sensitivity performance.

3.9 Alternative approaches

We have focused our attention so far on the characterization and computation of state variable realizations that achieve either a minimal L_1/L_2 sensitivity with respect to parameter errors or a minimal round-off noise gain to arithmetic computation errors. This way of addressing the FWL filter design problem as an optimal state variable realization problem has two consequences:

- The minimization of the FWL performance criteria defined in this chapter generically leads to fully parametrized solutions for A, B, C, d. This means that the computation of one output sample of an $n-th$ order filter typically requires $n^2 + 2n + 1$ scalar multiplications, whereas a direct form realization requires only $2n + 1$ multiplications. An improvement in FWL performance is therefore achieved at the expense of a reduction of the computational speed of the filter. In situations where FWL errors are not critical, it may therefore be desirable to search for suboptimal solutions that attempt a compromise between accuracy and speed of computation.

- By posing the problem as an optimization problem over the set S_H of 'equivalent' (in the infinite precision sense) state space realizations defined in (3.6), we have limited our search to this set. Alternative solutions to the design of digital filter realizations with low FWL errors can be obtained by widening the search to sets of realizations that do not belong to S_H.

We now briefly discuss these two observations and examine alternative ways of obtaining low FWL error realizations.

Accuracy versus speed of computations.

Since speed of computation is often a critical issue, a considerable effort has been made to obtain sparse realizations, i.e. realizations with a small number of real coefficients which, therefore, require a small number of multiplications per output sample computation. The sparsest realizations are of course the canonical or direct forms: they contain exactly the $2n$ parameters that appear in the transfer function and no more. However, we have seen that they have very poor FWL performance.

First it must be observed that, in both optimization problems discussed in this chapter, the optimal solutions are a *set* of realizations, thereby leaving some degrees of freedom that can possibly be used to force some coefficients to take zero, one or integer values. Thus optimal realizations can often be obtained that have fewer than $n^2 + 2n + 1$ real coefficients. In particular, this is always the case for realizations that minimize the L_1/L_2 sensitivity measure, as we shall show in Chapter 8. These sparse optimal realizations consist of Hessenberg or Schur forms which, typically, have a triangular or block triangular structure.

Further reductions in the number of real coefficients can be achieved by considering *suboptimal state space realizations*, where the term suboptimal refers to their FWL performance. Probably the easiest of such structures are the parallel and cascade realizations, in which a transfer function is decomposed, respectively, as a sum or a product of first and/or second order transfer functions. These individual first and second order transfer functions can then be realized either as direct forms (i.e. controllable or observable realizations), or, better, as optimal FWL realizations.

The implementation of an n-th order filter as a cascade or parallel connection of second and - possibly - first order sections does lead to a reduction of the roundoff noise gain compared to a straight direct form implementation of the n-th order filter even when the individual sections are realized in direct form. This is because closely spaced poles and zeros (which cause high roundoff noise gain and high sensitivity) can then be separated into different sections of the filter realization. A cascade or parallel implementation using direct forms in the individual sections leads to $2.5n + 1$ multiplications per output sample computation. From a FWL error performance point of view, a relatively cheap improvement can be obtained by optimizing the realizations of the second order filters for roundoff noise minimization, say. The cost in computations per output sample is then slightly increased: it is of the order of $4n$ for both parallel and cascade realizations. The reduction in roundoff noise gain, say, obtained by optimizing the individual second order realizations vis-à-vis a direct form implementation of these second order filters can be very significant. A detailed example is provided in [RM87]. In Chapter 8 we provide a further analysis of cascade and parallel realizations.

Besides parallel and cascade realizations, which are very easy to compute, other more sophisticated suboptimal realizations have been proposed and analysed, such as lattice filters and orthogonal filters. The number of multiplications per output sample computation is of the order of $4n$ for lattice filters, and $8n$ for orthogonal filters. Details about lattice filters and their noise properties can be found in [MG75], while [RM87] provides a good analysis of orthogonal filters.

Realizations outside the set S_H.

One way to extend the set of low FWL error realizations beyond the set S_H of realizations that have transfer function $H(z)$ is to replace the input-output description $H(z)$ of the digital filter by another equivalent representation. This can be done by replacing the shift operator z^{-1} that is used in the definition (3.3) of the input-output relation by some other operator. This has the effect of replacing the transfer function

$H(z)$ by an equivalent description, say $H_\delta(\delta)$, where δ is a polynomial function of z. The search for optimal realizations can then be performed over a new set of realizations that do not belong to S_H, but to a new set, say S_{H_δ}, of equivalent realizations. These new input-output descriptions can be thought of as generalized polynomial parametrizations, in which the basis for the description of polynomial operators is not z and the powers of z, but some polynomial function of z and its powers.

In the simplest case of a first order polynomial, this modification of the basis functions amounts to translating the origin of the complex plane from $z = 0$ to some other position. This is particularly useful from a FWL implementation point of view when the poles or zeros of a filter are not clustered around zero (as they seldom are in practice) but around some other value in the complex plane. By choosing this value as the new origin for the description of transfer functions one reduces the dynamic range that is necessary to describe the poles and zeros of the filter, thereby allowing for a greater accuracy with a given coefficient wordlength.

Generalized polynomial parametrizations will be discussed in Chapter 11 and 12. A well known special case of such generalized parametrizations is obtained by replacing z^{-1} by $z^{-1} = \frac{\delta^{-1}}{\Delta + \delta^{-1}}$ in (3.3), where Δ is some fixed integer number and δ is the new independent variable. This transformation can also be written as $\delta = \frac{z-1}{\Delta}$. With this substitution one can then write

$$H(z) = \frac{\sum_{i=0}^n b_i z^{-i}}{1 + \sum_{i=1}^n a_i z^{-i}} = \frac{\sum_{i=0}^n \beta_i \delta^{-i}}{\sum_{i=0}^n \alpha_i \delta^{-i}} \triangleq H_\delta(\delta). \qquad (3.66)$$

A pole at $z = 1$ in the transfer function $H(z)$ is replaced by a pole at $\delta = 0$ in the new transfer function $H_\delta(\delta)$. The search for optimal realizations can then be performed over the set S_δ of state space realizations of $H_\delta(\delta)$: these have been called δ-operator realizations [MG86]. With the special choice $\Delta = 1$, a δ-operator realization of a state equation effectively corresponds to replacing the classical computation

$$x(t+1) = Ax(t) + Bu(t)$$

by

$$x(t+1) = x(t) + (A - I)x(t) + Bu(t). \qquad (3.67)$$

When the poles of a filter are clustered around $z = 1$, as is typically the case for narrow band low pass filters or when the filter has been obtained from a continuous time system using fast sampling, the use of the δ transformation and of the corresponding δ realizations will produce lower sensitivity to coefficient errors whether or not optimal δ-operator realizations are used. This observation was apparently first made by Agarwal and Burrus [AB75] who called such structures *delay replaced* structures and advocated the use of delay replaced direct forms[12] to reduce the sensitivity to coefficient errors for such filters. The idea of delay replacement was further developed by Orlandi and Martinelli [OM84] and by Williamson [Wil88], until Middleton and Goodwin became prime advocates of such scheme, not just for numerical reasons but for the purpose of a unified development of continuous time and discrete time theories of filtering and control: see [MG86] and [MG90]. They coined the term δ-operator. In Chapter 11 we shall give a detailed analysis of δ-operator realizations, derive optimal δ-operator realizations and compare their FWL properties with those of the more classical shift operator realizations.

One procedure advocated for the reduction of roundoff errors at arithmetic operations is that of *integer gain residue feedback*. Consider, for example, the model (2.8) obtained with roundoff before multiplication (RBM), and assume for simplicity that the input signal $u(t)$ has B_s bits so that $Q[u(t)] = u(t)$. The model (2.8) can then be rewritten as:

$$x^*(t+1) = \beta^* Q[x^*(t)] + \mu^* u(t). \qquad (3.68)$$

Recall that in an RBM implementation the signal $x^*(t)$ of $(B_s + B_c)$ bits is rounded off to a B_s-bit signal $Q[x^*(t)]$. Now, defining the error

$$\epsilon(t) \triangleq x^*(t) - Q[x^*(t)]$$

we can then replace the computation (3.68) by the following scheme involving state error feedback:

$$x^*(t+1) = \beta^* Q[x^*(t)] + \mu^* u(t) + \Gamma \epsilon(t), \qquad (3.69)$$

where Γ is chosen as an integer coefficient gain. This scheme has been proposed and extensively investigated by Williamson (see [Wil86] and [Wil91]). The most common implementation of this residue feedback scheme is with identity feedback, i.e. $\Gamma = I$. Now, in such case equation (3.69) is replaced by

$$x^*(t+1) = x^*(t) + (\beta^* - 1)Q[x^*(t)] + \mu^* u(t). \qquad (3.70)$$

---

[12]Optimal realizations had not yet been developed in 1975.

Comparing with the expression (3.67), we note that identity residue feedback is identical to replacing the classical shift operator realization by a δ-operator realization. We shall return to these issues in Chapter 11.

3.10 Conclusions

In this chapter we have provided an analysis of FWL errors in state variable descriptions of a filter and we have reviewed the main 'classical' results on ways to minimize these errors. The minimization of FWL errors can essentially be reduced to an optimization problem over the set of similar (that is, equivalent) state space realizations. This way of attacking the problem is able to encompass several FWL performance criteria, such as transfer function sensitivity minimization and roundoff noise gain minimization. In this introductory chapter covering 'classical' material our attention has been focused on one particular L_1/L_2 sensitivity measure, and on one commonly accepted roundoff noise gain measure valid for l_2-constrained state space realizations. In the remainder of this book we extend these optimization techniques to a range of other FWL performance criteria and consider other sets of admissible state space realizations. We also address the FWL design problem not just for filter design but for control and identification problems as well.

We have illustrated the effectiveness of the optimal realizations through several simulations. Our simulations and calculations clearly establish that the direct (or canonical) forms, which are by far the simplest to implement and yield the greatest computational speed, exhibit a particularly poor performance with respect to FWL errors, whether it be errors in the coefficients or roundoff errors. This is particularly true when poles or zeros are clustered together, as is the case in narrow band filters or filters obtained from sampling continuous systems at high sampling rates. For a filter bandwidth f_m, say, the ratio $\frac{f_m}{f_s}$ is a good measure of the clustering of the poles, where f_s is the sampling frequency. Thus the FWL effects are particularly severe for narrow band filters, that is filters for which the ratio $\frac{f_m}{f_s}$ is small. In Chapter 6 we shall again observe the role of the pole locations and the adverse effect of closely clustered poles when we examine the sensitivity of poles and zeros with respect to parameter errors.

The time is ripe now to venture into some new material.

82   3. Parametrizations in digital system design

Appendix 3: Proof of Theorem 3.2.

Let W_c and W_o be the Gramians of some realization (A, B, C) in S_H and let T be a nonsingular similarity transformation matrix. Denote

$$W_c(T) \triangleq T^{-1}W_c T^{-T}, \quad W_o(T) \triangleq T^T W_o T$$
$$S(T) \triangleq tr(W_c(T))tr(W_o(T)). \qquad (3.A.1)$$

Evidently, the following inequality holds

$$tr(W_c)tr(W_o) \geq \min_{T:detT\neq 0} S(T).$$

We note that $S(T) = tr(W_c P^{-1})tr(W_o P) \triangleq R(P)$ with $P = TT^T$. Clearly, one has

$$\min_{T:detT\neq 0} S(T) \iff \min_{P>0} R(P).$$

Using the same techniques as in the proof of Lemma 4.A in Appendix 4.A, one can show that the minimum of $R(P)$ exists and that it can be achieved only by nonsingular P. So, such an optimal P must satisfy the following necessary condition:

$$\frac{dR(P)}{dP} = tr(W_c P^{-1})W_o - tr(W_o P)P^{-1}W_c P^{-1} = 0.$$

That is,

$$PW_o P = \frac{tr(W_o P)}{tr(W_c P^{-1})} W_c, \qquad (3.A.2)$$

which means

$$W_o(T) = \frac{tr(W_o(T))}{tr(W_c(T))} W_c(T). \qquad (3.A.3)$$

Lemma 3.1 shows the existence of an internally balanced realization, that is obtained by a similarity transformation T_b and that satisfies $W_c(T_b) = W_o(T_b) = \Sigma$, which is evidently a solution of (3.A.3). We note that $S(T_b) = (\sum_{i=1}^n \sigma_i)^2$. We now show that any $W_o(T)$ and $W_c(T)$ that satisfy (3.31) necessarily satisfy (3.A.3) and that the corresponding $S(T)$ are all identical and equal to $S(T_b)$. This will complete the proof of the first part of Theorem 3.2.

Let $W_o(T) = \rho^2 W_c(T)$ with $\rho^2 = tr(W_o)/tr(W_c)$. It then follows that $W_o(\rho^{-1/2}T) = W_c(\rho^{-1/2}T)$. Consider the singular value decomposition $W_o(\rho^{-1/2}T) = W_c(\rho^{-1/2}T) = QDQ^T$. Since

$$\lambda_i\{W_c(\rho^{-1/2}T)W_o(\rho^{-1/2}T)\} = \lambda_i\{W_c W_o\} = \sigma_i^2, i = 1,\ldots,n,$$

it follows that $D = \Sigma$. Finally, since $S(T)$ is unaffected by a scalar transformation, we have $S(T) = S(\rho^{-1/2}T) = (\sum_{i=1}^{n}\sigma_i)^2 = S(T_b)$. We conclude in particular that all solutions T of (3.A.3) yield the same value $S(T)$ which is therefore minimal.

To prove the second part, note that

$$tr(W_c) + tr(W_o) \geq 2\sqrt{tr(W_c)tr(W_o)} \geq 2\sum_{i=1}^{n}\sigma_i. \qquad (3.A.4)$$

The second inequality has been proved in the first part of this proof; equality holds if and only if (3.A.3) is satisfied. The first inequality is a geometric inequality; equality holds if and only if $tr(W_c) = tr(W_o)$. Therefore $tr(W_c) + tr(W_o) = 2\sum_{i=1}^{n}\sigma_i$ if and only if $W_c = W_o$. ∎

4

Frequency weighted optimal design

4.1 Introduction

In Chapter 3 we have reviewed 'classical' results on Finite Word Length state space digital filter design. Somewhat surprisingly, classicism in this area effectively refers to a period of development that only started in the mid-seventies for the results on minimizing the filter's round-off noise gain, while the 'classical' results on the minimization of the transfer function sensitivity date back only to the mid-eighties. The sensitivity minimization nut was cracked in [Thi84] because the author adopted a rather specific sensitivity measure, first suggested in [TT84], that lends itself to a tractable minimization problem. This sensitivity measure combines an L_1 norm on the sensitivity of the transfer function with respect to the parameters of the A matrix, with an L_2 norm on the sensitivity of the transfer function with respect to the parameters of the B and C matrices. We therefore called this an L_1/L_2 sensitivity measure. As is often the case in the development of scientific progress, Thiele's ground breaking contribution set the tone and his L_1/L_2 measure was initially adopted without much questioning by all other researchers in the field in their subsequent developments, including the authors of the present opus.

The L_1/L_2 sensitivity measure of Tavsanoglu and Thiele has several drawbacks.

- The sensitivity of the transfer function with respect to the parameters is considered to be equally important at all frequency points. Clearly, in most practical applications one is interested in the performance of the filter in some frequency bands more than in others.

- The rather illogical combination of an L_1 norm with respect to some parameters and an L_2 norm with respect to others is opportunistic rather than intuitive and reasonable. The L_1/L_2 sensitivity measure derives most of its bonus points from the fact that it can be easily optimized with respect to the choice of coordinates.

- The L_1/L_2-optimal realizations minimize a measure of the effects of coefficient errors on the transfer function, whereas in some applications the effects on other filter characteristics - such as, say, the pole locations - are more relevant.

- The result of this optimization yields a set of optimizing realizations that are typically fully parametrized, as we have observed in the examples of Chapter 3. This is because the measure itself assumes that all parameters in the matrices A, B and C are potentially implemented with error. If these matrices could be made to contain fixed numbers, such as 0, 1, or exact powers of the bit resolution, then this would not only reduce the computation time, but these numbers would be implemented without error, thereby improving the performance of the actually implemented filter.

Our motive in this and the next chapter will be to overcome some of these drawbacks. In this chapter we shall first extend the 'classical' L_1/L_2 sensitivity measure to a frequency weighted L_1/L_2 measure, thereby enabling us to take better account of practical design criteria. The optimal state space realization problem is rendered significantly more complicated by this frequency weighting and will require a more elaborate optimization algorithm.

Our efforts with respect to the other drawbacks mentioned above will be displayed in Chapters 5, 6, 7 and 8. In Chapter 5 we shall present a new sensitivity measure that uses the same L_2 norm for the sensitivities of the transfer function with respect to the parameters of the three matrices A, B and C. We shall compare it with the 'classical' L_1/L_2 measure. In Chapter 6 we shall discuss the sensitivity of the poles and zeros of a filter with respect to errors in its coefficients, while in Chapter 7 we shall present a new design method in which coefficient errors are treated as random noise, whereby the effects of coefficient errors and of roundoff errors can be combined into a synthetic design criterion that leads to a global optimization procedure. It will turn out that the sensitivity part of this criterion is precisely our new L_2 sensitivity norm introduced in Chapter 5. Finally, Chapter 8 will be devoted to various methods of obtaining sparse optimal and suboptimal realizations that have as many error-free coefficients as possible.

We conclude this introduction by mentioning that the results of this chapter were for the most part obtained in joint work with Brian Anderson and Jane Perkins, whose contribution is gratefully acknowledged: see [LAGP92].

4.2 Minimization of a frequency weighted sensitivity measure

We first note that the L_p norm in Definition 3.2 is a frequency independent mean value of a matrix function over the whole frequency range. Therefore, the sensitivity measure $M_{L_{12}}$ defined in (3.14) considers the sensitivity behaviour of the transfer function at one frequency point to be as important as at another frequency point. From a practical point of view, one is usually interested in the performance of the transfer function within a specified frequency range, the bandwidth of the transfer function for example, or even at some discrete frequency points. More precisely, one wants the transfer function to be less sensitive to the variations of the parameters in a certain frequency interval (the bandwidth, for example), and one can allow a greater sensitivity in a frequency domain that one is not interested in. This type of requirement can be met by using a weighted sensitivity function, and hence a corresponding weighted sensitivity measure.

A frequency weighted measure has already been introduced by Thiele [Thi86], but with a specific relationship between the weightings of the various terms of the measure. Under those constraints on the weightings, Thiele solved the sensitivity minimization problem using methods that are essentially the same as those explained in Chapter 3 for the unweighted problem. Here we address the case of general unconstrained frequency weightings, and we solve the corresponding optimal state variable design problem.

We start off by defining a frequency weighted sensitivity measure.

4.2.1 Weighted sensitivity measure of a realization

As in Chapter 3, we consider a state variable realization

$$\begin{aligned} x(t+1) &= Ax(t) + Bu(t) \\ y(t) &= Cx(t) + du(t), \end{aligned} \qquad (4.1)$$

4. Frequency weighted optimal design

and its transfer function

$$H(z) = C(zI - A)^{-1}B + d. \qquad (4.2)$$

Now let $W_A(z)$, $W_B(z)$ and $W_C(z)$ be three scalar functions of the complex variable z. Then the weighted sensitivity functions corresponding to those given in (3.10) are defined as

$$\begin{aligned}\frac{\delta H(z)}{\delta A} &\triangleq W_A(z)\frac{\partial H(z)}{\partial A} \\ \frac{\delta H(z)}{\delta B} &\triangleq W_B(z)\frac{\partial H(z)}{\partial B} \\ \frac{\delta H(z)}{\delta C^T} &\triangleq W_C(z)\frac{\partial H(z)}{\partial C^T}.\end{aligned} \qquad (4.3)$$

Note that the notation is not meant to suggest that δ is a derivative operator. Now let

$$W_A(z) = W_1(z)W_2(z) \qquad (4.4)$$

be any factorization of $W_A(z)$. Using (3.10)-(3.11), one then has the following explicit expressions

$$\frac{\delta H(z)}{\delta A} = [W_1(z)G(z)][W_2(z)F(z)]^T \qquad (4.5)$$

$$\frac{\delta H(z)}{\delta B} = W_B(z)G(z) \qquad (4.6)$$

$$\frac{\delta H(z)}{\delta C^T} = W_C(z)F(z), \qquad (4.7)$$

where $F(z)$ and $G(z)$ are as defined in (3.11):

$$\begin{aligned}F(z) &\triangleq (zI - A)^{-1}B = [f_1(z)\ldots f_n(z)]^T \\ G^T(z) &\triangleq C(zI - A)^{-1} = [g_1(z)\ldots g_n(z)]\end{aligned} \qquad (4.8)$$

The over-all weighted L_1/L_2 sensitivity measure is defined as

$$M^*_{L_{12}} \triangleq \|\frac{\delta H(z)}{\delta A}\|^2_1 + \|\frac{\delta H(z)}{\delta B}\|^2_2 + \|\frac{\delta H(z)}{\delta C^T}\|^2_2, \qquad (4.9)$$

where the norms have been defined in Definition 3.2. This measure can be rewritten as

$$\begin{aligned}M^*_{L_{12}} =\ & \|W_1(z)G(z)(W_2(z)F(z))^T\|^2_1 \\ & + \|W_B(z)G(z)\|^2_2 + \|W_C(z)F(z)\|^2_2.\end{aligned} \qquad (4.10)$$

A similarity transformation $x = Tz$ transforms $(A, B, C, F(z), G(z))$ into $(T^{-1}AT, T^{-1}B, CT, T^{-1}F(z), T^T G(z))$. This means that different realizations yield different sensitivity measures $M^*_{L_{12}}$. So an interesting problem is to find those realizations that minimize this sensitivity measure. The optimal FWL frequency weighted state space design can then be formulated as follows:

$$\min_{(A,B,C) \in S_H} M^*_{L_{12}} \qquad (4.11)$$

Comment: In [Thi86] Thiele introduced a frequency weighted sensitivity measure similar to (4.9) but with

$$W_1(z) = W_B(z) \text{ and } W_2(z) = W_C(z). \qquad (4.12)$$

For this special choice of weighting functions, Thiele showed how to compute the weighted Gramians and solved a number of sensitivity minimization problems to which we shall return later. The choice (4.12) is justified for the case where coloured noise with input spectrum $|W_C(z)|^2$ enters into the filter and where roundoff noise on the states is considered with a coloured spectrum $|W_B(z)|^2$: see [Thi86] for details. However, other choices can be motivated by other applications. To give but one example, in the optimal FWL controller realization problem, one minimizes a sensitivity measure of the closed loop transfer function with respect to all controller realizations. We shall show in Chapter 10 that the sensitivity functions of the closed loop transfer function with respect to the matrices (A, B, C) of the controller controller realization are the product of the sensitivity functions of the feedback controller transfer function with respect to these same matrices times a weighting function that is the same for all three sensitivity functions

$$W_A(z) = W_B(z) = W_C(z) = \frac{H_0(z)}{[1 + C(z)H_0(z)]^2}, \qquad (4.13)$$

where $H_0(z)$ is the transfer function of the open loop plant. We note that in this application Thiele's relationship (4.12) between the weightings is not satisfied.

In the next subsection, we discuss how to solve the optimal frequency weighted FWL state space design problem.

4.2.2 OPTIMAL FWL REALIZATIONS

The difficulty in solving (4.11) is due to the fact that the first term on the right of (4.9) is a complicated function of the realization (A, B, C). To overcome this, note that by the Cauchy-Schwartz inequality

$$\begin{aligned} \| \frac{\delta H(z)}{\delta A} \|_1^2 &= \| W_1(z)G(z)(W_2(z)F(z))^T \|_1^2 \\ &\leq \| W_1(z)G(z) \|_2^2 \| W_2(z)F(z) \|_2^2 \end{aligned} \quad (4.14)$$

where equality holds if and only if

$$\rho^2 G^H(z)G(z)|W_1(z)|^2 = F^H(z)F(z)|W_2(z)|^2 \quad \forall z \in \{|z|=1\}, \quad (4.15)$$

for some $\rho \neq 0 \in \mathbb{R}$. We will study the following upper bound of the measure $M_{L_{12}}^*$:

$$\begin{aligned} M_{L_{12}}^* \leq \bar{M}_{L_{12}}^* &\triangleq \| W_1(z)G(z) \|_2^2 \| W_2(z)F(z) \|_2^2 \\ &+ \| \frac{\delta H(z)}{\delta B} \|_2^2 + \| \frac{\delta H(z)}{\delta C^T} \|_2^2 . \end{aligned} \quad (4.16)$$

We shall present methods for minimizing $\bar{M}_{L_{12}}^*$ and examine under which conditions realizations that minimize the upper bound $\bar{M}_{L_{12}}^*$ also minimize the measure itself, $M_{L_{12}}^*$. It is easy to show with (3.23)-(3.24) that

$$\bar{M}_{L_{12}}^* = tr(K_{o1})tr(K_{c2}) + tr(K_{oB}) + tr(K_{cC}), \quad (4.17)$$

where K_{o1}, K_{c2}, K_{oB} and K_{cC} can be obtained by the following general expression:

$$K = \frac{1}{2\pi j} \oint_{|z|=1} X(z)X^H(z)z^{-1}dz \quad (4.18)$$

with $X(z) = G(z)W_1(z)$, $F(z)W_2(z)$, $G(z)W_B(z)$ and $F(z)W_C(z)$, respectively. We call these four matrices K_{o1}, K_{c2}, K_{oB}, K_{cC} weighted Gramians. An algorithm for computing a weighted Gramian can be found in Appendix 4.B at the end of this chapter.

A similarity transformation $x = Tz$ transforms $(A, B, C, K_{cC}, K_{c2}, K_{oB}, K_{o1})$ into $(T^{-1}AT, T^{-1}B, CT, T^{-1}K_{cC}T^{-T}, T^{-1}K_{c2}T^{-T}, T^T K_{oB}T, T^T K_{o1}T)$. So, the optimal FWL design problem of (4.11) is replaced by the following upper bound minimization:

$$\begin{aligned} \min_{T:\det T \neq 0} \{\bar{M}_{L_{12}}^* &= tr(T^T K_{o1}T)tr(T^{-1}K_{c2}T^{-T}) \\ &+ tr(T^T K_{oB}T) + tr(T^{-1}K_{cC}T^{-T})\}. \end{aligned} \quad (4.19)$$

Now, it is easy to see that

$$\bar{M}^*_{L_{12}} = tr(K_{o1}P)tr(K_{c2}P^{-1}) + tr(K_{oB}P) + tr(K_{cC}P^{-1}) \triangleq R(P) \quad (4.20)$$

where $P = TT^T$. Therefore,

$$\boxed{\min_{T:\det T \neq 0} \bar{M}^*_{L_{12}} \quad \Longleftrightarrow \quad \min_{P:P\triangleq TT^T, \det P \neq 0} R(P). \quad (4.21)}$$

In the remainder of this and the next section we study the minimization problem (4.21). Our developments will proceed as follows:

- First we show that a minimum of $R(P)$ exists and that it can only be achieved by nonsingular matrices P.

- We then show that this minimum is unique.

- We then proceed to the computation of the optimal solution P^{opt} by considering two different cases.

    We first consider the easier case where a proportionality relation exists between the weighted observability Gramians K_{o1} and K_{oB}, and similarly between the weighted controllability Gramians K_{c2} and K_{cC}. In this case, we shall compute an explicit solution of P^{opt}, and hence produce an explicit characterisation of the optimal realization set.

    We then consider the general situation where no such relation exists. In such case, no explicit expression of the optimal solution can be given and an iterative algorithm is required for its computation.

- We shall also show that in the first case considered above, and for the special choice $W_1(z) = W_2(z)$, the optimal realization set minimizes not just the upper bound $\bar{M}^*_{L_{12}}$, but the sensitivity measure $M^*_{L_{12}}$ itself.

4.2.3 EXISTENCE AND UNIQUENESS

Our first lemma shows that the minimum of $R(P)$ exists, and that it can be achieved by nonsingular P only. This means that (4.21) has solutions.

Lemma 4.1 : *With K_{oB} and K_{cC} nonsingular, the minimum of $R(P)$ defined in (4.20) exists and can be achieved only for nonsingular P.*

Proof: Since $tr(K_{o1}P)tr(K_{c2}P^{-1})$ is a positive differentiable function of P, and since K_{oB} and K_{cC} are positive definite, the result follows immediately from Lemma 4.A in Appendix 4.A. ∎

We have proved that (4.21) has solutions if K_{oB} and K_{cC} are both nonsingular. This condition is satisfied if the weighting functions $W_B(z)$ and $W_C(z)$ have no pole-zero cancellations with the system $H(z)$, i.e. if the system (A, B, C) is minimal and if the scalar weighting functions have no zeros at the poles of $H(z)$. In the sequel, this condition is assumed to be satisfied.

We shall now prove the uniqueness of the minimizing solution. To do so, we need the following lemma.

Lemma 4.2 : *Let M and X be two matrices of appropriate dimensions, then*

$$1)\ \frac{d[tr(MX)]}{dX} = M^T,\ 2)\ \frac{d[tr(MX^{-1})]}{dX} = -(X^{-1}MX^{-1})^T.$$

Proof: Denote $X \triangleq \{x_{ij}\}$ and $e_i \triangleq \begin{pmatrix} 0 & 0 & \ldots & 0 & 1 & 0 & \ldots & 0 \end{pmatrix}^T$ with the 1 in the i-th position. We then have

$$\frac{\partial tr(MX)}{\partial x_{ij}} = tr(\frac{\partial(MX)}{\partial x_{ij}}) = tr(Me_i e_j^T) = e_j^T M e_i = m_{ji}$$

from which the first part of the lemma follows directly.

Since $XX^{-1} = I$, one has $\partial(XX^{-1})/\partial x_{ij} = 0$. It then follows that $(\partial X/\partial x_{ij})X^{-1} + X(\partial X^{-1}/\partial x_{ij}) = 0$, and hence

$$\partial X^{-1}/\partial x_{ij} = -X^{-1}(\partial X/\partial x_{ij})X^{-1}.$$

Now,

$$\begin{aligned}\frac{\partial tr(MX^{-1})}{\partial x_{ij}} &= tr(\frac{\partial(MX^{-1})}{\partial x_{ij}}) = tr[M(-X^{-1}\frac{\partial X}{\partial x_{ij}}X^{-1})] \\ &= tr(-X^{-1}MX^{-1}e_i e_j^T) = e_j^T(-X^{-1}MX^{-1})e_i.\end{aligned}$$
(4.22)

The second part of the lemma follows directly from the above expression. ∎

4. Frequency weighted optimal design

With Lemma 4.2, one has

$$\frac{\partial R(P)}{\partial P} = -tr(K_{o1}P)P^{-1}K_{c2}P^{-1} - P^{-1}K_{cC}P^{-1}$$
$$+ tr(K_{c2}P^{-1})K_{o1} + K_{oB}. \quad (4.23)$$

By letting $\partial R(P)/\partial P = 0$, one gets a necessary condition which the solution of (4.21) must satisfy:

$$P[tr(K_{c2}P^{-1})K_{o1} + K_{oB}]P = tr(K_{o1}P)K_{c2} + K_{cC}. \quad (4.24)$$

Our next result shows the uniqueness of the solution of (4.24).

Theorem 4.1 : *With the four symmetric positive definite matrices K_{o1}, K_{c2}, K_{oB} and K_{cC} defined by (4.18), (4.24) has a unique solution, and hence so does (4.21).*

Proof: Let P_0 and P be two solutions of (4.24). From

$$P_0[tr(K_{c2}P_0^{-1})K_{o1} + K_{oB}]P_0 = tr(K_{o1}P_0)K_{c2} + K_{cC},$$

it follows that

$$I[tr(\tilde{K}_{c2}I^{-1})\tilde{K}_{o1} + \tilde{K}_{oB}]I = tr(\tilde{K}_{o1}I)\tilde{K}_{c2} + \tilde{K}_{cC},$$

where

$$\tilde{K}_{o1} = P_0^{1/2}K_{o1}P_0^{1/2}, \quad \tilde{K}_{c2} = P_0^{-1/2}K_{c2}P_0^{-1/2},$$
$$\tilde{K}_{oB} = P_0^{1/2}K_{oB}P_0^{1/2}, \quad \tilde{K}_{cC} = P_0^{-1/2}K_{cC}P_0^{-1/2}.$$

This means that by proper choice of the initial realization, (4.24) has the unit matrix as a solution. So, without loss of generality, one can assume $P_0 = I$. Therefore, one only needs to prove that if P is a solution of (4.24), then $P = I$ under the following constraint

$$tr(K_{c2})K_{o1} + K_{oB} = tr(K_{o1})K_{c2} + K_{cC} \quad (4.25)$$

or equivalently,

$$tr(\tilde{K}_{c2})\tilde{K}_{o1} + \tilde{K}_{oB} = tr(\tilde{K}_{o1})\tilde{K}_{c2} + \tilde{K}_{cC}, \quad (4.26)$$

where

$$\tilde{K}_{o1} = U^T K_{o1} U, \quad \tilde{K}_{c2} = U^T K_{c2} U,$$
$$\tilde{K}_{oB} = U^T K_{oB} U, \quad \tilde{K}_{cC} = U^T K_{cC} U \quad (4.27)$$

for an arbitrary orthogonal matrix U. In particular, one has

$$\left(\sum_{i=1}^{n}\tilde{K}_{c2}(i,i)\right)\tilde{K}_{o1}(j,j)+\tilde{K}_{oB}(j,j)$$
$$=\left(\sum_{i=1}^{n}\tilde{K}_{o1}(i,i)\right)\tilde{K}_{c2}(j,j)+\tilde{K}_{cC}(j,j) \qquad (4.28)$$

for all $j = 1, 2, ..., n$. Now, by SVD one has $P = UX^2U^T$ where U is some orthogonal matrix and $X^2 = diag(x_1^2, x_2^2, ..., x_n^2)$ with $x_1^2 \geq x_2^2 \geq .... \geq x_n^2 > 0$. Inserting $P = UX^2U^T$ in (4.24) and using (4.27) yields

$$\left\{\left(\sum_{i=1}^{n}\tilde{K}_{c2}(i,i)x_i^{-2}\right)\tilde{K}_{o1}(j,j)+\tilde{K}_{oB}(j,j)\right\}x_j^4$$
$$=\left(\sum_{i=1}^{n}\tilde{K}_{o1}(i,i)x_i^2\right)\tilde{K}_{c2}(j,j)+\tilde{K}_{cC}(j,j)$$

or

$$\left\{\left(\sum_{i=1}^{n}\tilde{K}_{c2}(i,i)x_i^{-2}\right)\tilde{K}_{o1}(j,j)+\tilde{K}_{oB}(j,j)\right\}x_j^2$$
$$=\left\{\left(\sum_{i=1}^{n}\tilde{K}_{o1}(i,i)x_i^2\right)\tilde{K}_{c2}(j,j)+\tilde{K}_{cC}(j,j)\right\}x_j^{-2} \qquad (4.29)$$

for all $j = 1, 2, ..., n$. On the one hand, for $j = 1$, it follows from (4.29) that

$$\left(\sum_{i=1}^{n}\tilde{K}_{c2}(i,i)\right)\tilde{K}_{o1}(1,1)+\tilde{K}_{oB}(1,1)x_1^2$$
$$\leq \left(\sum_{i=1}^{n}\tilde{K}_{o1}(i,i)\right)\tilde{K}_{c2}(1,1)+\tilde{K}_{cC}(1,1)x_1^{-2}$$

since $x_1^2 \geq x_2^2 \geq ... \geq x_n^2 > 0$. On the other hand, taking $j = 1$ and the same U as in (4.27), it follows from (4.28) that

$$\left(\sum_{i=1}^{n}\tilde{K}_{c2}(i,i)\right)\tilde{K}_{o1}(1,1)+\tilde{K}_{oB}(1,1)$$
$$=\left(\sum_{i=1}^{n}\tilde{K}_{o1}(i,i)\right)\tilde{K}_{c2}(1,1)+\tilde{K}_{cC}(1,1).$$

One concludes that $x_1^2 \leq 1$. Similarly, for $j = n$ one can obtain

$$\left(\sum_{i=1}^{n} \tilde{K}_{c2}(i,i)\right) \tilde{K}_{o1}(n,n) + \tilde{K}_{oB}(n,n)x_n^2$$
$$\geq \left(\sum_{i=1}^{n} \tilde{K}_{o1}(i,i)\right) \tilde{K}_{c2}(n,n) + \tilde{K}_{cC}(n,n)x_n^{-2}$$

and (4.28) yields

$$\left(\sum_{i=1}^{n} \tilde{K}_{c2}(i,i)\right) \tilde{K}_{o1}(n,n) + \tilde{K}_{oB}(n,n)$$
$$= \left(\sum_{i=1}^{n} \tilde{K}_{o1}(i,i)\right) \tilde{K}_{c2}(n,n) + \tilde{K}_{cC}(n,n).$$

This implies that $x_n^2 \geq 1$. Since $x_1^2 \geq x_n^2$, one has $x_1^2 = x_2^2 = \ldots = x_n^2 = 1$ which leads to $P = I$. This completes the proof. ∎

4.3 Computation of the optimal realization set

It appears difficult to find an explicit expression of the solution P of (4.24) when no particular relation exists between the frequency weightings. We will show later that in this general case P^{opt} can be computed as the limiting solution of a gradient algorithm. But first we show that an analytic solution of (4.24) can be computed for the special case where

$$K_{oB} = \varrho_1 K_{o1}, \quad K_{cC} = \varrho_2 K_{c2}, \qquad (4.30)$$

with ϱ_1 and ϱ_2 two positive constants. We note that one particular case where this relationship holds is when $W_1(z) = W_B(z)$ and $W_2(z) = W_C(z)$ as in [Thi86]; in that particular case $\varrho_1 = \varrho_2 = 1$. To compute P^{opt} when (4.30) holds we need the following lemma.

Lemma 4.3 : *Let $W > 0$ and $M \geq 0$ be square symmetric matrices of the same dimension. The equation $PWP = M$ has a unique solution $P = P^T \geq 0$ and the solution is given by*

$$P = W^{-1/2}[W^{1/2}MW^{1/2}]^{1/2}W^{-1/2},$$

where for any $X \geq 0$, $X^{1/2}$ denotes the unique symmetric matrix satisfying $X^{1/2} \geq 0$ and $X^{1/2}X^{1/2} = X$.

4. Frequency weighted optimal design

Proof: Let $W^{1/2}$ be a positive definite square root of W. Clearly, this square root is unique [Bel70]. Then

$$PWP = M \iff W^{1/2}PW^{1/2}W^{1/2}PW^{1/2} = W^{1/2}MW^{1/2}$$
$$\iff W^{1/2}PW^{1/2} = [W^{1/2}MW^{1/2}]^{1/2}$$

which leads to

$$P = W^{-1/2}[W^{1/2}MW^{1/2}]^{1/2}W^{-1/2}.$$

Evidently, P is unique. ∎

Theorem 4.2 : *With four symmetric positive definite matrices K_{o1}, K_{oB}, K_{c2} and K_{cC} satisfying (4.30), there exists a unique solution P^{opt} of (4.21) which is given by*

$$P^{opt} = \rho K_{o1}^{-1/2}[K_{o1}^{1/2}K_{c2}K_{o1}^{1/2}]^{1/2}K_{o1}^{-1/2} \qquad (4.31)$$

where

$$\varrho = (\varrho_2/\varrho_1)^{1/2}. \qquad (4.32)$$

In addition, the optimal solutions of the optimization problem (4.19) are given by

$$T^{opt} = \varrho^{1/2}K_{o1}^{-1/2}[K_{o1}^{1/2}K_{c2}K_{o1}^{1/2}]^{1/4}V, \qquad (4.33)$$

where V is an arbitrary orthogonal matrix. The Gramians of the optimal realizations are characterized by

$$\tilde{K}_{o1} = \varrho^2 \tilde{K}_{c2}. \qquad (4.34)$$

Proof: With (4.24) and (4.30), it follows from Lemma 4.3 that P^{opt} is given by (4.31) where ϱ is given by

$$\varrho = \sqrt{\frac{tr(K_{o1}P) + \varrho_2}{tr(K_{c2}P^{-1}) + \varrho_1}} = \sqrt{\frac{\varrho tr[(K_{o1}^{1/2}K_{c2}K_{o1}^{1/2})^{1/2}] + \varrho_2}{\varrho^{-1}tr[(K_{o1}^{1/2}K_{c2}K_{o1}^{1/2})^{1/2}] + \varrho_1}}$$

which yields (4.32). The uniqueness is evident. Since the minimum of $R(P)$ exists and since (4.24) is a necessary condition for achieving this minimum, (4.31)-(4.32) is the unique solution of (4.21) for the case

(4.30). To obtain the set of optimal transformations T^{opt}, note that P^{opt} can be factored as

$$\begin{aligned} P^{opt} &= \varrho K_{o1}^{-1/2}[K_{o1}^{1/2}K_{c2}K_{o1}^{1/2}]^{1/2}K_{o1}^{-1/2} \\ &= \{\varrho^{1/2}K_{o1}^{-1/2}[K_{o1}^{1/2}K_{c2}K_{o1}^{1/2}]^{1/4}V\} \\ &\quad \times\{\varrho^{1/2}K_{o1}^{-1/2}[K_{o1}^{1/2}K_{c2}K_{o1}^{1/2}]^{1/4}V\}^T \end{aligned}$$

where V is an arbitrary orthogonal matrix. Expression (4.33) then follows from $P = TT^T$. We can now compute the Gramians of the optimal realizations:

$$\begin{aligned} \tilde{K}_{o1} &= (T^{opt})^T K_{o1} T^{opt} = \varrho V^T (K_{o1}^{1/2}K_{c2}K_{o1}^{1/2})^{1/2}V \\ \tilde{K}_{c2} &= (T^{opt})^{-1} K_{c2} (T^{opt})^{-T} = \varrho^{-1}V^T (K_{o1}^{1/2}K_{c2}K_{o1}^{1/2})^{1/2}V \end{aligned}$$

and hence

$$\tilde{K}_{o1} = \varrho^2 \tilde{K}_{c2}. \qquad (4.35)$$

Of course, by the relation (4.30), we also have

$$\tilde{K}_{oB} = \varrho_1 \tilde{K}_{o1} \text{ and } \tilde{K}_{cC} = \varrho_2 \tilde{K}_{c2}. \qquad (4.36)$$

∎

Next we show that if $W_1(z) = W_2(z)$ holds in addition to (4.30), then the sensitivity measure itself is also minimized by the optimal realization set characterized by (4.35).

Corollary 4.1 : *Assume that, in addition to the relation (4.30) between the respective weighted Gramians, the two factors of the frequency weighting $W_A(z)$ are identical, i.e. $W_1(z) = W_2(z) \; \forall z$. Then the realizations characterized by (4.35) minimize the frequency weighted sensitivity measure (4.9).*

Proof: With the additional condition $W_1(z) = W_2(z)$, the optimality condition (4.35) can be written

$$\begin{aligned} &\frac{1}{2\pi j}\oint_{|z|=1} \tilde{G}(z)\tilde{G}^H(z)|W_1(z)|^2 z^{-1} dz = \\ &\frac{1}{2\pi j}\oint_{|z|=1} \varrho^2 \tilde{F}(z)\tilde{F}^H(z)|W_1(z)|^2 z^{-1} dz. \end{aligned} \qquad (4.37)$$

It then immediately follows from Lemma 2 in [Thi86] that

$$\tilde{G}^H(z)\tilde{G}(z) = \varrho^2 \tilde{F}^H(z)\tilde{F}(z) \; \forall z \in \{|z|=1\}. \qquad (4.38)$$

4. Frequency weighted optimal design

Therefore the Cauchy-Schwartz inequality is satisfied with equality in (4.14) and hence the result follows by the same argument as in [Thi86]. ∎

We now turn to the general case where the relation (4.30) does not hold. In such case, an explicit expression of the solution of (4.24) does not appear to be at hand. However, P^{opt} can be obtained by an iterative procedure using a gradient algorithm:

$$P(k+1) = P(k) - \mu \frac{\partial R(P)}{\partial P}\Big|_{P=P(k)}, \qquad (4.39)$$

where $\partial R(P)/\partial P$ is given by (4.23) and μ is a positive step size. We have proved above that the function $R(P)$ has a unique (and hence global) minimum achieved by a nonsingular P. Therefore the above algorithm will converge to P^{opt} for any positive definite initial condition.

As with almost any numerical minimization algorithm applied to a problem for which no analytic solution is available, no fast rules on the step size choice can be given. Our arguments have shown that the surface near the minimum is approximately quadratic, which gives a slight insight into the rate of convergence.

The choice of an appropriate initial condition will improve the convergence of the algorithm (4.39). We note that a necessary condition for (4.24) is $tr(K_{oB}P) = tr(K_{cC}P^{-1})$. So we use as initial condition a P_0 that minimizes $tr(K_{o1}P)tr(K_{c2}P^{-1})$ and ensures $tr(K_{oB}P) = tr(K_{cC}P^{-1})$ at the same time. We know that the P_0 minimizing the above trace product is not unique. In fact, suppose a P_1 has been found that minimizes this product, then so does kP_1 for any $k > 0$. Clearly, $P_0 = kP_1$ with

$$k^2 = tr(K_{cC}P_1^{-1})/tr(K_{oB}P_1) \qquad (4.40)$$

will minimize $tr(K_{o1}P)tr(K_{c2}P^{-1})$ while at the same time producing the equality $tr(K_{oB}P) = tr(K_{cC}P^{-1})$. It can be shown [Li90] that such a P_1 can be chosen to be $T_b T_b^T$ where T_b internally balances K_{o1} and K_{c2}, that is

$$T_b^{-1} K_{c2} T_b^{-T} = T_b^T K_{o1} T_b \triangleq diag(\eta_1, \eta_2, ..., \eta_n).$$

This transformation matrix T_b can be obtained using a numerically well-conditioned algorithm due to Laub [Lau80].

Finally we note that for any optimal $P^{opt} = T^{opt}(T^{opt})^T$, the corresponding optimal transformation matrices can be constructed as

$$T^{opt} = (P^{opt})^{1/2} V \qquad (4.41)$$

for any orthogonal matrix V. All the arguments above can be summarized by the following theorem.

Theorem 4.3 : *The optimal transformation matrices, that is the solutions of the frequency weighted L_1/L_2 transfer function sensitivity minimization problem (4.19), are not unique and can be characterized by (4.41) where P^{opt} is determined by the system and the weighting functions (it is the unique solution of (4.24)) while V is an arbitrary orthogonal matrix.*

This means that there is a degree of freedom characterized by the set of orthogonal matrices V in this optimal transformation set. In the next section we show how to exploit this freedom in order to simplify the implementation and to improve the computational performance of the system.

Comment: As we stated above, Thiele [Thi86] has also considered the frequency weighted sensitivity measure (4.9) but only in the special case where $W_1(z) = W_B(z)$ and $W_2(z) = W_C(z)$. He showed how to minimize the upper bound $\bar{M}^*_{L_{12}}$ for this case. He also showed that if, in addition, $W_1(z) = W_2(z) = W_B(z) = W_C(z)$, then the realizations that minimize the upper bound also minimize the weighted sensitivity measure itself. Our results extend Thiele's result in several ways. First we have explicitly characterized the realizations that minimize the upper bound in the case where the relation (4.12) is replaced by the weaker relation (4.30). Secondly we have shown that if, in addition to (4.30), the factors of $W_A(z)$ are chosen to be identical, $W_1(z) = W_2(z)$, then this optimal realization set also mimimizes the sensitivity measure $M^*_{L_{12}}$ itself. Finally, we have shown how to compute a realisation set that minimizes the upper bound even in the most general case where no special constraints hold between any of the frequency weightings or frequency weighted Gramians.

4.4 Numerical example

We now illustrate our previous theoretical results using for $H(z)$ the sixth-order narrow band low pass filter already presented in Example 3.2. Recall that this filter has a normalized sampling frequency $f_s = 1$ and filter design parameters $f_p = 0.03125$ (passband frequency), $F_s = 0.0390625$ (stopband frequency), and $\epsilon_p = 1 dB$ (passband ripple).

4. Frequency weighted optimal design

The upper limit of the frequency response is $0dB$, while the stopband attenuation is $46.68dB$. Recall that the controllable realization R_c of $H(z)$ is given by:

$\boxed{R_c:}$

$$A_c = \begin{pmatrix} 5.6526 & -13.3818 & 16.9792 & -12.1765 & 4.6789 & -0.7526 \\ 1 & 0 & 0 & 0 & 0 & 0 \\ 0 & 1 & 0 & 0 & 0 & 0 \\ 0 & 0 & 1 & 0 & 0 & 0 \\ 0 & 0 & 0 & 1 & 0 & 0 \\ 0 & 0 & 0 & 0 & 1 & 0 \end{pmatrix}$$

$$B_c = (1\ 0\ 0\ 0\ 0\ 0)^T, \qquad d = 0.4708 \times 10^{-2}$$

$$C_c = (0.1511\ -0.4558\ 0.3855\ 0.116\ -0.3074\ 0.1165) \times 10^{-2}.$$

Taking this realization as the initial realization, we can compute the corresponding Gramians without weighting (that is $W_1(z) = W_2(z) = W_B(z) = W_C(z) = 1$). In this case, one optimal realization that minimizes the sensitivity measure is the balanced form R_b characterized by $K_{c2} = K_{cC} = K_{o1} = K_{oB} = diagonal$. This realization has been given in Example 3.2 in Chapter 3.

As stated before, the basic idea in using the weighting functions is to emphasize the behaviour of the transfer function in some frequency domain of interest. So the choice of weighting functions depends completely on the specifications imposed on the implemented filter.

Take $W_1(z) = W_B(z) = W_C(z) = H_w(z)$ and $W_2(z) = 1$ with $H_w(z)$ a low pass narrow band filter of order 10, whose frequency response is shown in Figure 4.1.

We can get the unique solution P^{opt} of (4.24) by using the algorithm (4.39) and hence the optimal similarity transformation matrices T^{opt} constructed by (4.41). We denote by $R^{opt}(M^*_{L_{12}})$ the realization determined by T^{opt} in (4.41) with $V = I$:

4. Frequency weighted optimal design

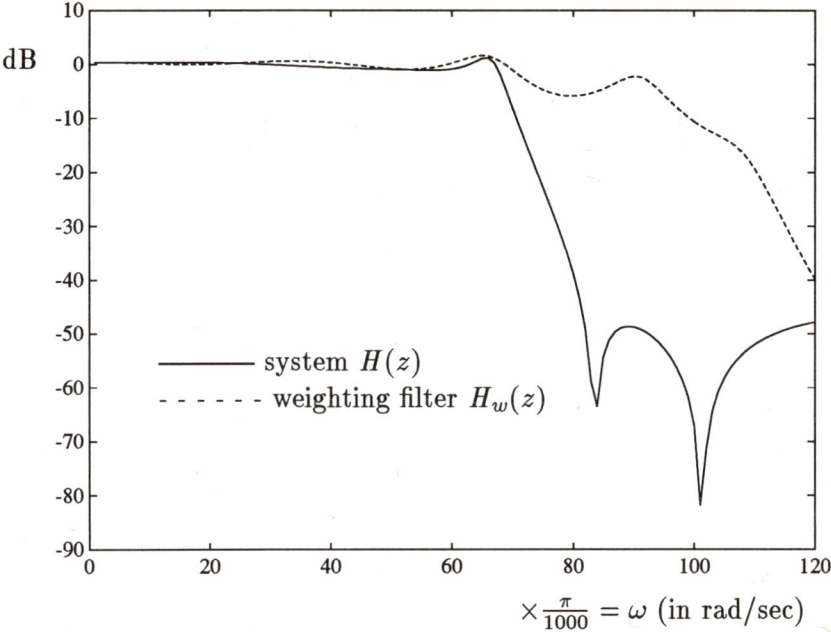

FIGURE 4.1. Magnitude frequency responses of the system and of the weighting filter

$\boxed{R^{opt}(M^*_{L_{12}}):}$

$$A^{opt}(M^*_{L_{12}}) =$$

$$\begin{pmatrix} 0.9813 & -0.0962 & -0.0280 & -0.0364 & -0.0540 & -0.1045 \\ 0.0849 & 0.9769 & -0.1155 & -0.0352 & -0.0415 & -0.0466 \\ 0.0247 & 0.1341 & 0.9454 & -0.0975 & -0.0370 & -0.0099 \\ 0.0302 & -0.0010 & 0.1545 & 0.9127 & -0.0805 & -0.0010 \\ 0.0478 & 0.0563 & 0.0070 & 0.1416 & 0.9511 & -0.0645 \\ 0.0831 & 0.0529 & 0.0477 & 0.0175 & 0.0499 & 0.8853 \end{pmatrix}$$

$$B^{opt}(M^*_{L_{12}}) =$$

$$(0.0346 \; -0.1496 \; 0.2620 \; -0.2316 \; 0.1032 \; -0.0185)^T$$

$$C^{opt}(M^*_{L_{12}}) =$$

$$(0.1394 \; 0.0036 \; -0.0178 \; -0.0050 \; 0.0783 \; 0.3971).$$

We truncate the fractional part of every coefficient of R_b and $R^{opt}(M^*_{L_{12}})$, which are of infinite precision, to 8 bits. We then compute the actual frequency response of these two FWL realizations and compare them with the ideal frequency response of the filter. The magnitudes of these frequency responses are given in Figure 4.2.

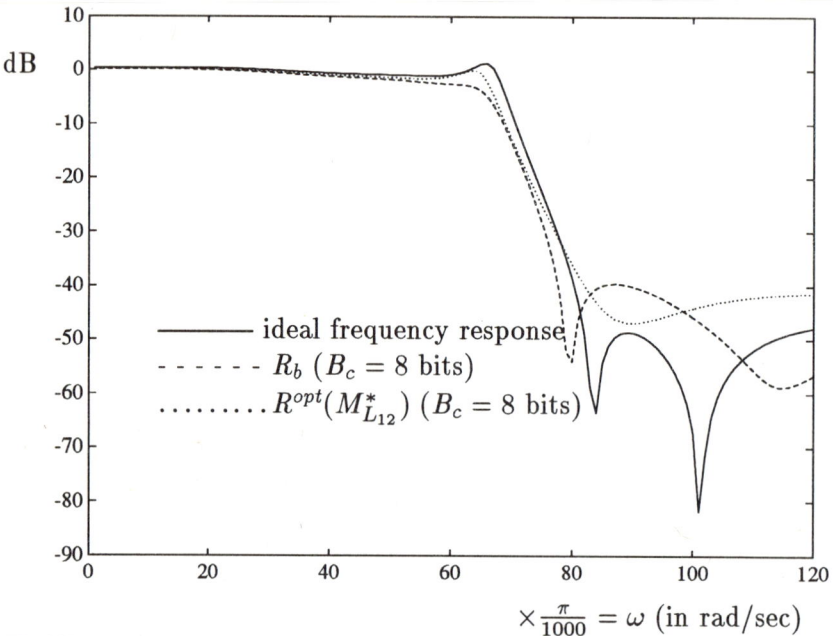

FIGURE 4.2. Magnitude frequency responses of FWL weighted and non weighted optimal realizations

Comments:

1. The balanced form R_b yields a nice performance, but its behaviour is not quite satisfactory in the frequency range between $\frac{50}{1000}\pi$ and $\frac{70}{1000}\pi$ (in rad/sec). We note that this interval belongs to the passband of the filter. Therefore, more effort should be made to improve the performance of the actually implemented filter in this frequency interval.

2. $H_w(z)$ is a low pass filter (see Figure 4.1) and its passband covers that of $H(z)$. Taking the weighting functions equal to $H_w(z)$ means that we weigh the sensitivity of the filter within the filter bandwidth. So, it is expected that a better performance of the actually implemented filter with the weighted optimal realization $R^{opt}(M^*_{L_{12}})$ will be achieved within this bandwidth compared with the optimal realization R_b without weighting. This is confirmed by Figure 4.2.

4.5 Conclusions

The optimal FWL state space digital system design problem becomes more practically relevant when the classical sensitivity measure, which weighs the sensitivity functions evenly over all frequencies, is replaced by a frequency weighted measure. This allows one to search for state space realizations that minimize the FWL error effects in frequency ranges where the application requires a high accuracy, while allowing for larger errors in frequency ranges where such errors do not seriously degrade the actual performance of the filter. The price to be paid for the introduction of such frequency weighting is that the optimal solution sets are no longer characterized by simple relationships between observability and controllability Gramians, as they were for the unweighted L_1/L_2 measure.

Just as in the initial derivation of the optimal solutions for the unweighted measure, our approach in this chapter has been to replace the minimization of the weighted L_1/L_2 measure by the minimization of an upper bound, obtained from the Cauchy-Schwartz inequality. The frequency weightings make the computation of the expression for the optimal solution set significantly harder, as the rather lengthy derivations of the present chapter have shown. However, one should bear in mind that even though the analytical effort required to establish the existence and uniqueness of the optimal solution was far from negligible, the actual computation of a realization minimizing the upper bound of the frequency weighted measure for a given transfer function and with given weightings using the gradient algorithm (4.39) is not particularly hard.

We have characterized the set of all optimal realizations via a set of optimal similarity transformations, and our simulations have shown the effect of different frequency weightings on the FWL performance of the corresponding optimal realizations.

Appendix 4.A: Proof of existence of a minimum.

Lemma 4.A: *Let $P \in \mathbb{R}^{n \times n}$ belong to the set of nonnegative definite matrices and let*

$$R(P) \triangleq f(P) + tr(PM) + tr(P^{-1}W), \qquad (4.A.1)$$

where $f(P)$ is a scalar positive differentiable function of P, and $M \in \mathbb{R}^{n \times n}$ and $W \in \mathbb{R}^{n \times n}$ are two positive definite matrices. Then the minimum of $R(P)$ exists and can be achieved only for nonsingular P.

Proof: By SVD, any nonnegative definite matrix P can be decomposed into $P = \{p_{ij}\} = U^T \Sigma^2 U$, where $U = \{u_{ij}\}$ is some orthogonal matrix and $\Sigma = diag(\sigma_1, \sigma_2, \ldots, \sigma_n)$, $\sigma_i \geq \sigma_{i+1} \geq 0$. So one has

$$p_{ij} = \sum_{k=1}^{n} u_{ki} u_{kj} \sigma_k^2 \qquad \forall i, j. \qquad (4.A.2)$$

It is well known [Nei67] that orthogonal matrices belong to a differentiable manifold of dimension $n(n-1)/2$. Therefore, locally the elements u_{ij} of an orthogonal matrix U in $\mathbb{R}^{n \times n}$ can be reparametrized as continuous functions of $n(n-1)/2$ parameters, i.e. of a vector

$$\bar{\theta} \triangleq (\theta_1 \theta_2 \ldots \theta_N)^T \text{ with } N = n(n-1)/2 \text{ for } |\theta_l| \leq \pi \; \forall l. \quad (4.A.3)$$

Hence, p_{ij} is a continuous function of $\{\theta_l\}$ and $\{\sigma_k^2\}$ for all i and j. Now, we note that

$$R(P) = f(P) + \sum_{i=1}^{n} \bar{m}_{ii} \sigma_i^2 + \sum_{i=1}^{n} \bar{w}_{ii} \sigma_i^{-2},$$

where

$$\bar{M} = UMU^T \triangleq \{\bar{m}_{ij}\}, \quad \bar{W} = UWU^T \triangleq \{\bar{w}_{ij}\}.$$

For any nonnegative definite matrix $M = \{m_{ij}\} \geq 0$, it is well known [Bel70] that $\lambda_{\min}(M) \leq m_{ii}$ for $i = 1, 2, \ldots, n$, where $\lambda_{\min}(M)$ denotes the minimal eigenvalue of M. Clearly therefore,

$$R(P) \geq \bar{m}_{11} \sigma_1^2 \geq \lambda_{\min}(\bar{M}) \sigma_1^2 = \lambda_{\min}(M) \sigma_1^2 \qquad (4.A.4)$$

and

$$R(P) \geq \bar{w}_{nn} \sigma_n^{-2} \geq \lambda_{\min}(\bar{W}) \sigma_n^{-2} = \lambda_{\min}(W) \sigma_n^{-2}. \qquad (4.A.5)$$

Also, for $P = I$, we get

$$R(I) = f(I) + tr(M) + tr(W) \triangleq E_0 > 0.$$

Now choose any E in $\mathbb{R}_+$ such that $E \geq max\{E_0, \lambda_{\min}(M), \lambda_{\min}(W)\}$ and define

$$C_-^2(E) \triangleq \lambda_{min}(W)/E \leq 1 \text{ and } C_+^2(E) \triangleq E/\{\lambda_{min}(M)\} \geq 1.$$

It then follows that $R(P) \geq E$ if either $\sigma_n^2 \leq C_-^2(E)$ or $\sigma_1^2 \geq C_+^2(E)$.

Since M and W are nonsingular by assumption and since E is finite, $C_-^2(E)$ and $C_+^2(E)$ as defined above satisfy $0 < C_-^2(E)$ and $C_+^2(E) < +\infty$. We now define the following closed set:

$$P_E \triangleq \{P : |\theta_i| \leq \pi, 0 < C_-^2(E) \leq \sigma_i^2 \leq C_+^2(E) < +\infty \; \forall i\}. \quad (4.A.6)$$

For any P outside P_E (i.e. any P such that $\sigma_n^2 < C_-^2(E)$ or $\sigma_1^2 > C_+^2(E)$), we have $R(P) > E$. On the other hand, $P = I$ is in P_E with $R(I) \leq E$ and, using (4.A.5) and $\sigma_n^2 \leq C_+^2(E)$ together with the definition of $C_+^2(E)$, yields

$$R(P) \geq \frac{\lambda_{\min}(M)\lambda_{\min}(W)}{E} > 0 \; \forall P \text{ in } P_E. \quad (4.A.7)$$

Since $R(P)$ is a continuous function of P in the closed set P_E with a strictly positive lower bound, it follows that $R(P)$ has its global minimum within this set, and since all the elements of P_E are nonsingular and bounded by the assumption on M and W, any element P for which $R(P)$ achieves the global minimum is nonsingular. ∎

Appendix 4.B: Computation of weighted Gramians.

The weighted Gramians defined in (4.18) can be computed by solving a series of integrals with the help of the method of residues or using an iterative procedure [Åst70]. If the weighting functions are proper rational functions, these Gramians can also be obtained by solving Lyapunov equations as we now show.

Suppose that the weighting function, say $W(z)$, is a proper rational function. It can then be written in the following general form: $W(z) = \alpha + \Psi(zI - \Phi)^{-1}\Omega$. The integrands $X(z)$ of the weighted Gramians (4.18) are all of the form:

$$X(z) = (zI - J)^{-1} L W(z) \triangleq Y(z) W(z) \qquad (4.B.1)$$

where $Y(z) = F(z), G(z)$. It can then be shown [Thi86] and [AM89] that the corresponding weighted Gramian is the (1,1) block matrix of the matrix

$$M = \begin{pmatrix} M_{11} & M_{12} \\ M_{12}^T & M_{22} \end{pmatrix}$$

satisfying the following Lyapunov equation:

$$M = \begin{pmatrix} J & L\Psi \\ 0 & \Phi \end{pmatrix} M \begin{pmatrix} J & L\Psi \\ 0 & \Phi \end{pmatrix}^T + \begin{pmatrix} \alpha L \\ \Omega \end{pmatrix} \begin{pmatrix} \alpha L \\ \Omega \end{pmatrix}^T. \qquad (4.B.2)$$

Thus every weighted Gramian in (4.18) can be expressed as $K = M_{11}$ with the appropriate definitions for $\Phi, \Omega, \Psi, \alpha, L$ and J. So the computation of the weighted Gramians is obtained by solving the following general Lyapunov equation:

$$M = QMQ^T + qq^T \qquad (4.B.3)$$

where Q and q are the block matrices corresponding to (4.B.2). The solution of this equation can be expressed by

$$M = \sum_{i=0}^{\infty} Q^i qq^T (Q^i)^T. \qquad (4.B.4)$$

Using the identity $Vec(SUV) = (V^T \otimes S)Vec(U)$ where $Vec(.)$ is the column stacking operator and $\otimes$ denotes the *Kronecker* product, it is easy to show that M satisfies the following linear equation (see e.g. [VB89]):

$$(I - Q \otimes Q)Vec(M) = Vec(qq^T). \qquad (4.B.5)$$

It should be pointed out that from a numerical point of view it is desirable that the matrix Q be in Schur form for computing M [BS72]. This can be done by transforming the matrices Φ and J into Schur form [GL83] by applying an orthogonal pre-transformation to both of them. Very good discussions about the numerical aspects of systems and control algorithms can be found in [Doo89].

5

A new transfer function sensitivity measure

5.1 Introduction

In Chapters 3 and 4 we have studied the optimal FWL state space design problem of a digital linear, time invariant system in terms of minimizing a global sensitivity measure, with and without frequency weighting, respectively. The initial unweighted sensitivity measure of the transfer function with respect to the parameters of the state space model was first proposed by Tavsanoglu and Thiele [TT84], and it was extended in Chapter 4 to a weighted version. In both cases, the measure contained a mixture of an L_1 norm on the sensitivity function with respect to the matrix A and an L_2 norm with respect to the vectors B and C.

The combination of an L_1 norm with respect to some parameters and an L_2 norm with respect to others is opportunistic rather than logical. The choice of an L_1/L_2 sensitivity measure is essentially motivated by the fact that it can easily be optimized with respect to the choice of coordinates. This, in turn, is caused by the analytic properties of the first term, which admits an easily computable upper bound, compatible with the other two terms. The optimal realizations obtained with this optimal design procedure show very nice performance in a FWL implementation, as we have seen earlier in Chapters 3 and 4. However, it would be more natural to use a sensitivity measure that is based on an L_2 norm for all three sensitivity functions instead of a mixture of L_1 and L_2 norms.

The object of the present chapter is precisely to show that such a coherent L_2 measure can indeed be optimized with respect to the set S_H of equivalent state space realizations. In fact, we show in this chapter that the existence of an optimal solution set can be proved by essentially the same technique as was used in Chapter 4 to show the existence of optimal solutions for the frequency weighted sensitivity minimization problem. A positive definite function $R(P)$ is minimized with respect to the set of symmetric nonnegative definite matrices P, and it is shown that the solution P is unique and positive definite. The set of square roots of P then define the set of optimal similarity transformations. A constructive procedure for the computation of the optimal P is again obtained by an iterative gradient algorithm as in Chapter 4.

The solution of the L_2 sensitivity minimization problem was obtained independently by Perkins, Helmke and Moore (see [PHM90] and [HM91]). They show that the optimal solution P is the unique positive definite solution of a gradient flow, which must then be solved using an analog computation.

An outline of this chapter is as follows. In Section 5.2, we define the L_2 sensitivity measure, compute its coordinate dependence, and formulate the optimal realization problem in terms of the minimization of this measure. The optimal realization problem is then shown to have a solution and a computational algorithm is given. The relationship between the L_2 and the more classical L_1/L_2 sensitivity measure are discussed in Section 5.3, and we illustrate our theoretical results by an example in Section 5.4.

5.2 Minimization of an L_2 sensitivity measure

5.2.1 THE L_2 SENSITIVITY MEASURE OF A REALIZATION

We recall that in Chapter 3 we have studied the 'classical' L_1/L_2 sensitivity measure initially introduced by Tavsanoglu and Thiele [TT84]:

$$M_{L_{12}} \triangleq \| \frac{\partial H(z)}{\partial A} \|_1^2 + \| \frac{\partial H(z)}{\partial B} \|_2^2 + \| \frac{\partial H(z)}{\partial C^T} \|_2^2, \qquad (5.1)$$

where the sensitivity functions and the L_1 and L_2 norms have been defined in Definitions 3.1 and 3.2. Because of its mixture of L_1 and L_2 norms we have called this an L_1/L_2 sensitivity measure. With the same definitions and notations as in Chapter 3, we now define the L_2 sensitivity measure of a realization (A, B, C) as follows:

$$M_{L_2} \triangleq \| \frac{\partial H(z)}{\partial A} \|_2^2 + \| \frac{\partial H(z)}{\partial B} \|_2^2 + \| \frac{\partial H(z)}{\partial C^T} \|_2^2. \quad (5.2)$$

The only difference between these two definitions lies in the first term, and in Section 5.3 we shall study the relationship between these two measures by computing the difference between $\|\frac{\partial H}{\partial A}\|_1^2$ and $\|\frac{\partial H}{\partial A}\|_2^2$.

In addition to its rather more intrinsic appeal as a coherent transfer function sensitivity measure, we shall see in Chapter 7 that this L_2 measure can also be obtained when the parameter errors are considered as noisy perturbations and when their effect on the transfer function is interpreted as a noise gain. This will be part of a new filter design procedure in which the coefficient errors and the roundoff noise errors are both treated as noise, thereby allowing for a synthetic design.

5.2.2 Optimal L_2 sensitivity realizations

In this subsection we examine the L_2-sensitivity measure minimization problem, and show how to compute the optimal FWL realizations that minimize this measure.

We first compute the new term $\| \frac{\partial H(z)}{\partial A} \|_2^2$ as a function of the system matrices (A, B, C) and examine its dependence with respect to similarity transformation matrices.

$$\begin{aligned}
\| \frac{\partial H(z)}{\partial A} \|_2^2 &= \frac{1}{2\pi j} \oint_{|z|=1} tr\{(\frac{\partial H(z)}{\partial A})(\frac{\partial H(z)}{\partial A})^H\} z^{-1} dz \\
&= tr\{\frac{1}{2\pi j} \oint_{|z|=1} (\frac{\partial H(z)}{\partial A})(\frac{\partial H(z)}{\partial A})^H z^{-1} dz\}) \\
&\triangleq tr(W_A) \quad (5.3)
\end{aligned}$$

where

$$W_A \triangleq \frac{1}{2\pi j} \oint_{|z|=1} (\frac{\partial H(z)}{\partial A})(\frac{\partial H(z)}{\partial A})^H z^{-1} dz \quad (5.4)$$

From (3.10) and (3.11), one has

$$\begin{aligned}
\frac{\partial H(z)}{\partial A} &= G(z) F^T(z) = \sum_{i=0}^{\infty} g(i) z^{-(i+1)} \sum_{k=0}^{\infty} f^T(j) z^{-(j+1)} \\
&= \sum_{i,j \geq 0} g(i) f^T(j) z^{-(j+i+2)} \triangleq \sum_{k=0}^{\infty} h(k) z^{-(k+2)} \quad (5.5)
\end{aligned}$$

where $F(z)$ and $G(z)$, defined in (3.11), have been rewritten as

$$F(z) \triangleq (zI - A)^{-1}B = \sum_{i=0}^{\infty} A^i B z^{-(i+1)}$$
$$\triangleq \sum_{i=0}^{\infty} f(i) z^{-(i+1)}, \qquad (5.6)$$

$$G(z) \triangleq (zI - A^T)^{-1}C^T = \sum_{i=0}^{\infty} (CA^i)^T z^{-(i+1)}$$
$$\triangleq \sum_{i=0}^{\infty} g(i) z^{-(i+1)}, \qquad (5.7)$$

and where $h(k)$ is defined as :

$$h(k) \triangleq \sum_{i+j=k} g(i) f^T(j). \qquad (5.8)$$

It follows from (5.4) and (5.5) that

$$W_A = \sum_{k=0}^{\infty} h(k) h^T(k). \qquad (5.9)$$

We now examine how W_A depends upon the choice of the coordinate system matrix T. If the realization (A, B, C) is obtained from an initial realization (A_0, B_0, C_0) by a similarity transformation T, the impulse response sequences $\{f(i)\}$ and $\{g(i)\}$ can be shown to have the following expressions:

$$f(i) = T^{-1} f_0(i), \quad g(i) = T^T g_0(i) \quad \forall i,$$

where $\{f_0(i)\}$ and $\{g_0(i)\}$ are the impulse responses of the vectors $F_0(z) = (zI - A_0)^{-1} B_0$ and $G_0(z) = (zI - A_0^T)^{-1} C_0^T$, respectively. So, we have $h(k) = T^T h_0(k) T^{-T}$ with $h_0(k)$ given by (5.8) but with $f(i)$ and $g(i)$ replaced by $f_0(i)$ and $g_0(i)$, respectively. Therefore,

$$W_A = \sum_{k=0}^{\infty} T^T h_0(k) T^{-T} T^{-1} h_0^T(k) T. \qquad (5.10)$$

With the expressions (5.3) and (5.10) for $\|\frac{\partial H(z)}{\partial A}\|_2^2$ and the earlier expressions (3.23) and (3.24) for $\|\frac{\partial H(z)}{\partial B}\|_2^2$ and $\|\frac{\partial H(z)}{\partial C^T}\|_2^2$, we can now express our new sensitivity measure M_{L_2} explicitly as a function of the transformation matrix T.

$$\begin{aligned} M_{L_2}(T) &= \|\frac{\partial H(z)}{\partial A}\|_2^2 + \|\frac{\partial H(z)}{\partial B}\|_2^2 + \|\frac{\partial H(z)}{\partial C^T}\|_2^2 \\ &= tr(W_A) + tr(W_o) + tr(W_c) \\ &= tr\left(\sum_{k=0}^{\infty} T^T h_0(k) T^{-T} T^{-1} h_0^T(k) T\right) + tr(T^T W_o^0 T) \\ &+ tr(T^{-1} W_c^0 T^{-T}), \end{aligned} \quad (5.11)$$

where (W_c^0, W_o^0) is the Gramian pair, as defined in Chapter 3, corresponding to the initial realization (A_0, B_0, C_0). Since $tr(AB) = tr(BA)$ for any A and B, we note that the three terms in (5.11) depend only on TT^T. Defining $P \triangleq TT^T$, we can rewrite $M_{L_2}(T)$ as

$$\begin{aligned} M_{L_2}(T) &= tr\left(\sum_{k=0}^{\infty} P h_0(k) P^{-1} h_0^T(k)\right) + tr(P W_o^0) + tr(P^{-1} W_c^0) \\ &\triangleq R(P) \end{aligned} \quad (5.12)$$

The optimal FWL design problem is to identify those realizations that minimize M_{L_2}:

$$\boxed{min_{(A,B,C) \in S_H} M_{L_2} \Longrightarrow (A, B, C)^{opt}.} \quad (5.13)$$

This is equivalent to

$$\min_{T: det T \neq 0} M_{L_2}(T) \Longrightarrow T^{opt} \quad (5.14)$$

or

$$\min_{P > 0} R(P) \Longrightarrow P^{opt}, \quad (5.15)$$

where $M_{L_2}(T)$ and $R(P)$ are expressed in terms of an initial realization (A_0, B_0, C_0).

In the next subsection, we first show that the optimization problem (5.15) has a unique solution and we give an algorithm for computing an optimal realization.

5.2.3 Solution of the optimal realization problem

The following lemma shows that the minimum of $R(P)$ exists, and that it is achieved by nonsingular P only.

114   5. A new transfer function sensitivity measure

Lemma 5.1 : *For a minimal realization of an asymptotically stable system $H(z)$, the minimum of $R(P)$ defined in (5.12) exists and is achieved by a nonsingular P.*

Proof: We note that the first term of $R(P)$ is positive and differentiable for $P > 0$. Since W_o^0 and W_c^0 are positive definite if the system is asymptotically stable and minimal, the lemma follows directly from Lemma 4.A in Appendix 4.A. ∎

Having proved the existence of a solution to the problem (5.15), we now show how to construct such a solution.

Theorem 5.1 : *For a minimal realization of an asymptotically stable minimal system $H(z)$, $R(P)$ has a unique global minimum. The solution of (5.15) is the unique solution of*

$$P\left[W_o^0 + \sum_{k=0}^{\infty} h_0(k) P^{-1} h_0^T(k)\right] P = W_c^0 + \sum_{k=0}^{\infty} h_0^T(k) P h_0(k). \quad (5.16)$$

Proof: Using Lemma 4.2, one can get the following expression

$$\frac{\partial R(P)}{\partial P} = \sum_{k=0}^{\infty} \{h_0(k) P^{-1} h_0^T(k) - P^{-1} h_0^T(k) P h_0(k) P^{-1}\}$$
$$+ W_o^0 - P^{-1} W_c^0 P^{-1}. \quad (5.17)$$

Any solution P of (5.15) must necessarily satisfy $\frac{\partial R(P)}{\partial P} = 0$. Since P must be nonsingular, (5.17) is equivalent with (5.16). It remains to be shown that the solution of (5.16) is unique. Let P_0 and P be two solutions of (5.16). We first show that, without loss of generality, one can assume $P_0 = I$. If P_0 is a solution of (5.16), corresponding to the triple $\{W_c^0, W_o^0, h_0(k)\}$ obtained from an initial realization $R_0 = (A_0, B_0, C_0)$, then the transformed realization

$$R_T = (P_0^{-1/2} A_0 P_0^{1/2}, P_0^{-1/2} B_0, C_0 P_0^{1/2})$$

produces a corresponding triple

$$\{P_0^{-1/2} W_c^0 P_0^{-1/2}, P_0^{1/2} W_o^0 P_0^{1/2}, P_0^{1/2} h_0(k) P_0^{-1/2}\}.$$

Clearly, the equation (5.16) determined by this triple has the identity matrix I as a solution. One then takes R_T as the initial realization with corresponding solution $P_o = I$. Now we consider any other solution, P, of (5.16). By Singular Value Decomposition (SVD), P can be decomposed as $P = VX^2V^T$, where V is some orthogonal matrix and $X = diag(x_1, x_2, \ldots, x_n)$, $x_i \geq x_{i+1} > 0$. So, (5.16) can be rewritten as

$$X^2 \left[\bar{W}_o^0 + \sum_{i=0}^{\infty} \bar{h}_0(i) X^{-2} \bar{h}_0^T(i) \right] X^2 = \bar{W}_c^0 + \sum_{i=0}^{\infty} \bar{h}_0^T(i) X^2 \bar{h}_0(i), \quad (5.18)$$

where $\bar{W}_c^0 = V^T W_c^0 V$, $\bar{W}_o^0 = V^T W_o^0 V$ and $\bar{h}_0(i) = V^T h_0(i) V \quad \forall i$. Similarly, for the solution $P_0 = I$, (5.16) yields

$$I \left[\bar{W}_o^0 + \sum_{i=0}^{\infty} \bar{h}_0(i) I \bar{h}_0^T(i) \right] I = \bar{W}_c^0 + \sum_{i=0}^{\infty} \bar{h}_0^T(i) I \bar{h}_0(i). \quad (5.19)$$

Indeed, if $P = I$ is the solution of (5.16) for some triple $\{W_c^0, W_o^0, h_0(k)\}$, then it is also solution of (5.16) for any other triple

$$\{U^T W_c^0 U, U^T W_o^0 U, U^T h_0(k) U\},$$

with U an orthogonal matrix. By taking $U = V$, (5.19) is obtained.

The (k,k)-th equations of (5.18) are given by

$$x_k^4 \left[\bar{W}_o^0(k,k) + \sum_{i=0}^{\infty} \left(\sum_{j=1}^{n} h_{kj}^2(i) x_j^{-2} \right) \right] = \bar{W}_c^0(k,k)$$

$$+ \sum_{i=0}^{\infty} \left(\sum_{j=1}^{n} h_{jk}^2(i) x_j^2 \right) \quad (5.20)$$

or

$$x_k^4 \left[\bar{W}_o^0(k,k) + \sum_{j=1}^{n} \left(\sum_{i=0}^{\infty} h_{kj}^2(i) \right) x_j^{-2} \right] = \bar{W}_c^0(k,k)$$

$$+ \sum_{j=1}^{n} \left(\sum_{i=0}^{\infty} h_{jk}^2(i) \right) x_j^2, \quad (5.21)$$

where $h_{jk}(i)$ is the (j,k)-th element of $\bar{h}_0(i)$. Similarly, for (5.19):

$$\bar{W}_o^0(k,k) + \sum_{j=1}^{n} \left(\sum_{i=0}^{\infty} h_{kj}^2(i) \right) = \bar{W}_c^0(k,k) + \sum_{j=1}^{n} \left(\sum_{i=0}^{\infty} h_{jk}^2(i) \right) \quad \forall k. \quad (5.22)$$

Letting $k = n$ in (5.21), one obtains

$$x_n^4 \left[\bar{W}_o^0(n,n) + \sum_{j=1}^{n} \left(\sum_{i=0}^{\infty} h_{nj}^2(i) \right) x_j^{-2} \right] = \bar{W}_c^0(n,n) + \sum_{j=1}^{n} \left(\sum_{i=0}^{\infty} h_{jn}^2(i) \right) x_j^2.$$

This leads to the following inequality by noting that $x_n \leq x_i$ for $i = 1, 2, \ldots, n-1$:

$$x_n^4 \left[\bar{W}_o^0(n,n) + x_n^{-2} \sum_{j=1}^{n} \left(\sum_{i=0}^{\infty} h_{nj}^2(i) \right) \right] \geq \bar{W}_c^0(n,n) + x_n^2 \sum_{j=1}^{n} \left(\sum_{i=0}^{\infty} h_{jn}^2(i) \right).$$

Equivalently,

$$\bar{W}_o^0(n,n) x_n^2 + \sum_{j=1}^{n} \left(\sum_{i=0}^{\infty} h_{nj}^2(i) \right) \geq \bar{W}_c^0(n,n) x_n^{-2} + \sum_{j=1}^{n} \left(\sum_{i=0}^{\infty} h_{jn}^2(i) \right).$$
(5.23)

Clearly, equality in (5.23) holds if and only if $x_i = x_n$ for all $i = 1, 2, \ldots, n-1$. Letting $k = n$ in (5.22) and comparing with (5.23) yields $x_n \geq 1$.

We now turn to the case $k = 1$. Expression (5.21) specializes to

$$x_1^4 \left[\bar{W}_o^0(1,1) + \sum_{j=1}^{n} \left(\sum_{i=0}^{\infty} h_{1j}^2(i) \right) x_j^{-2} \right] = \bar{W}_c^0(1,1) + \sum_{j=1}^{n} \left(\sum_{i=0}^{\infty} h_{j1}^2(i) \right) x_j^2.$$

This leads to the following inequality by noting that $x_1 \geq x_i$ for $i = 2, 3, \ldots, n$:

$$x_1^4 \left[\bar{W}_o^0(1,1) + x_1^{-2} \sum_{j=1}^{n} \left(\sum_{i=0}^{\infty} h_{1j}^2(i) \right) \right] \leq \bar{W}_c^0(1,1) + x_1^2 \sum_{j=1}^{n} \left(\sum_{i=0}^{\infty} h_{j1}^2(i) \right).$$

Equivalently,

$$\bar{W}_o^0(1,1) x_1^2 + \sum_{j=1}^{n} \left(\sum_{i=0}^{\infty} h_{1j}^2(i) \right) \leq \bar{W}_c^0(1,1) x_1^{-2} + \sum_{j=1}^{n} \left(\sum_{i=0}^{\infty} h_{j1}^2(i) \right).$$
(5.24)

Letting $k = 1$ in (5.22), one has

$$\bar{W}_o^0(1,1) + \sum_{j=1}^{n}\left(\sum_{i=0}^{\infty} h_{1j}^2(i)\right) = \bar{W}_c^0(1,1) + \sum_{j=1}^{n}\left(\sum_{i=0}^{\infty} h_{j1}^2(i)\right). \quad (5.25)$$

A comparison of (5.24) and (5.25) suggests that $x_1 \leq 1$. By combining the conclusions $x_1 \leq 1, x_n \geq 1$, and the prior assumption $x_1 \geq x_2 \geq \ldots \geq x_n$, it follows that $x_i = 1$ for $i = 1, \ldots, n$. This implies $X = I$ and $P = P_o$. ∎

It seems impossible to derive an explicit expression of the optimal P from (5.16). Just as in Chapter 4, one can always obtain this optimal P by an iterative procedure using a gradient algorithm:

$$P(k+1) = P(k) - \mu \frac{\partial R(P)}{\partial P}\bigg|_{P=P(k)} \quad (5.26)$$

where $\frac{\partial R(P)}{\partial P}$ is given by (5.17), and μ is a positive step size.

Since $R(P)$ has no local minimum (or maximum), this algorithm converges to the unique solution P^{opt} of (5.15) for any positive definite initial condition as long as μ is small enough. Just as in Chapter 4, the optimal realization set is then obtained from the initial realization by any similarity transformation T such that $T = P^{1/2}V$, where $P^{1/2}$ is any square root of P and V is any orthogonal matrix.

Comments :

1. Any square root of the optimal P^{opt} defines an optimal transformation: $P^{opt} = T^{opt}(T^{opt})^T$. The set of optimal transformations $\{T^{opt}\}$ is obtained by multiplying any arbitrary T^{opt} by the set of orthogonal matrices. This degree of freedom can be used to obtain some additional performance improvements by the use of special forms, such as Schur or Hessenberg forms. We shall discuss this in detail in Chapter 8.

2. The same minimization results are obtained by Helmke and Moore using Topology and Morse theories. It is shown [HM91] that the optimal realization subset can be obtained by using Gradient Flow techniques whose solution converges exponentially fast to the solution P^{opt}:

$$P^{opt} = \lim_{t \to \infty} P(t) \quad (5.27)$$

with $P(t)$ being updated by

$$\dot{P}(t) = \frac{\partial R[P(t)]}{\partial P(t)}. \tag{5.28}$$

Clearly, our algorithm (5.26) is a discrete version of the above Gradient Flows.

3. In an alternative approach Helmke and Moore also consider Gradient Flows evolving on the manifold of all realizations (A, B, C) of $H(z)$ [HM91]. This approach has the advantage of yielding a direct method for computing the subset of sensitivity optimal realizations by avoiding the intermediate step of computing coordinate transformation matrices. They show that their solutions converge exponentially fast to the optimal realization subset. Clearly, the Gradient Flow techniques require an analog computer to perform the computational work.

5.3 Relationship between L_1/L_2 and L_2 sensitivity measures

In the previous section, we have established the existence of optimal realizations that minimize the L_2 sensitivity measure. Now we show that this optimal realization subset and the one that minimizes the L_1/L_2 sensitivity measure are generally not identical.

In Chapter 3, we have seen that balanced realizations (A_b, B_b, C_b) belong to the set of L_1/L_2-optimal realizations. Such realizations are not unique, but one of them satisfies the following relationship [Kun78] and [WK83]:

$$A_b = Q A_b^T Q, \quad B_b = Q C_b^T, \tag{5.29}$$

where Q is a signature matrix, determined by the system itself.

According to Theorem 3.3, the L_1/L_2 optimal realization set is characterized by:

$$A^{opt}(M_{L_{12}}) = U^T A_b U, \quad B^{opt}(M_{L_{12}}) = U^T B_b, \quad C^{opt}(M_{L_{12}}) = C_b U, \tag{5.30}$$

where U is an arbitrary orthogonal matrix. It follows from (5.29) and (5.30) that

$$A^{opt}(M_{L_{12}}) = V^T (A^{opt}(M_{L_{12}}))^T V, \quad B^{opt}(M_{L_{12}}) = V^T (C^{opt}(M_{L_{12}}))^T, \tag{5.31}$$

where
$$V = U^T Q U = V^T. \tag{5.32}$$

Clearly, for any realization $(A^{opt}(M_{L_{12}}), B^{opt}(M_{L_{12}}), C^{opt}(M_{L_{12}}))$, which is L_1/L_2-optimal, one has

$$\begin{aligned}
F_{opt}(z) &\triangleq \sum_{i=0}^{\infty} f_{opt}(i) z^{-(i+1)} = (zI - A^{opt}(M_{L_{12}}))^{-1} B^{opt}(M_{L_{12}}) \\
&= (zI - V^T (A^{opt}(M_{L_{12}}))^T V)^{-1} (V^T (C^{opt}(M_{L_{12}}))^T) \\
&= V^T (zI - (A^{opt}(M_{L_{12}}))^T)^{-1} (C^{opt}(M_{L_{12}}))^T \\
&\triangleq V^T G_{opt}(z) = V^T \sum_{i=0}^{\infty} g_{opt}(i) z^{-(i+1)},
\end{aligned} \tag{5.33}$$

and hence

$$h_{opt}(k) \triangleq \sum_{i+j=k} g_{opt}(i) f_{opt}^T(j) = \sum_{i+j=k} g_{opt}(i) g_{opt}^T(j) V. \tag{5.34}$$

Therefore,

$$\begin{aligned}
h_{opt}(k) h_{opt}^T(k) &= \left\{ \sum_{i+j=k} g_{opt}(i) g_{opt}^T(j) \right\}^2 \quad \text{and} \\
h_{opt}^T(k) h_{opt}(k) &= V^T \left\{ \sum_{i+j=k} g_{opt}(i) g_{opt}^T(j) \right\}^2 V.
\end{aligned} \tag{5.35}$$

We now consider equation (5.16) that characterizes the L_2-optimal realizations, and we take the initial realization (A_0, B_0, C_0) in (5.16) to be any L_1/L_2-optimal realization $(A^{opt}(M_{L_{12}}), B^{opt}(M_{L_{12}}), C^{opt}(M_{L_{12}}))$, for which $W_o^0 = W_c^0$. We examine whether $P = I$ is a solution of this equation. A positive answer would indicate that $(A^{opt}(M_{L_{12}}), B^{opt}(M_{L_{12}}), C^{opt}(M_{L_{12}}))$ is also an L_2-optimal realization, since $P = I$ implies $T = I$.

Consider thus $P = I$ in (5.16), together with (5.35) and $W_c^0 = W_o^0$. It then reduces to

$$\sum_{k=0}^{\infty} \left\{ \sum_{i+j=k} g_{opt}(i) g_{opt}^T(j) \right\}^2 = V^T \left[\sum_{k=0}^{\infty} \left\{ \sum_{i+j=k} g_{opt}(i) g_{opt}^T(j) \right\}^2 \right] V. \tag{5.36}$$

120   5. A new transfer function sensitivity measure

With an obvious definition for M, this can be rewritten as

$$M = U^T Q U M U^T Q U, \qquad (5.37)$$

where M is a given symmetric positive definite matrix, Q is a given signature matrix, and U is an arbitrary orthogonal matrix. Let $M = W^T D^2 W$ be the SVD of M. Two cases can be distinguished :

1. if $D^2 = I$, i.e. $M = I$, then (5.37) is satisfied for any orthogonal U, and hence the two optimal realization sets are identical. However, there is no indication that a balanced realization should produce $M = I$.

2. if $D^2 \neq I$, then (5.37) is satisfied for any orthogonal U if and only if

$$Q = \pm I. \qquad (5.38)$$

Since Q is determined by the system $H(z)$ itself, it does not hold in general that the L_1/L_2-optimal realization subset and the L_2-optimal realization subset are identical.

Comments:

1. It follows from (5.29)-(5.30) that if $Q = \pm I$, then $A^{opt}(M_{L_{12}})$ is normal.[1] We shall see in Chapter 6 that normal matrices have a minimal pole sensitivity. Hence, if a system has the property that $Q = \pm I$, then the optimal realization sets minimizing the L_1/L_2 sensitivity measure and the L_2 sensitivity measure are identical, and these optimal realizations also minimize the pole sensitivity measure.

2. Even in the special case where the two optimal realization subsets are identical, this does not imply that the two measures take the same minimum value. In our next theorem we show that, for every realization, the L_2 measure is larger than the L_1/L_2 measure.

Theorem 5.2 : *For every minimal realization of an asymptotically stable system $H(z)$, the L_2 sensitivity measure is larger than the L_1/L_2 sensitivity measure.*

---

[1] We recall that a matrix A is normal if $AA^T = A^T A$.

Proof: For a given realization (A, B, C) the difference between the two measures is

$$M_{L_2} - M_{L_{12}} = \| \frac{\partial H(z)}{\partial A} \|_2^2 - \| \frac{\partial H(z)}{\partial A} \|_1^2 \qquad (5.39)$$

It follows from Definition 3.2 that

$$\| \frac{\partial H(z)}{\partial A} \|_1^2 = \left\{ \frac{1}{2\pi} \int_0^{2\pi} \| \frac{\partial H(e^{j\omega})}{\partial A} \|_F^1 d\omega \right\}^{(1/1) \times 2}$$

$$= \left\{ \frac{1}{2\pi} \int_0^{2\pi} \| \frac{\partial H(e^{j\omega})}{\partial A} \|_F^1 d\omega \right\}^2$$

and

$$\| \frac{\partial H(z)}{\partial A} \|_2^2 = \left\{ \frac{1}{2\pi} \int_0^{2\pi} \| \frac{\partial H(e^{j\omega})}{\partial A} \|_F^2 d\omega \right\}^{(1/2) \times 2}$$

$$= \frac{1}{2\pi} \int_0^{2\pi} \| \frac{\partial H(e^{j\omega})}{\partial A} \|_F^2 d\omega.$$

Denote $f(\omega) \triangleq \| \frac{\partial H(e^{j\omega})}{\partial A} \|_F$. Then :

$$M_{L_2} - M_{L_{12}} = \frac{1}{2\pi} \int_0^{2\pi} f^2(\omega) d\omega - \left[\frac{1}{2\pi} \int_0^{2\pi} f(\omega) d\omega \right]^2. \qquad (5.40)$$

For those readers familiar with probability theory the right hand side is clearly nonnegative, and for those who have to admit they have never heard of a mean square error and a variance, we offer the following argument:

$$f(\omega) = \frac{1}{2\pi} \int_0^{2\pi} f(\omega) d\omega + \left[f(\omega) - \frac{1}{2\pi} \int_0^{2\pi} f(\omega) d\omega \right].$$

Taking the square on both sides and integrating yields :

$$\frac{1}{2\pi} \int_0^{2\pi} f^2(\omega) d\omega = \left[\frac{1}{2\pi} \int_0^{2\pi} f(\omega) d\omega \right]^2$$
$$+ \frac{1}{2\pi} \int_0^{2\pi} \left[f(\omega) - \frac{1}{2\pi} \int_0^{2\pi} f(\omega) d\omega \right]^2 d\omega.$$

Comparing with (5.40), this shows that $M_{L_2} \geq M_{L_{12}}$ with equality if and only if $\| \frac{\partial H(e^{j\omega})}{\partial A} \|_F$ is constant over all ω. This will never happen except in pathological cases. ∎

As a direct result of Theorem 5.2, we have the following corollary:

Corollary 5.1 : *The minimum of the L_2 sensitivity measure is larger than that of the L_1/L_2 sensitivity measure.*

Remark: if we consider that the L_2 sensitivity measure is the real sensitivity measure of the transfer function $H(z)$, then the L_1/L_2 sensitivity measure is a lower bound of this measure and this lower bound is never achieved.

5.4 An example

Consider a narrow band low pass filter of order four given by its controllable realization R_c.

$\boxed{R_c :}$

$$A_c = \begin{pmatrix} 3.5906 & -4.8535 & 2.9260 & -0.6636 \\ 1.0000 & 0 & 0 & 0 \\ 0 & 1.0000 & 0 & 0 \\ 0 & 0 & 1.0000 & 0 \end{pmatrix}$$

$$B_c = \begin{pmatrix} 1 & 0 & 0 & 0 \end{pmatrix}^T$$

$$C_c = 10^{-3} \times \begin{pmatrix} 0.2353 & 0.0355 & 0.2147 & 0.0104 \end{pmatrix}$$

$$d = 3.0996 \times 10^{-5}. \tag{5.41}$$

The corresponding Gramians are:

$$W_{cc} = 10^5 \times \begin{pmatrix} 2.1022 & 2.0914 & 2.0593 & 2.0067 \\ 2.0914 & 2.1022 & 2.0914 & 2.0593 \\ 2.0593 & 2.0914 & 2.1022 & 2.0914 \\ 2.0067 & 2.0593 & 2.0914 & 2.1022 \end{pmatrix}$$

$$W_{co} = \begin{pmatrix} 0.0512 & -0.1328 & 0.1157 & -0.0338 \\ -0.1328 & 0.3453 & -0.3012 & 0.0880 \\ 0.1157 & -0.3012 & 0.2630 & -0.0769 \\ -0.0338 & 0.0880 & -0.0769 & 0.0225 \end{pmatrix}. \tag{5.42}$$

A balanced realization that minimizes the L_1/L_2 sensitivity measure is:

5. A new transfer function sensitivity measure

$\boxed{R^{opt}(M_{L_{12}}):}$

$$A^{opt}(M_{L_{12}}) = \begin{pmatrix} 0.9813 & 0.0927 & -0.0222 & 0.0098 \\ -0.0927 & 0.9378 & 0.1076 & -0.0247 \\ -0.0222 & -0.1076 & 0.8734 & 0.0963 \\ -0.0098 & -0.0247 & -0.0963 & 0.7982 \end{pmatrix}$$

$$B^{opt}(M_{L_{12}}) = \begin{pmatrix} 0.1670 & 0.2221 & 0.1568 & 0.0540 \end{pmatrix}^T$$

$$C^{opt}(M_{L_{12}}) = \begin{pmatrix} 0.1670 & -0.2221 & 0.1568 & -0.0540 \end{pmatrix} \quad (5.43)$$

The realization $R^{opt}(M_{L_{12}})$ is obtained from the controllable realization R_c through the transformation

$$T^{opt}(M_{L_{12}}) = \begin{pmatrix} 366.8884 & -424.3290 & 227.3725 & -31.2340 \\ 337.3366 & -452.2991 & 314.1244 & -95.1682 \\ 306.5954 & -471.0780 & 404.9193 & -186.4350 \\ 275.1450 & -480.9063 & 495.4803 & -311.6325 \end{pmatrix}.$$

By using the algorithm (5.26) proposed in Section 5.2 with $R^{opt}(M_{L_{12}})$ as initial realization, one obtains the following optimal realization that minimizes the L_2 sensitivity measure:

$\boxed{R^{opt}(M_{L_2}):}$

$$A^{opt}(M_{L_2}) = \begin{pmatrix} 0.9718 & 0.0935 & -0.0290 & 0.0110 \\ -0.0939 & 0.9197 & 0.1002 & -0.0187 \\ -0.0295 & -0.1025 & 0.8903 & 0.0910 \\ -0.0113 & -0.0179 & -0.0865 & 0.8087 \end{pmatrix}$$

$$B^{opt}(M_{L_2}) = \begin{pmatrix} 0.1817 & 0.2585 & 0.1967 & 0.0642 \end{pmatrix}^T$$

$$C^{opt}(M_{L_2}) = \begin{pmatrix} 0.1761 & -0.2496 & 0.1869 & -0.0623 \end{pmatrix}. \quad (5.44)$$

The corresponding transformation from the controllable realization R_c is the product of $T^{opt}(M_{L_{12}})$ and of the square root of the optimal P obtained as the solution of the recursive algorithm. It is given by:

$$T^{opt}(M_{L_2}) = \begin{pmatrix} 383.8532 & -469.9862 & 282.1280 & -42.8333 \\ 355.9921 & -507.6227 & 374.7432 & -111.7574 \\ 326.3468 & -536.6054 & 471.6920 & -208.1652 \\ 295.3544 & -556.6425 & 569.1714 & -338.3927 \end{pmatrix}.$$

5. A new transfer function sensitivity measure

The L_2 sensitivity measures for the three realizations R_c, $R^{opt}(M_{L_{12}})$ and $R^{opt}(M_{L_2})$ are given in Table 5.1. For the sake of comparison, the L_1/L_2 sensitivity measures are also given for the same three realizations.

	Sensitivity measure	
	L_2	L_1/L_2
R_c	1.0613×10^7	1.4144×10^6
$R^{opt}(M_{L_{12}})$	27.7458	5.2035
$R^{opt}(M_{L_2})$	26.6919	5.3557

TABLE 5.1. L_2 and L_1/L_2 measures for three different realizations.

Comments:

- We first note that the optimal realizations (whether with respect to the L_2 measure or with respect to the L_1/L_2 measure) have sensitivities that are several orders of magnitude smaller than the controllable realization, whatever the norm that is chosen to measure these sensitivities. This confirms our earlier examples and findings that all indicate poor sensitivity performance for the canonical forms.

- The L_2-optimal realization has a slightly smaller L_2 measure than the L_1/L_2-optimal realization, and the converse holds with respect to the L_1/L_2 measure. This is probably an indication that a realization that is optimal with respect to one particular transfer function sensitivity measure is near optimal with respect to another such measure.

- The computations confirm our theoretical results that the L_2 measure of a realization is always larger than the L_1/L_2 measure of the same realization.

5.5 Conclusions

We have proposed a new transfer function sensitivity measure that has the merit of being more logically coherent in that it uses the same norms for the sensitivities of the transfer function with respect to the three matrices A, B, and C. This has led us to a measure, called L_2 measure, whose coordinate dependence is significantly more complicated than that of the earlier L_1/L_2 measure. However, we have observed that the optimization techniques laid out in Chapter 4 - and in particular the proofs of existence and uniqueness of the solution - can be transported *mutatis mutandis* to the solution of this new problem. The efforts extended in the present chapter to crack down the minimization of our new L_2 measure will become fully vindicated in Chapter 7 where we shall propose a synthetic sensitivity plus roundoff noise minimizing strategy, which will lead to a criterion that looks strikingly similar to the L_2 sensitivity measure of this chapter.

But before we address the roundoff noise minimization problem (in a synthetic 'sensitivity plus roundoff noise' design) we shall address another sensitivity minimizing realization problem, namely the design of state space realizations that minimize the sensitivity of pole and zero locations with respect to coefficients of the realization.

6

Pole and zero sensitivity minimization

6.1 Introduction

The classical results reviewed in Chapter 3 on minimum sensitivity realizations and our extensions of Chapters 4 and 5 all dealt with the problem of minimizing some sensitivity measure of the transfer function with respect to the coefficients of all equivalent state space realizations. In practice it is often the case that it is more important to minimize the errors in the location of certain poles and zeros of a system with respect to coefficient quantization than the errors in the transfer function. Examples of such situations are the design of notch filters, where the zero locations need to be very precise, or the design of controllers where a high precision might be desired on the dominant pole of the closed loop system. This chapter therefore looks at various ways of defining pole and/or zero sensitivities, and of solving the corresponding optimal filter design problems.

In [Kai66], Kaiser considered the pole and zero sensitivity with respect to the coefficients of the denominator and numerator of the transfer function $H(z)$. Clearly, a parametrization in terms of numerator and denominator coefficients corresponds to a special case of state space realizations of a system, namely the *direct form* discussed in Chapter 3. Mantey [Man68] was probably the first to study the pole sensitivity problem with respect to realizations. He argued that in order to minimize the pole sensitivity measure the poles have to be implemented directly (in block diagonal forms). But this conclusion was arrived at by comparing a few realizations and hence it falls short of a global analysis. A global minimization of a pole sensitivity measure with respect to the whole realization set is evidently of importance.

A transfer function can be entirely characterized by its poles and zeros to within a constant multiplier. When a system is parametrized by a state space model, it can be shown (and will be in this chapter) that the pole and zero sensitivities depend on the choice of the particular realization. It is therefore interesting to find realizations that minimize the sensitivity of the poles, the zeros, or some poles and/or some zeros with respect to the coefficients of these realizations.

Classically, the pole and zero sensitivities are studied separately. Since pole and zero perturbations both affect the same system, it is desirable to have a measure that unifies these two sensitivities. In addition, there are practical situations where errors on some particular poles (or zeros) can be more damaging than errors on others, and where one therefore seeks a measure that allows one to impose a tighter precision on some poles and/or zeros than on others. In this chapter, we introduce such a general pole-zero sensitivity measure, where a particular weight is allowed on each pole and each zero. It is a generalization of the separate pole sensitivity and zero sensitivity measures already introduced in the literature (see e.g. [SW84] and [Wil86]). We show how to compute this measure for a particular state space realization and we then study its dependence with respect to the class of similarity transformations. We are then in a position to pose and solve the pole-zero sensitivity minimization problem, where the minimization is again performed with respect to the class of all equivalent state space realizations. We note that pole and zero sensitivities exist only in the case of single poles (or single zeros); the state variable realization optimization games can therefore only be played in this particular case.

The outline of the chapter is as follows. In Section 6.2 we define our global pole-zero sensitivity measure and compare it briefly with more classical measures. As a function of a realization (A, B, C, d), the poles and zeros can be considered as eigenvalues of matrices, at least - for the zeros - if $d \neq 0$. We limit our analysis to this case of direct feedthrough systems, and we carefully derive expressions for the general eigenvalue sensitivity problem in Section 6.3. In Section 6.4 we give the optimal solution for the rather classical case of an unweighted pole sensitivity measure which leads to normal matrices. The computation of the zero sensitivity measure is detailed in Section 6.5, while Section 6.6 is devoted to computing the dependence of the pole-zero measure with respect to the coordinates of the state realization. Optimal realizations minimizing the pole-zero sensitivity measure are computed in Section 6.7, and simulations are presented in Section 6.8.

6.2 A pole-zero sensitivity measure

Consider a discrete-time linear time-invariant system characterized by its transfer function $H(z)$ and let (A, B, C, d) be a state space realization of this system, with $H(z) = d + C(zI - A)^{-1}B$. If we denote by $\lambda(M)$ the eigenvalue set of a matrix M, then the poles $\{\lambda_k\}$ of $H(z)$ are the eigenvalues of A, $\{\lambda_k\} = \lambda(A)$, while the zeros $\{v_k\}$ of $H(z)$ are the eigenvalues of

$$Z \triangleq A - d^{-1}BC \tag{6.1}$$

provided $d \neq 0$. That is $\{v_k\} = \lambda(Z)$ (see [DW78]). Throughout this chapter, our zero sensitivity analysis will be limited to the case of systems without delay, i.e. $d \neq 0$.[1]

When the parameters in the matrices A, B, C, d of the state space realization are implemented with error (as is the case in a FWL implementation of the system) this produces an error in the poles and zeros of the system. We note that the poles are affected only by the parameters of the matrix A, while the zeros are affected by all four elements of the realization. In computing the effect of parameter errors on the poles and zeros of the system, one must clearly consider each of these poles and zeros individually. Indeed, there is no a priori reason to believe that errors on the parameters of the matrix A, say, have the same effect on a pole that is close to zero and on one that is close to the unit circle, or on one that is close to another pole. Indeed, simulations show these effects to be different. One of our tasks in this section, therefore, will be to compute the *pole sensitivity functions* $\{\frac{\partial \lambda_k}{\partial A}\}$ and the *zero sensitivity functions* $\{\frac{\partial v_k}{\partial A}, \frac{\partial v_k}{\partial B}, \frac{\partial v_k}{\partial C}, \frac{\partial v_k}{\partial d}\}$ of individual poles and zeros with respect to the matrices A, B, C, d. Assuming that these sensitivity functions can be computed, we shall then adopt the following global *pole-zero sensitivity measure*, denoted M_{pz}.

Pole-zero sensitivity measure

$$\begin{aligned} M_{pz} = \sum_{k=1}^{n} \{ & w_{\lambda_k} \|\frac{\partial \lambda_k}{\partial A}\|_F^2 + w_{v_k} [\|\frac{\partial v_k}{\partial A}\|_F^2 + \|\frac{\partial v_k}{\partial B}\|_F^2 \\ & + \|\frac{\partial v_k}{\partial C}\|_F^2 + \|\frac{\partial v_k}{\partial d}\|_F^2] \} \end{aligned} \tag{6.2}$$

---

[1] All our results on pole sensitivity hold without this assumption, however.

130   6. Pole and zero sensitivity minimization

where $\{w_{\lambda_k}, k = 1, \ldots, n\}$ and $\{w_{v_k}, k = 1, \ldots, n\}$ are nonnegative weightings that reflect the relative importance that the designer may care to attach to the sensitivity of individual poles and/or zeros in a sensitivity minimization problem. For example, in an optimal FWL realization design for a filter one may often be much less worried about errors on fast time constants (corresponding to poles well within the unit circle) than about errors on slow time constants (corresponding to poles close to the unit circle). From a practical point of view, it therefore makes more sense to minimize a weighted pole sensitivity measure, or even a weighted pole-zero measure, and we shall return to this problem in the following sections.

Comment: Classically, separate measures have been defined for the sensitivities of the poles and of the zeros with respect to parameter errors. For example, in [SW84] the following pole sensitivity measure was adopted and studied

$$\Psi_p = \sum_{k=1}^{n} \|\frac{\partial \lambda_k}{\partial A}\|_F^2, \qquad (6.3)$$

while in [Wil86] the zero sensitivity measure was defined as

$$\Psi_z = \sum_{k=1}^{n} \|\frac{\partial v_k}{\partial Z}\|_F^2 \qquad (6.4)$$

with $Z = A - d^{-1}BC$. We note that the pole sensitivity measure Ψ_p is a special case of our measure M_{pz} in which the weightings w_{λ_k} have all been set to 1 while the weightings w_{v_k} have been set to 0. As for the zero sensitivity measure Ψ_z, it seems to have been chosen more for its ease of computation than for its usefulness. Indeed, Ψ_z reflects the effect on the zero locations of errors in Z rather than errors in A, B, C, d. We shall define a more appropriate version of (6.4) in which the sensitivities are taken with respect to the elements of A, B, C, d.

Our global pole-zero sensitivity measure (6.3) allows much greater flexibility (through the choice of appropriate weightings w_{λ_k} and w_{v_k}) than the separate pole and zero measures and includes them as special cases. Our main contribution in this chapter will be to solve the optimal realization problem for the minimization of the pole-zero sensitivity measure M_{pz}. Before we can embark on this optimization exercise, we need to obtain a computable expression of this measure as a function of system parameters and to display its dependence on the particular choice of coordinate space.

Since the weighting factors w_{λ_k} and w_{v_k} are the designer's choice and reflect the particular application, the computation of M_{pz} in (6.2) requires that we obtain expressions for the sensitivity of the poles $\{\lambda_k\}$ and of the zeros $\{v_k\}$ with respect to the elements in a realization (A,B,C,d) of $H(z)$, namely $\{\frac{\partial \lambda_k}{\partial A}\}$ and $\{\frac{\partial v_k}{\partial B}, \frac{\partial v_k}{\partial B}, \frac{\partial v_k}{\partial C}, \frac{\partial v_k}{\partial d}\}$. We note that the question of pole sensitivities appears much simpler since the poles are just the eigenvalues of A. We therefore start our analysis with a derivation of the sensitivities of the eigenvalues of a matrix with respect to changes in the elements of this matrix.

6.3 The eigenvalue sensitivity problem

Given the all pervasive role that eigenvalues play in daily life (at least in the lives of mathematicians and engineers), questions about numerical procedures for the computation of eigenvalues and about the sensitivity of the result to errors in the entries of the matrix have haunted the nights of many mathematicians and numerical analysts. The sensitivity analysis - more often called perturbation analysis in mathematical circles - is rendered even more relevant by the fact that algorithms for the computation of eigenvalues are iterative in nature, and the eventual result is therefore at best an approximation of the exact eigenvalues.

It is therefore not surprising that there is an extensive literature on the *perturbation of eigenvalues*: see e.g. [SS90], [GL83] and [Ste73]. Many of these results produce bounds on the distance between the eigenvalues of a matrix M and those of a perturbed matrix $M + E$ in terms of the norm of this perturbation E. Here we are interested in sensitivity results, because our ultimate aim is to formulate and solve sensitivity optimization problems with respect to the set of equivalent realizations. Results on the sensitivity of an eigenvalue are rather more scarce; they are also, of necessity, limited to the case where the eigenvalue is a differentiable function of the matrix. We shall see that this requires that the eigenvalue be simple (as opposed to a repeated or multiple eigenvalue). A rigorous derivation of the eigenvalue sensitivity can be found in [SS90] using the powerful theory of Gerschgorin disks. A much simpler derivation can be found in [SW84]; however, it is incorrect because it fails to take into account the fact that the eigenvectors (and not just the eigenvalues) of a perturbed matrix are also perturbed.

6. Pole and zero sensitivity minimization

At the risk of boring some readers with our appeal to such high level concepts as the implicit function theorem, we give here a self-contained derivation of the computation of the sensitivity of an eigenvalue. The impatient reader can jump straight to the result in expressions (6.24) and (6.25).

Let $M \in \mathbb{R}^{n \times n}$, and let $\{\lambda_k\}$ be the eigenvalue set of M: $\{\lambda_k = \lambda(M)\}$. Define $P \triangleq (\lambda, M)$ where $\lambda \in \mathbb{C}$ is some scalar complex variable and define a scalar function f of P as:

$$f(P) = det(\lambda I - M). \tag{6.5}$$

Clearly,

$$f(P) = \prod_{k=1}^{n}(\lambda - \lambda_k), \tag{6.6}$$

and hence

$$\frac{\partial f(P)}{\partial \lambda} = \sum_{k=1}^{n} \prod_{l \neq k}(\lambda - \lambda_l). \tag{6.7}$$

Now, let M^* be a particular matrix in $\mathbb{R}^{n \times n}$ whose eigenvalue set is $\{\lambda_k^*\}$. The matrix M^*, together with each of its eigenvalues λ_k^*, form a particular point set $\{P_k^* = (\lambda_k^*, M^*), k = 1, \ldots, n\}$. We now derive expressions for the sensitivity of one of these eigenvalues, λ_k^*, with respect to changes in M^*. We first note that

$$\frac{\partial f(P_k^*)}{\partial \lambda_k^*} = \prod_{l \neq k}(\lambda_k^* - \lambda_l^*). \tag{6.8}$$

Therefore $\frac{\partial f(P_k^*)}{\partial \lambda_k^*} \neq 0$ if and only if λ_k^* is not a repeated eigenvalue of M^*. In the sequel, we shall assume that M^* has no repeated eigenvalues; *thus our eigenvalue sensitivity analysis will be limited to the case of distinct eigenvalues.*

Since $f(P_k^*) = 0$ and $\frac{\partial f(P_k^*)}{\partial \lambda_k^*} \neq 0$, it follows by the implicit function theorem [Spi65] that there exists a unique function

$$\lambda_k = g_k(M), \tag{6.9}$$

which satisfies $f(g_k(M), M) = 0$ in some neighbourhood of M^*. In addition, in this neighbourhood we have:

$$\frac{\partial f}{\partial m_{ij}} + \frac{\partial f}{\partial g_k}\frac{\partial g_k}{\partial m_{ij}} = 0. \tag{6.10}$$

Therefore, in this neighbourhood $\frac{\partial \lambda_k}{\partial m_{ij}}$ exists and is given by

$$\frac{\partial \lambda_k}{\partial m_{ij}} \triangleq \frac{\partial g_k(M)}{\partial m_{ij}} = -\frac{\frac{\partial f(P)}{\partial m_{ij}}}{\frac{\partial f(P)}{\partial \lambda_k}}. \tag{6.11}$$

We have just shown that the expression (6.11) for $\frac{\partial \lambda_k}{\partial m_{ij}}$ is valid if M has no repeated eigenvalues, and in the process we have derived the expression (6.7) for the denominator. In order to compute the numerator, we need to invoke some well known properties of matrix algebra.

Denote $W(P) \triangleq \lambda I - M = \{w_{ij}\}$. Then the following are two well known properties of matrices:

$$f(P) \triangleq det W = \sum_{j=1}^{n} w_{ij} det W_{ij} \quad \text{for } i = 1, \ldots, n, \tag{6.12}$$

and

$$(det W)I = W\, adj W, \tag{6.13}$$

where W_{ij} is the (i,j)-th minor of W, i.e. it is the $(n-1) \times (n-1)$ matrix obtained by deleting the i-th row and j-th column of $(-1)^{i+j}W$, and $adj W$ is the adjoint matrix defined as

$$(adj W)_{ij} = det W_{ji}, \quad i,j = 1, \ldots, n. \tag{6.14}$$

We now show that, for a minimal realization (A, B, C, d) of a system, if M is replaced by either A or $Z \triangleq A - d^{-1}BC$, then the matrix W has rank at least equal to $n-1$, and the adjoint is not a zero matrix.

Lemma 6.1 : *Let (A, B, C, d) be any minimal realization of a transfer function $H(z)$. Then, with $M = A$ or $M = A - d^{-1}BC$, with $W = \lambda I - M$ and for any $\lambda \in \mathbb{C}$, the rank of W is at least equal to $n-1$ and the matrix $adj W$ is not the zero matrix, i.e. there exists at least one element of $\{det W_{ij}, i, j = 1, \ldots, n\}$ that is nonzero.*

Proof: Consider first a controllable realization (A_c, B_c, C_c, d) of $H(z)$ with A_c in top companion form. Then $Z_c = A_c - d^{-1}B_cC_c$ is also in top companion form. Now, if $W_c = \lambda I - M_c$ with $M_c = A_c$ or $M_c = Z_c$, then one can see that $det W_{c1n} = (-1)^{n-1}$ for any λ. This means that W_c has a rank no smaller than $(n-1)$ for any λ. Consider now an arbitrary realization (A, B, C, d). Then there exists a nonsingular matrix T such that $A = T^{-1}A_cT$. But then, for the two possible choices of M and M_c

above, we have $W \triangleq \lambda I - M = T^{-1}W_c T$, and hence W also has a rank at least equal to $(n-1)$ for any λ, which implies that W has at least one $(n-1) \times (n-1)$ full rank submatrix. Thus there exists at least a $detW_{ij} \neq 0$ for some i, j. ∎

We are now able to compute $\frac{\partial f(P)}{\partial M}$ and hence $\frac{\partial \lambda}{\partial M}$. Noting that $w_{ij} = -m_{ij}$ if $i \neq j$, $w_{ii} = \lambda - m_{ii}$ and that W_{ij} is independent of m_{ij} for all $i, j = 1, \ldots, n$, we have

$$\frac{\partial f(P)}{\partial m_{ij}} = -detW_{ij} \quad \forall i, j, \tag{6.15}$$

and hence

$$\frac{\partial f(P)}{\partial M} = -\{adjW(P)\}^T. \tag{6.16}$$

Combining (6.11) and (6.16) finally yields the following expression:

$$\frac{\partial \lambda}{\partial M} = \{adjW(P)\}^T \{\frac{\partial f(P)}{\partial \lambda}\}^{-1}, \tag{6.17}$$

where $\frac{\partial f(P)}{\partial \lambda}$ is given by (6.7). Clearly, when M goes to M^* (and thus λ_l goes to λ_l^* for $l = 1, \ldots, n$) and when λ tends to λ_k^*, expression (6.17) yields the sensitivity of λ_k^* with respect to M^*.

Comments:

- We note that when P goes to P_k^* for some $k \in \{1, \ldots, n\}$, then $\frac{\partial f(P)}{\partial \lambda}$ tends to $\frac{\partial f(P_k^*)}{\partial \lambda_k^*} = \prod_{l \neq k}(\lambda_k^* - \lambda_l^*)$, while $adjW(P)$ tends to a matrix that is different from the zero matrix as shown in Lemma 6.1 and that is actually of rank 1 as shown from (6.13). The result (6.17) then shows that the closer the eigenvalue λ_k^* is to some other eigenvalue, the higher its sensitivity to M^*. Eigenvalues will be close together when a discrete time system is obtained from a continuous time system by fast sampling, in which case the poles of the discrete time system are clustered around $z = +1$. Another example are low pass narrow band digital filters, as illustrated in Chapter 3.

- The formula (6.17) has been derived under the assumption that λ is a differentiable function of M. In the case of multiple eigenvalues, this is no longer true and one should therefore not use the above analysis to infer that the sensitivity of a multiple eigenvalue is infinite, as incorrectly asserted by some authors (see e.g. [Wil91]). In fact, the sensitivity is undefined when M^* has an eigenvalue λ_k of algebraic multiplicity $m > 1$. In such case, a different analysis, using higher order Taylor series expansions, can be used to compute the finite displacement of multiple eigenvalues under the effect of matrix perturbations.[2] We repeat that our subsequent discussion is limited to the case of distinct eigenvalues, that is M^* and hence M (which is a matrix in the neighbourhood of M^*) is assumed to have simple eigenvalues.

Using (6.17) as a starting point, we now derive an expression for the sensitivity $\frac{\partial \lambda_k^*}{\partial M^*}$ directly in terms of left and right eigenvectors of M^*. Let $X = (x_1, x_2, ..., x_n)$ be a matrix formed by n right eigenvectors of M corresponding to the eigenvalues $(\lambda_1, ..., \lambda_n)$, i.e. $Mx_k = \lambda_k x_k$ for $k = 1, ..., n$. Let $D \triangleq diag(\lambda_1, \lambda_2, ..., \lambda_n)$; hence $MX = XD$. Since the $\{\lambda_k\}$ are distinct, X is nonsingular. It then follows that $M = XDX^{-1}$ and hence

$$W(P) = \lambda I - M = X(\lambda I - D)X^{-1}. \qquad (6.18)$$

As long as $\lambda \neq \lambda_i, i = 1, ..., n$, the matrix $\lambda I - D$ is invertible. It follows from (6.6) and (6.13) that

$$\begin{aligned} adjW(P) &= X(\lambda I - D)^{-1} X^{-1} detW \\ &= X[(\lambda I - D)^{-1} \prod_{i=1}^{n}(\lambda - \lambda_i)]X^{-1} \\ &= X[diag(\prod_{i \neq 1}(\lambda - \lambda_i), \prod_{i \neq 2}(\lambda - \lambda_i), ..., \prod_{i \neq n}(\lambda - \lambda_i))]X^{-1}. \end{aligned}$$
$$(6.19)$$

If we now let P tend to P_k^* (i.e. $\lambda \to \lambda_k^*$ and $M \to M^*$), then all the diagonal elements except the k-th one tend to zero in the above expression, which converges to

$$adjW(P_k^*) = X^* E_k X^{*-1} \prod_{l \neq k}(\lambda_k^* - \lambda_l^*), \qquad (6.20)$$

---

[2]The authors would like to thank Professor G.H. Allen for bringing this fact to their attention.

136   6. Pole and zero sensitivity minimization

where $E_k = diag(0, ..., 0, 1, 0, ..., 0)$ with the 'one' in the k-th position. The expression for the sensitivity now follows from (6.17) and (6.20):

$$\left(\frac{\partial \lambda_k^*}{\partial M^*}\right)^T = X^* E_k X^{*-1}, \qquad (6.21)$$

where X^* is the matrix of right eigenvectors of M^*. From now on, we drop the superscripts "*", and we rewrite (6.21) as

$$\left(\frac{\partial \lambda_k}{\partial M}\right)^T = X E_k X^{-1} \triangleq X E_k Y^H = x_k y_k^H \qquad (6.22)$$

where X is the matrix of right eigenvectors of M and $Y = (y_1 \; y_2 \; \cdots \; y_n) \triangleq X^{-H}$ with "H" denoting the transpose and conjugate operation. Since $M = XDX^{-1}$, we have $X^{-1}M = DX^{-1}$. Therefore, with $Y^H \triangleq X^{-1}$, it follows that the vectors $\{y_k\}$ are the reciprocal left eigenvectors of M.

The expression (6.22) has been obtained without any particular normalization on the right eigenvectors x_k, but with the constraint $y_k^H x_k = 1$ on y_k. A special case of (6.21) is obtained when the right eigenvectors are normalized, $x_k^H x_k = \|x_k\|_2^2 = 1$. An insightful variant is obtained if, in addition, the left eigenvectors are normalized in the same way as the right ones. More precisely, define the matrix $U \triangleq Y diag(\|y_1\|_2^{-1}, ..., \|y_n\|_2^{-1})$ and note that $\|u_k\|_2 = 1, k = 1, ..., n$. Then $U^H M = DU^H$, $y_k = u_k \|y_k\|_2$ and $x_k^H u_k = \frac{x_k^H y_k}{\|y_k\|_2} = \frac{1}{\|y_k\|_2}$. Therefore $x_k y_k^H = \frac{x_k u_k^H}{x_k^H u_k}$ and expression (6.22) is replaced by the following alternative expression:

$$\left(\frac{\partial \lambda_k}{\partial M}\right)^T = X E_k X^{-1} = \frac{x_k u_k^H}{x_k^H u_k} \qquad (6.23)$$

where $\|x_k\|_2 = \|u_k\|_2 = 1$.

Expressions (6.22) and (6.23) are very useful because they give the sensitivities of the n eigenvalues of the matrix M as a function of its n pairs of reciprocal eigenvectors.

If we define $\Psi_k \triangleq \|\frac{\partial \lambda_k}{\partial M}\|_F^2$ as the *sensitivity measure* for $\lambda_k(M)$, then we have

$$\begin{aligned} \Psi_k &= \|y_k\|_2^2 \|x_k\|_2^2 = tr(y_k y_k^H) tr(x_k x_k^H) \\ &= tr\{(y_k x_k^H)(y_k x_k^H)^H\}. \end{aligned} \qquad (6.24)$$

If the right eigenvectors are normalized, $||x_k||_2 = 1$, we get the following alternative expressions:

$$\Psi_k = ||y_k||_2^2 = \frac{1}{|x_k^H u_k|^2}, \qquad (6.25)$$

where the y_k and u_k are as defined above.

In the next Lemma we show that the sensitivity measure of an eigenvalue is always larger than or equal to one.

Lemma 6.2 : *Let $\Psi_k \triangleq ||\frac{\partial \lambda_k}{\partial M}||_F^2$ be the sensitivity measure of $\lambda_k(M)$ with respect to M. Then*

$$\Psi_k \geqslant 1, \qquad (6.26)$$

with equality if and only if the reciprocal eigenvectors x_k and y_k are identical.

Proof: Consider, for example, the expression (6.25) with x_k normalized. By the Cauchy-Schwartz inequality, $|x_k^H u_k| \leqslant ||x_k||_2 ||u_k||_2 = 1$, and equality holds if and only if $x_k = c u_k$ for some $c \in \mathbb{C}$. But since x_k and u_k have unit norm, $|c| = 1$. Also, since y_k is colinear with u_k, and $y_k^H x_k = 1$, it follows that $\Psi_k = 1$ if and only if $x_k = y_k = u_k$. ∎

It can also be shown [SS90], [Ste73] that if λ_k is a simple eigenvalue of M and if $\bar{\lambda}_k$ is the corresponding eigenvalue of the perturbed matrix $M + E$, then

$$|\lambda_k - \bar{\lambda}_k| \leq ||E||_2 ||y_k||_2$$

or, alternatively,

$$|\lambda_k - \bar{\lambda}_k| \leq ||E||_2 \frac{1}{|x_k^H u_k|},$$

where x_k and y_k are obtained here on the basis of normalized right eigenvectors. Hence, $||y_k||_2$ and $\frac{1}{|x_k^H u_k|}$ can be seen as condition numbers for the eigenvalue λ_k. Roughly speaking, if an order ϵ perturbation is made on the $||\cdot||_2$-norm of the matrix M, then the eigenvalue λ_k may be perturbed by up to $\frac{\epsilon}{|x_k^H u_k|}$ if the second expression is used. Thus, if $|x_k^H u_k|$ is small, then λ_k can be regarded as ill-conditioned. Note that $|x_k^H u_k|$ is the cosine of the angle between left and right eigenvectors associated with λ_k.

Example [Ste73] : Let

$$M = \begin{pmatrix} 1 & 100 \\ 0 & 2 \end{pmatrix}$$

Then M has the simple eigenvalue $\lambda_1 = 1$ with a normalized right eigenvector $x_1 = (1\ 0)^T$ and, correspondingly, $y_1^H = (1\ -100)^T$ or $u_1^H = (0.1\ -1)^T$. Since $||y_1||_2 = \frac{1}{x_1^H u_1} \simeq 100$, we expect a perturbation in M to result in a perturbation in λ_1 that is amplified by a factor 100. Indeed, the perturbed matrix

$$M + E = \begin{pmatrix} 1 & 100 \\ -10^{-5} & 2 \end{pmatrix}$$

has an eigenvalue $\lambda_1 \simeq 1.001$.

In line with the definition (6.3) of pole sensitivity measure we define the *overall eigenvalue sensitivity measure of a matrix* M as follows:

$$\Psi(M) = \sum_{k=1}^{n} \Psi_k. \qquad (6.27)$$

The eigenvalue sensitivity measure as just defined has the following properties:

Theorem 6.1 : *Let $M \in \mathbb{R}^{n \times n}$ have distinct eigenvalues; then the overall eigenvalue sensitivity measure is not smaller than n, that is $\Psi(M) \geqslant n$, and equality holds if and only if M is normal (that is $M^H M = M M^H$).*

Proof: By Lemma 6.2 above it follows that $\Psi(M) \geqslant n$ with equality if and only if $y_k = x_k, k = 1, \ldots, n$, assuming that $\{x_k\}$ are the normalized right eigenvectors. But then it follows from $Y^H X = I$ that $X^H X = I$. This implies that the eigenvectors of M are orthogonal. This is a defining property of normal matrices, and hence M is normal (see, e.g. [Bel70] p. 226). ■

We have thus completed our examination of the sensitivity of simple eigenvalues of a matrix with respect to perturbations in the elements of that matrix. Since the pole sensitivities of a filter are precisely the sensitivities of the eigenvalues of the state transition matrix A, we now exploit our results for the separate issue of the minimization of an unweighted pole sensitivity measure with respect to all similar realizations.

6.4 Pole sensitivity minimization and normal matrices

We shall denote by Ψ_{pk} the sensitivity measure of the k-th pole of the realization (A, B, C, d) of a system:

$$\Psi_k \triangleq \|\frac{\partial \lambda_k}{\partial A}\|_F^2.$$

The unweighted pole sensitivity measure Ψ_p of the system (A, B, C, d), defined in (6.3), can now be written

$$\Psi_p \triangleq \sum_{k=1}^{n} \Psi_{pk} = \Psi(A), \qquad (6.28)$$

where $\Psi(A)$ is the eigenvalue sensitivity measure computed in the last subsection. It is obtained from (6.27) with M replaced by the real state transition matrix A.

Our last result shows that *normal matrices* have the property that their eigenvalues are globally least sensitive to errors in the matrix entries, where the term *globally* refers to the fact that all the eigenvalues are being given equal weight in the sensitivity measure $\Psi(M)$. If the unweighted pole sensitivity measure $\Psi_p \triangleq \sum_{k=1}^{n} \|\frac{\partial \lambda_k}{\partial A}\|_F^2$ is adopted, then Theorem 6.1 tells us that this measure is minimal (and equal to n) if and only if A is normal. For a real matrix, this means that $AA^T = A^T A$.

Comment: We have defined the pole sensitivity measure as an unweighted measure and have shown that the optimal realizations are those that have A in normal form. It follows directly from Lemma 6.2 and the proof of Theorem 6.1 that, even if a weighted pole sensitivity measure is adopted, with different weights for different poles, the optimum is still achieved for realizations that have the transition matrix A in normal form.

Normality has often been used as a design goal. MacFarlane and Hung [MH82] have stated, "... an approximation to normality is something which one strives to achieve in the feedback design process", while Patel and Toda [PT82] have shown that the robustness bound for a certain class of parameter errors is maximized when the plant matrix A is normal. In [SW84], Skelton and Wagie have proposed a new optimal control problem performance index by adding an 'abnormality' penalty to the classical performance index. The idea is to achieve a normal or quasi-normal closed loop state transition matrix by appropriate choice of the feedback gain matrix. Since the achievement of an exactly normal closed loop matrix may not be achievable, their technique is to add an "abnormality" penalty $\|AA^T - A^TA\|_F^2$ to the criterion.

Since our objective here is one of minimal pole sensitivity realization, the question arises as to whether a system $H(z)$ can always be represented in a coordinate space in which the matrix A is in normal form. Stated more simply, can any matrix $A \in \mathbb{R}^{n\times n}$ be transformed to normal form by a real similarity transformation? The answer is yes if and only if A has a full set of independent eigenvectors; this is the case, in particular, if A has simple eigenvalues. To prove this result we need the following preliminary lemma.

Lemma 6.3 : *Let $M \in \mathbb{R}^{n\times n}$ and let $X \triangleq (x_1,\ldots,x_n)$ be a matrix of right eigenvectors of M corresponding to the eigenvalues $(\lambda_1,\ldots,\lambda_n)$. Moreover, let us assume that M has k real eigenvalues which are $\{\lambda_i, i=1,2,\cdots,k\}$. Then $XD^{-2}X^H$ is a real nonnegative definite matrix for any matrix D of the form $D \triangleq \text{diag}(D_R, D_C, D_C)$ where $D_R \triangleq \text{diag}(r_1,\ldots,r_k)$, $D_C \triangleq \text{diag}(c_1, c_2,\ldots,c_l)$, $k+2l = n$, and the numbers $r_1,\ldots,r_k$ and $c_1,\ldots,c_l$ are real and arbitrary.*

Proof: If λ_i is real, then $x_i \in \mathbb{R}^n$; if λ_i is complex with eigenvector x_i, then $\bar{\lambda}_i \in \lambda(M)$ with eigenvector $\bar{x}_i$, where "—" denotes complex conjugate. Therefore, without loss of generality X can be written as

$$X = \begin{pmatrix} X_R & X_C & \bar{X}_C \end{pmatrix},$$

where X_R consists of all the real eigenvectors and X_C consists of the complex eigenvectors which, together with their complex conjugates and X_R, make up the matrix X. Then:

$$\begin{aligned}XD^{-2}X^H &= \begin{pmatrix} X_R D_R^{-2} & X_C D_C^{-2} & \bar{X}_C D_C^{-2} \end{pmatrix} \begin{pmatrix} X_R & X_C & \bar{X}_C \end{pmatrix}^H \\ &= X_R D_R^{-2} X_R^T + 2\text{Re}(X_C D_C^{-2} X_C^H) \end{aligned} \quad (6.29)$$

where 'Re' means the real part. It follows that $XD^{-2}X^H$ is real and nonnegative definite. ∎

We can now prove the announced result and characterize the set of similarity transformations.

Theorem 6.2 : *Let $A_0 \in \mathbb{R}^{n \times n}$. Then A_0 can be transformed to normal form by a real similarity transformation if and only if it has a full set of linearly independent eigenvectors. If so, the set of real similarity matrices that transform A_0 to normal form are completely characterized by*

$$T = (X_0 D^{-2} X_0^H)^{1/2} Q, \qquad (6.30)$$

where $X_0 \triangleq (x_1^0, \ldots, x_n^0)$ is a nonsingular matrix of right eigenvectors of A_0, D is any diagonal positive definite matrix and Q is an arbitrary orthogonal matrix.

Proof: Let $A = T^{-1} A_0 T$ for some arbitrary nonsingular T, and let X be the matrix of right eigenvectors of A corresponding to X_0. Then $X = T^{-1} X_0$. Now A is normal if and only if the $\{x_i\}$ form an orthogonal set (see e.g. [Bel70]), that is

$$X^H X = D^2 \qquad (6.31)$$

where D is any diagonal positive definite matrix. We show that this can hold if and only if X_0 is nonsingular. Indeed if (6.31) holds, then one has $X_0^H T^{-T} T^{-1} X_0 = D^2$ and X_0 is clearly nonsingular. Conversely, if X_0 is nonsingular, then by the previous Lemma, $X_0 D^{-2} X_0^H$ is real and positive definite. Therefore with T defined by (6.30) and $X = T^{-1} X_0$, it is easy to see that X satisfies (6.31). Hence A is normal.

The above argument shows that the matrices T that yield the property (6.31) must satisfy $X_0^H T^{-T} T^{-1} X_0 = D^2$ or, equivalently, $TT^T = X_0 D^{-2} X_0$. These are all characterized by (6.30). ∎

The set of matrices that have independent eigenvectors are called *non-defective*. They include all matrices with distinct eigenvalues but also those with possibly multiple eigenvalues provided their corresponding eigenvectors are independent. An alternative characterization is that defective matrices can be diagonalized by similarity transformations. We conclude by noting that if, in particular, a system $H(z)$ has distinct eigenvalues it can always be realized with A in normal form, and hence the unweighted pole sensitivity measure Ψ_p can be minimized within the class of equivalent realizations. Furthermore, these pole sensitivity minimizing realizations are not unique: they are all related by similarity transformations of the form (6.30).

142   6. Pole and zero sensitivity minimization

6.5 Zero sensitivity measure

We recall that for a realization (A, B, C, d) with $d \neq 0$ the zeros of the system are the eigenvalues of $Z \triangleq A - d^{-1}BC$. Hence they depend not only on the matrix A, but also on B, C and d. Therefore, one needs to study the sensitivity of the zeros of Z with respect to all the parameters in (A, B, C, d), and this is of course not identical to computing the sensitivity of these zeros with respect to Z, as suggested in [Wil86] and [Wil91]. Rather, we need to compute $\frac{\partial \lambda(Z)}{\partial Q}$ for $Q = A, B, C, d$ with $Z = A - d^{-1}BC$. We now embark on this exercise.

Denote $F \triangleq d^{-1}B$, which leads to $Z = A - FC = \{z_{ij}\}$ with $z_{ij} = a_{ij} - f_i c_j$. Let $\{v_i, i = 1, \ldots, n\}$ be the zeros of the system, i.e. the eigenvalues of Z, and let v_k be any one of these eigenvalues. Then

$$\frac{\partial v_k}{\partial a_{ij}} = \sum_{l,p} \frac{\partial v_k}{\partial z_{lp}} \frac{\partial z_{lp}}{\partial a_{ij}} = \frac{\partial v_k}{\partial z_{ij}},$$

which leads to

$$\frac{\partial v_k}{\partial A} = \frac{\partial v_k}{\partial Z}, \qquad (6.32)$$

where $\frac{\partial v_k}{\partial Z}$ is given by (6.22) or (6.23) with M replaced by Z and λ_k by v_k.

Now consider $\frac{\partial v_k}{\partial C}$:

$$\frac{\partial v_k}{\partial c_j} = \sum_{l,p} \frac{\partial v_k}{\partial z_{lp}} \frac{\partial z_{lp}}{\partial c_j} = \sum_{l=1}^{n} \frac{\partial v_k}{\partial z_{lj}}(-f_l) = -F^T \frac{\partial v_k}{\partial Z^{(j)}}$$

with $\frac{\partial v_k}{\partial Z^{(j)}}$ denoting the j-th column of $\frac{\partial v_k}{\partial Z}$. It then follows that

$$\frac{\partial v_k}{\partial C} = -F^T \frac{\partial v_k}{\partial Z} = -d^{-1} B^T \frac{\partial v_k}{\partial Z}. \qquad (6.33)$$

An identical computation yields $\frac{\partial v_k}{\partial F} = -\frac{\partial v_k}{\partial Z} C^T$. Since

$$\frac{\partial v_k}{\partial b_i} = \frac{\partial v_k}{\partial f_i} \frac{\partial f_i}{\partial b_i} = d^{-1} \frac{\partial v_k}{\partial f_i},$$

one therefore gets

$$\frac{\partial v_k}{\partial B} = -\frac{\partial v_k}{\partial Z}(d^{-1} C^T). \qquad (6.34)$$

Finally,

$$\frac{\partial v_k}{\partial d} = \sum_{i=1}^{n} \frac{\partial v_k}{\partial f_i}\frac{\partial f_i}{\partial d} = \sum_{i=1}^{n} \frac{\partial v_k}{\partial f_i}(-d^{-2}b_i) = d^{-2}B^T \frac{\partial v_k}{\partial Z}C^T. \quad (6.35)$$

We have thus expressed the sensitivities of an arbitrary zero, v_k, of $H(z)$ with respect to the entries of (A, B, C, d) as functions of the sensitivity, $\frac{\partial v_k}{\partial Z}$, of this zero with respect to Z. Since v_k is an eigenvalue of Z, we can now use the expressions of the eigenvalue sensitivity measures computed previously to obtain the zero sensitivity measure Ψ_{zk}. We assume again that Z has a complete set of independent eigenvectors[3] and we introduce the full rank matrix $X_z \triangleq [x_z(1), \ldots, x_z(n)]$ of right eigenvectors of Z and the matrix $Y_z \triangleq X_z^{-H} = [y_z(1), \ldots, y_z(n)]$ of reciprocal left eigenvectors.

We define the sensitivity measure of an arbitrary zero v_k of the system (A, B, C, d) as

$$\Psi_{zk} = \|\frac{\partial v_k}{\partial A}\|_F^2 + \|\frac{\partial v_k}{\partial B}\|_F^2 + \|\frac{\partial v_k}{\partial C}\|_F^2 + \|\frac{\partial v_k}{\partial d}\|_F^2. \quad (6.36)$$

Using expressions (6.22) and (6.24) we obtain

$$\begin{aligned}\Psi_{zk} &= tr\{(y_z(k)x_z^H(k))(y_z(k)x_z^H(k))^H\} + tr(\alpha_k^2 y_z(k)y_z^H(k)) \\ &+ tr(\beta_k^2 x_z(k)x_z^H(k)) + \alpha_k^2\beta_k^2,\end{aligned} \quad (6.37)$$

where

$$\begin{aligned}\alpha_k^2 &\triangleq |d^{-1}x_z^H(k)C^T|^2 = |d^{-1}Cx_z(k)|^2, \\ \beta_k^2 &\triangleq |d^{-1}B^T y_z(k)|^2.\end{aligned} \quad (6.38)$$

If, instead, expressions (6.23) and (6.25) are used with normalized right eigenvectors $\{x_z(k)\}$ and normalized left eigenvectors $\{u_z(k)\}$, then the following alternative expression is obtained for the zero sensitivity measure:

$$\Psi_{zk} = \frac{1}{|x_z^H(k)u_z(k)|^2}(1 + \alpha_k^2) + \beta_k^2 + \alpha_k^2\beta_k^2. \quad (6.39)$$

The overall unweighted zero sensitivity measure is then given by

$$\Psi_z = \sum_{k=1}^{n} \Psi_{zk} \quad (6.40)$$

---

[3] Recall again that this is the case, in particular, if all the zeros are different, but it is not limited to this case.

while the weighted measure M_{pz} introduced in (6.2) can now be written as:

$$M_{pz} = \sum_{k=1}^{n}\{w_{\lambda_k}\Psi_{pk} + w_{v_k}\Psi_{zk}\}. \tag{6.41}$$

Before we can handle the minimization of Ψ_z or M_{pz} we need to examine the dependence of Ψ_{pk} and Ψ_{zk} on the coordinates of the realization.

6.6 Pole-zero sensitivity coordinate dependence

Two different realizations of the same system $H(z)$ have the same poles and the same zeros, but their pole and zero sensitivity measures may be quite different. We now examine the dependence of the pole and zero sensitivity measures, Ψ_{pk} and Ψ_{zk}, with respect to a coordinate change. That is, we study the effect of a similarity transformation by a nonsingular transformation matrix T on these measures. First we examine the effect of a real similarity transformation T on the eigenvalue sensitivity measure.

Consider a matrix M^0 that has a complete set of independent eigenvectors and a nonsingular matrix T, and denote $M = T^{-1}M^0T$. Clearly, $\lambda(M) = \lambda(M^0)$. Let x_k^0 be a right eigenvector of M^0 corresponding to the eigenvalue λ_k and let y_k^0 be its reciprocal left eigenvector, i.e. $y_k^{0H}x_k^0 = 1$. The corresponding eigenvectors of M for the same eigenvalue λ_k are given as follows:

$$x_k = T^{-1}x_k^0, \quad y_k = T^T y_k^0. \tag{6.42}$$

It then follows from (6.24) that the eigenvalue sensitivity measure in the new coordinate system is given by

$$\begin{aligned}\Psi_k(T) &= tr(T^T y_k^0 y_k^{0H} T) tr(T^{-1} x_k^0 x_k^{0H} T^{-T}) \\ &= tr\{T^T y_k^0 x_k^{0H} T^{-T}(T^T y_k^0 x_k^{0H} T^{-T})^H\}. \end{aligned} \tag{6.43}$$

Now, let (A_0, B_0, C_0, d) be some initial realization in S_H with $Z_0 = A_0 - d^{-1}B_0C_0$ and let $(A, B, C, d) \in S_H$ be obtained from (A_0, B_0, C_0, d) through a similarity transformation T. Let $\{x_p^0(k)\}$ and $\{y_p^0(k)\}$, respectively $\{x_z^0(k)\}$ and $\{y_z^0(k)\}$, be the left and right eigenvectors of A_0, respectively Z_0, for this initial realization. The pole sensitivity measure $\Psi_{pk}(T)$ in the new realization is obtained directly from (6.43):

$$\begin{aligned}\Psi_{pk}(T) &= tr\{T^T y_p^0(k) x_p^{0H}(k) T^{-T}(T^T y_p^0(k) x_p^{0H}(k) T^{-T})^H\} \\ &= tr\{P y_p^0(k) x_p^{0H}(k) P^{-1}(y_p^0(k) x_p^{0H}(k))^H\}, \end{aligned} \tag{6.44}$$

where $P \triangleq TT^T$. Similarly, substituting $x_z(k) = T^{-1}x_z^0(k)$ and $y_z(k) = T^T y_z^0(k)$ in (6.37), yields the following expression for the zero sensitivity measure:

$$\begin{aligned}
\Psi_{zk}(T) &= tr\{(T^T y_z^0(k) x_z^{0H}(k) T^{-T})(T^T y_z^0(k) x_z^{0H}(k) T^{-T})^H\} \\
&\quad + tr\{\alpha_k^2 T^T y_z^0(k) y_z^{0H}(k) T\} + tr\{\beta_k^2 T^{-1} x_z^0(k) x_z^{0H}(k) T^{-T}\} \\
&\quad + \alpha_k^2 \beta_k^2 \\
&= tr\{P y_z^0(k) x_z^{0H}(k) P^{-1} (y_z^0(k) x_z^{0H}(k))^H\} \\
&\quad + tr\{\alpha_k^2 P y_z^0(k) y_z^{0H}(k)\} + tr\{\beta_k^2 P^{-1} x_z^0(k) x_z^{0H}(k))\} \\
&\quad + \alpha_k^2 \beta_k^2.
\end{aligned}$$
(6.45)

The numbers $\{\alpha_k^2\}$ and $\{\beta_k^2\}$ have been defined in (6.38). It follows from $C = C_0 T$, $B = T^{-1} B_0$ and (6.42) that these numbers are coordinate independent.[4] In addition, we show in the next Lemma that these numbers are nonzero for any minimal realization of $H(z)$.

Lemma 6.4 : *Let (A, B, C, d) be a minimal realization of $H(z)$ with $d \neq 0$. Then the numbers $\{\alpha_k^2\}$ and $\{\beta_k^2\}$ defined in (6.38) are strictly positive for all $k = 1, \ldots, n$.*

Proof: Suppose that $x_z(k)$ is an eigenvector of the zero $v_k \in \lambda(Z)$ with $Z = A - d^{-1} BC$, then one has

$$(A - d^{-1} BC) x_z(k) = v_k x_z(k).$$

If $\alpha_k^2 = 0$, which means $C x_z(k) = 0$, then the above equation becomes

$$A x_z(k) = v_k x_z(k)$$

which means that v_k is also a pole of the system. Since (A, B, C, d) is a minimal realization, this is impossible. The same argument shows that $\beta_k^2 > 0 \ \forall k$. ∎

We have now derived expressions for the way the sensitivity measure of a particular pole or zero of a minimal realization depends on the coordinate system in which this realization is described. All the preliminary tools have been collected to address the ultimate objective of this section: the solution of optimal realization problems for the minimization of pole, zero, or pole-zero sensitivity measures.

---

[4]However, they do depend on the particular choice of eigenvectors $\{x_z^0(k), k = 1, \ldots, n\}$ of the matrix Z_0.

6.7 Optimal realizations for pole-zero sensitivity minimization

Even though the poles and zeros of a system are of course realization independent, we have seen so far in this section that their sensitivity with respect to the parameters of the state space realization depend very clearly on the coordinate space, as is evidenced by the expressions (6.44) and (6.45). Using these expressions we can then formulate a number of pole, zero, or pole-zero sensitivity minimization problems, where the minimization is carried out with respect to all equivalent realizations $\{A, B, C, d \in S_H\}$. We address here the minimization of three different criteria: the unweighted pole sensitivity measure Ψ_p defined in (6.28), the unweighted zero sensitivity measure Ψ_z defined in (6.40), and the weighted pole-zero measure M_{pz} defined in (6.41).

The optimal unweighted pole sensitivity realization problem has been completely solved in Section 6.4. The optimal solution is to have A in normal form as proved in Theorem 6.1. We have shown in Theorem 6.2 that any non-defective (or, equivalently, diagonalizable) matrix A_0 can be transformed to normal form and we have displayed the set of similarity transformation matrices that achieve this job. We recall that any system with distinct poles yields a non-defective transition matrix.

We now address the optimal realization problem for the unweighted zero sensitivity measure Ψ_z. Similarly to Theorem 6.1 for the unweighted pole sensitivity measure, our next result gives a lower bound for this measure.

Theorem 6.3 : *Let (A, B, C, d) be a minimal realization of a system $H(z)$ with $d \neq 0$ and distinct zeros. Then the zeros of $H(z)$ are the eigenvalues of $Z \triangleq A - d^{-1}BC$. The unweighted sensitivity measure (6.41) is lower bounded by*

$$\Psi_z \geq n + 2 \sum_{k=1}^{n} |\alpha_k \beta_k| + \sum_{k=1}^{n} \alpha_k^2 \beta_k^2. \tag{6.46}$$

This lower bound is achieved if and only if Z is normal with its right eigenvector matrix X_z satisfying

$$X_z^H X_z = diag(|\frac{\alpha_1}{\beta_1}|, \ldots, |\frac{\alpha_n}{\beta_n}|). \tag{6.47}$$

If (A_0, B_0, C_0, d) is an initial realization of $H(z)$ with $Z_0 = A_0 - d^{-1}B_0C_0$, and if $X_{z,0}$ is a corresponding matrix of independent right eigenvectors of Z_0, then (A_0, B_0, C_0, d) can be transformed to a form that achieves this lower bound by a similarity transformation

$$T^{opt} = (X_{z,0} D^{-2} X_{z,0}^H)^{1/2} Q, \tag{6.48}$$

where $D^2 = diag(|\frac{\alpha_1}{\beta_1}|, \ldots, |\frac{\alpha_n}{\beta_n}|).$[5]

Proof: To obtain the lower bound, let us consider expression (6.37). By Lemma 6.2 the first term on the right hand side is lower bounded by 1 with equality if and only if $\forall k$, $y_z(k) = \gamma_k x_z(k)$ for some γ_k. This last condition is equivalent with $Y_z = X_z \Gamma$ with $\Gamma \triangleq diag(\gamma_1, \ldots, \gamma_n)$. This, together with $Y_z^H X_z = I$, implies that Z is normal and that $X_z^H X_z = \Gamma^{-H}$. Since $X_z^H X_z$ is Hermitian it follows that the $\{\gamma_k\}$ must be positive.

Now consider the second and third terms of (6.37). We have

$$tr(\alpha_k^2 y_z(k) y_z^H(k)) + tr(\beta_k^2 x_z(k) x_z^H(k))$$
$$= \alpha_k^2 \|y_z(k)\|_2^2 + \beta_k^2 \|x_z(k)\|_2^2$$
$$\geq 2|\alpha_k \beta_k| \times \|y_z(k)\|_2 \|x_z(k)\|_2 \tag{6.49}$$

and equality holds if and only if

$$|\alpha_k| \times \|y_z(k)\|_2 = |\beta_k| \times \|x_z(k)\|_2. \tag{6.50}$$

Now by the Cauchy Schwartz inequality,

$$\|y_z(k)\|_2 \|x_z(k)\|_2 \geq |y_z^H(k) x_z(k)| = 1,$$

with equality if and only if $\forall k$, $y_z(k) = \gamma_k x_z(k)$ for some γ_k. Therefore the expression on the right hand side of (6.49) can be lower bounded by

$$2|\alpha_k \beta_k| \times \|y_z(k)\|_2 \|x_z(k)\|_2 \geq 2|\alpha_k \beta_k|. \tag{6.51}$$

To achieve both bounds with equality, one needs to satisfy both $y_z(k) = \gamma_k x_z(k)$ and (6.50). This, and the positivity of the $\{\gamma_k\}$, implies $\gamma_k = |\frac{\beta_k}{\alpha_k}|$ $\forall k$. It then follows that $Y_z = X_z \Gamma$, with $\Gamma = diag(|\frac{\beta_1}{\alpha_1}|, \ldots, |\frac{\beta_n}{\alpha_n}|)$. Now denote $D^{-2} \triangleq \Gamma$. Condition (6.47) then follows by noting that $Y_z^H X_z = I = D^{-2} X_z^H X_z$, while the expression (6.48) for T^{opt} follows from Theorem 6.2. ■

---

[5] We recall our earlier footnote that the $\{\alpha_k, \beta_k\}$ depend on the particular choice of eigenvectors in $X_{z,0}$. Since they are coordinate independent, they depend in the same way on the columns of X_z. For example, consider the simple case where all the eigenvectors in $X_{z,0}$ are scaled up by a same scalar factor μ, then all the $\{\alpha_k\}$ are multiplied by $|\mu|$, while all the $\{\beta_k\}$ are divided by $|\mu|$, and therefore in the new coordinate system, the columns of X_z are also multiplied by μ, which is consistent with (6.47).

148   6. Pole and zero sensitivity minimization

Often, the primary concern is the minimization of the pole sensitivity which is achieved by transforming A to normal form. In general this does not yield a normal Z, and one may then wish to use the additional degrees of freedom to make the zero sensitivity measure small. This objective has been formulated, but not solved, in [Wil91]. One particular case of interest is when the Cauchy index of $H(z)$ is equal to the order n of the system; in such case A and Z can be made normal simultaneously: see [Wil86] for details.

We turn finally to the optimal realization problem for the weighted measure M_{pz} defined in (6.41) or, equivalently, (6.2). The solution of the unweighted zero sensitivity measure problem, expressed in terms of a class of optimal $\{A, B, C, d\}$, as opposed to optimal $\{Z\}$, will be obtained as a special case of this general problem by setting $w_{\lambda_k} = 0, k = 1, \ldots, n$ and $w_{v_k} = 1, k = 1, \ldots, n$. In addition the measure M_{pz} allows for a tradeoff between pole and zero sensitivity minimization while at the same time allowing for a special emphasis on certain poles and/or zeros.

For a minimal realization with a set of distinct poles and zeros, it follows from (6.41), (6.44) and (6.45) that M_{pz} can be written as

$$M_{pz} = \sum_{k=1}^{2n} tr(PH_k P^{-1} H_k^H) + tr(PM_y) + tr(P^{-1} M_x) + c$$
$$\triangleq R(P), \qquad (6.52)$$

where $P = TT^T$ and

$$H_k = w_{\lambda_k}^{1/2} y_p^0(k) x_p^{0H}(k), \quad k = 1, \ldots, n$$
$$= w_{v_k}^{1/2} y_z^0(k-n) x_z^{0H}(k-n), \quad k = n+1, \ldots, 2n \quad (6.53)$$

$$M_y = \sum_{k=1}^{n} w_{v_k} \alpha_k^2 y_z^0(k) y_z^{0H}(k)$$
$$= Y_z^0 diag(w_{v_1} \alpha_1^2, \ldots, w_{v_n} \alpha_n^2) Y_z^{0H} \qquad (6.54)$$

$$M_x = \sum_{k=1}^{n} w_{v_k} \beta_k^2 x_z^0(k) x_z^{0H}(k)$$
$$= X_z^0 diag(w_{v_1} \beta_1^2, \ldots, w_{v_n} \beta_n^2) X_z^{0H} \qquad (6.55)$$

$$c = \sum_{k=1}^{n} w_{v_k} \alpha_k^2 \beta_k^2. \qquad (6.56)$$

So, the minimization problem can be formulated as

$$\min_{(A,B,C,d)\in S_H} M_{pz} \iff \min_{P>0} R(P). \qquad (6.57)$$

Our first result shows that the solution exists.

Theorem 6.4 : *Let $H(z)$ be a discrete time transfer function with a set of distinct poles and zeros, and let (A_0, B_0, C_0, d_0) be a minimal realization of $H(z)$. Furthermore, let $X_p^0 \triangleq [x_p^0(1) \;\ldots\; x_p^0(n)]$ and $X_z^0 \triangleq [x_z^0(1) \;\ldots\; x_z^0(n)]$ be right eigenvector matrices of A_0 and $Z_0 \triangleq A_0 - d_0^{-1} B_0 C_0$, respectively. Let $Y_p^0 \triangleq [y_p^0(1) \;\ldots\; y_p^0(n)]$ and $Y_z^0 \triangleq [y_z^0(1) \;\ldots\; y_z^0(n)]$ be the matrices of reciprocal left eigenvectors and let $\{w_{v_k} > 0\}$ for $k = 1, \ldots, n$. Then, with $P = TT^T$, the minimum of (6.52) exists and is achieved by a nonsingular matrix P.*

Proof: Since the realization is assumed minimal, the $\{\alpha_k\}$ and $\{\beta_k\}$ are all nonzero by Lemma 6.4. In addition, since the poles and zeros are all distinct, the matrices Y_z^0 and X_z^0 are nonsingular. Therefore M_y and M_x in (6.54) and (6.55) are strictly positive, and by Lemma 4.A of Appendix 4.A, $R(P)$ has a global minimum which can be achieved only for positive definite P. ∎

Our next result shows that the minimization of $R(P)$ leads to a unique solution.

Theorem 6.5 : *Under the same conditions as in Theorem 6.4, the minimum of $R(P)$ is unique and is obtained as the solution of the following equation:*

$$P\left[M_y + \sum_{k=1}^{2n} H_k P^{-1} H_k^H\right] P = M_x + \sum_{k=1}^{2n} H_k^H P H_k. \qquad (6.58)$$

Proof: Since the solution exists by the previous result, a necessary condition is that it be a solution of $\frac{\partial R(P)}{\partial P} = 0$:

$$\begin{aligned}\frac{\partial R(P)}{\partial P} &= \sum_{k=1}^{2n} \{H_k P^{-1} H_k^H - P^{-1} H_k^H P H_k P^{-1}\} \\ &\quad + M_y - P^{-1} M_x P^{-1},\end{aligned} \qquad (6.59)$$

where the nonsingularity of P is guaranteed by Theorem 6.4. Pre- and post-multiplying by P yields (6.58). Now we note that equation (6.58) has exactly the same form as equation (5.16) in Theorem 5.1. In addition it follows from the proof of Theorem 6.4 that the matrices M_y and M_x are positive definite. The proof of uniqueness is then identical to that of Theorem 5.1. ∎

Comments:

- The optimal solution to the minimization of M_{pz} is given as the unique solution P of (6.58). This means that the optimal realization problem has a set of optimal solutions characterized by

$$\{A, B, C, d = (T^{opt})^{-1} A_0 T^{opt}, (T^{opt})^{-1} B_0, C_0 T^{opt}, d_0\} \quad (6.60)$$

  where $T^{opt} = \bar{T}V$, $\bar{T}$ is any square root of P ($P = \bar{T}\bar{T}^T$, and V is any orthogonal matrix.

- In order to apply Lemma 4.A to our situation we have imposed in Theorems 6.4 and 6.5 that the weightings $\{w_{v_k}\}$ on the zero sensitivity measures should all be strictly positive. This is a sufficient condition for the existence of a unique minimum, but it is by no means necessary since we know that the minimization of the unweighted pole sensitivity measure (which is a special case of the minimization of M_{pz} with $\{w_{v_k} = 0\}$) has a well defined solution as shown in Theorem 6.1.

- As a special case, Theorem 6.5 gives a solution to the optimal realization problem for the unweighted zero sensitivity minimization, by choosing $\{w_{\lambda_k} = 0\}$ and $\{w_{v_k} = 1\}$. The solution is given in terms of a characterization of the optimal $\{A, B, C, d\}$, rather than in terms of the optimal $\{Z\}$, and is therefore a much more useful result.

- We have just discussed the minimization problem of M_{pz} without constraint. As mentioned before, an important practical consideration is to avoid overflow and this can be done by using an l_2 scaling technique. The minimization of M_{pz} can be done with an l_2 scaling constraint. This issue will be discussed and the corresponding algorithm will be given in the next chapter.

It seems difficult to have an explicit expression of the solutions to the matrix equation (6.58). Just as in the previous section, one can always obtain the solutions by an iterative procedure using a gradient algorithm:

$$P(k+1) = P(k) - \mu \frac{\partial R(P)}{\partial P}\bigg|_{P=P(k)} \quad (6.61)$$

where μ is the step size and $\frac{\partial R(P)}{\partial P}$ is computed with (6.59). Since there is no local minimum (or maximum either), this algorithm will converge to the unique solution to the equation (6.58) as long as μ is small enough.

6.8 Numerical example

We now illustrate our theoretical results with a numerical example. We consider a filter that is initially described in direct form. We then compute optimal realizations that minimize, respectively, the unweighted pole sensitivity measure Ψ_p, the unweighted zero sensitivity measure Ψ_z, and a weighted pole zero sensitivity measure M_{pz}. The coefficients of these different realizations are perturbed and the effects of these perturbations on the pole and zero locations are examined and compared. The filter is given in direct form, R_c, as follows:

$\boxed{R_c:}$

$$A_c = \begin{pmatrix} 3.5906 & -4.8535 & 2.9260 & -0.6636 \\ 1.0000 & 0 & 0 & 0 \\ 0 & 1.0000 & 0 & 0 \\ 0 & 0 & 1.0000 & 0 \end{pmatrix} \quad B_c = \begin{pmatrix} 1 \\ 0 \\ 0 \\ 0 \end{pmatrix}$$

$$C_c = \begin{pmatrix} 0.2353 & 0.0355 & 0.2147 & 0.0104 \end{pmatrix} \times 10^{-3} \quad d = 0.1425.$$

The poles and zeros of this filter are presented in vector form, respectively,

$$V_p = \begin{pmatrix} 0.9321 \pm j0.1361 \\ 0.8632 \pm j0.0522 \end{pmatrix}, \quad V_z = \begin{pmatrix} 1.0444 \pm j0.2130 \\ 0.7501 \pm j0.1467 \end{pmatrix}.$$

We note that the first two poles are close to the unit circle, with a modulus of 0.9420. It may therefore prove useful to pay particular attention to their precision by choosing a realization in which these poles are rather insensitive to coefficient errors, but to illustrate the uses and effects of the different measures, consider first the minimization of the unweighted pole sensitivity measure Ψ_p. By Theorem 6.2, the realization R_c can be transformed to normal form by a similarity transformation $T^{opt}(\Psi_p) = (X_c X_c^H)^{1/2}$ where X_c is a right eigenvector matrix

152   6. Pole and zero sensitivity minimization

corresponding to the direct form R_c. We get:

$$T^{opt}(\Psi_p) = \begin{pmatrix} 196.9896 & 187.7522 & 175.7374 & 161.1424 \\ 187.7522 & 199.4658 & 207.0774 & 209.9985 \\ 175.7374 & 207.0774 & 238.0392 & 267.4614 \\ 161.1424 & 209.9985 & 267.4614 & 334.5247 \end{pmatrix}.$$

The corresponding realization $R^{opt}(\Psi_p)$ is as follows:

$\boxed{R^{opt}(\Psi_p) :}$

$$A^{opt}(\Psi_p) = \begin{pmatrix} 0.9080 & -0.0959 & -0.0561 & -0.0418 \\ 0.1180 & 0.9249 & -0.0546 & -0.0209 \\ 0.0016 & 0.0893 & 0.8896 & -0.0440 \\ 0.0126 & 0.0132 & 0.0615 & 0.8680 \end{pmatrix}$$

$$B^{opt}(\Psi_p) = \begin{pmatrix} 0.6920 \\ -1.7542 \\ 1.4991 \\ -0.4307 \end{pmatrix} \quad C^{opt}(\Psi_p) = \begin{pmatrix} 0.0924 \\ 0.0979 \\ 0.1026 \\ 0.1063 \end{pmatrix}^T.$$

The unweighted pole and zero sensitivity measures for $R^{opt}(\Psi_p)$ are $\Psi_p(T^{opt}(\Psi_p)) = \Psi_p^{min} = 4$ and $\Psi_z(T^{opt}(\Psi_p)) = 1.7132 \times 10^2$.

Consider now the minimization of the unweighted zero sensitivity measure Ψ_z. According to Theorem 6.3, one can compute the optimal zero sensitivity transformation $T^{opt}(\Psi_z)$ using the expression (6.48) with $Q = I$. This yields:

$$T^{opt}(\Psi_z) = \begin{pmatrix} 237.6822 & 189.6767 & 141.7718 & 94.8774 \\ 189.6767 & 196.1607 & 185.8373 & 159.1512 \\ 141.7718 & 185.8373 & 223.3090 & 245.9469 \\ 94.8774 & 159.1512 & 245.9469 & 361.1948 \end{pmatrix}.$$

The corresponding realization $R^{opt}(\Psi_z)$ is given by:

$\boxed{R^{opt}(\Psi_z) :}$

$$A^{opt}(\Psi_z) = \begin{pmatrix} 1.0744 & -0.1548 & -0.0239 & -0.0388 \\ 0.1156 & 0.9440 & -0.1356 & -0.0196 \\ 0.0228 & 0.1510 & 0.8260 & -0.0681 \\ 0.0438 & 0.0364 & 0.1218 & 0.7462 \end{pmatrix}$$

$$B^{opt}(\Psi_z) = \begin{pmatrix} 0.0611 \\ -0.1343 \\ 0.1019 \\ -0.0263 \end{pmatrix} \quad C^{opt}(\Psi_z) = \begin{pmatrix} 0.0941 \\ 0.0932 \\ 0.0905 \\ 0.0845 \end{pmatrix}^T.$$

The realization $R^{opt}(\Psi_z)$ yields $\Psi_z(T^{opt}(\Psi_z)) = \Psi_z^{min} = 7.9019$ and $\Psi_p(T^{opt}(\Psi_z)) = 85.9928$.

For the sake of comparison, the pole and zero sensitivity measures of the direct form R_c have also been computed. They are $\Psi_p = 1.8860 \times 10^7$ and $\Psi_z = 3.3652 \times 10^5$. We summarize these results in Table 6.1.

	Unweighted sensitivity measures	
	Ψ_p	Ψ_z
R_c	1.8860×10^7	3.3652×10^5
$R^{opt}(\Psi_p)$	4	1.7132×10^2
$R^{opt}(\Psi_z)$	85.9928	7.9019

TABLE 6.1. Unweighted pole and zero sensitivity measures for three realizations.

The conclusions to be drawn from the Table are self explanatory. To confirm these theoretical results, we now present some simulations. We generate a succession of 5 different perturbations on all coefficients of (A, B, C, d) for each of the three realizations R_c, $R^{opt}(\Psi_p)$ and $R^{opt}(\Psi_z)$. To make things clear, let $R = (A, B, C, d)$ denote any one of the three realizations above. We then consider a series of 5 perturbations of R by $\{\Delta R(i) = (E_A(i), E_B(i), E_C(i), E_d(i)), i = 1, 2, ..., 5\}$, where $E_A(i), E_B(i), E_C(i)$ and $E_d(i)$ are the perturbation matrices added to A, B, C and d for the i-th perturbation. All the elements of $\Delta R(i)$ are generated with the $MATLAB$ command 'rand('normal')'. These perturbations are such that the ∞-norm of each matrix is bounded by some design parameter σ: $||E_x(i)||_\infty \leq \sigma \ \forall i$, and for $x = A, B, C, d$.

We have computed the poles and zeros for the initial realizations R_c, $R^{opt}(\Psi_p)$, $R^{opt}(\Psi_z)$ and for their perturbed realizations $R_E = R + \Delta R(i) = (A + E_A(i), B + E_B(i), C + E_C(i), d + E_d(i)), i = 1, 2, ..., 5$. Figures 6.1, 6.2 and 6.3 represent, respectively, the pole-zero distributions for the realizations R_c with $\sigma = 5 \times 10^{-5}$, $R^{opt}(\Psi_p)$ with $\sigma = 5 \times 10^{-3}$ and $R^{opt}(\Psi_z)$ again with $\sigma = 5 \times 10^{-3}$. In each figure, the poles of the exact realizations are represented by a '+' while

6. Pole and zero sensitivity minimization

the poles of the 5 perturbed realizations are represented by 'x'. For the zero locations, the exact (unperturbed) zeros are represented by a '*' while the zero locations of the 5 perturbed realizations are denoted 'o'. The following observations can be drawn from these simulations.

- First we note that the errors on the poles and zeros of the canonical realization, R_c, are roughly of the same magnitude as those of the two optimal realizations for perturbations on the coefficients that are two orders of magnitude smaller. This immediately shows the dramatic improvements that can be obtained on pole and/or zero errors using optimal realizations.

- The figures clearly indicate that the optimization procedures achieve their stated goal: the errors on the poles are exceedingly small in the realization $R^{opt}(\Psi_p)$, while the errors on the zeros are exceedingly small in the realization $R^{opt}(\Psi_z)$.

- The errors on the zeros are larger in the realization $R^{opt}(\Psi_p)$ than the errors on the poles in the realization $R^{opt}(\Psi_z)$. This is in accordance with the theoretical values of their respective sensitivities, which show that the zero sensitivity of $R^{opt}(\Psi_p)$ is about twice the pole sensitivity of $R^{opt}(\Psi_z)$: see Table 6.1.

We note that the zero sensitivity measure of $R^{opt}(\Psi_p)$, 1.7132×10^2, is far from the minimal value, 7.9019, while the pole sensitivity measure of $R^{opt}(\Psi_z)$, 85.9928, is far higher than the minimal value, 4. Thus, these two optimal realizations are significantly different. It therefore makes sense to use our combined measure M_{pz} in order to achieve a tradeoff between the two apparently conflicting design criteria Ψ_p and Ψ_z.

From a stability point of view, one would like to implement a realization for which the two poles closest to the unit circle, numbered 1 and 2, have a smaller sensitivity. Their sensitivities in the three realizations examined so far are: $\Psi_p(1) = \Psi_p(2) = 1.3651 \times 10^6$ for R_c, $\Psi_p(1) = \Psi_p(2) = 1$ for $R^{opt}(\Psi_p)$, and $\Psi_p(1) = \Psi_p(2) = 9.4566$ for $R^{opt}(\Psi_z)$.

We therefore choose the following weighting factors:

$$\begin{pmatrix} w_{\lambda_1} & w_{\lambda_2} & w_{\lambda_3} & w_{\lambda_4} \end{pmatrix} = \begin{pmatrix} 20 & 20 & 1 & 1 \end{pmatrix}$$

and

$$\begin{pmatrix} w_{v_1} & w_{v_2} & w_{v_3} & w_{v_4} \end{pmatrix} = \begin{pmatrix} 1 & 1 & 1 & 1 \end{pmatrix}.$$

6. Pole and zero sensitivity minimization

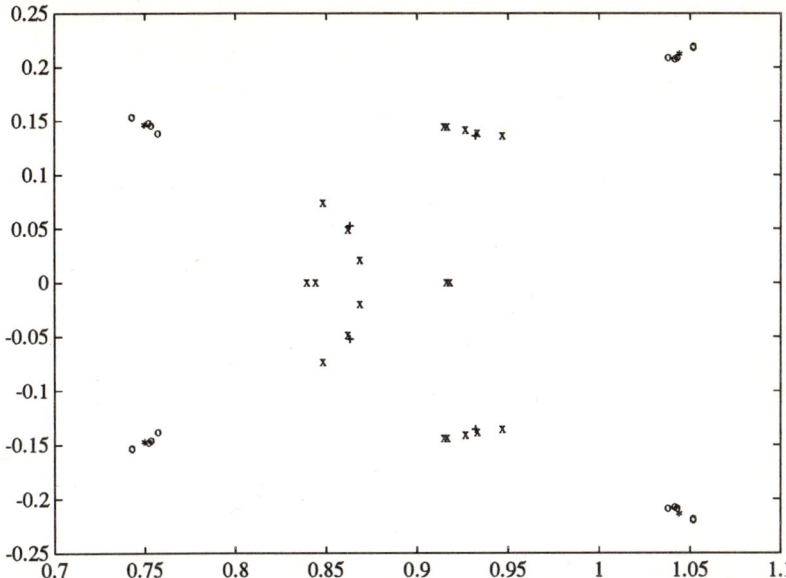

FIGURE 6.1. Poles and zeros for 5 different perturbations of R_c.

Starting with $R^{opt}(\Psi_z)$ as initial realization, one can find the optimal realization that minimizes M_{pz} using the recursive algorithm (6.61) (with $\mu = 0.001$). After convergence of P to its optimal value, the optimal similarity transformation $T^{opt}(\Psi_{pz})$ is obtained as the symmetric square root of P:

$$T^{opt}(\Psi_{pz}) = \begin{pmatrix} 237.6058 & 189.8789 & 141.5923 & 94.9307 \\ 189.8789 & 195.6219 & 186.3178 & 159.0079 \\ 141.5923 & 186.3178 & 222.8786 & 246.0758 \\ 94.9307 & 159.0079 & 246.0758 & 361.1559 \end{pmatrix}$$

with which the corresponding optimal realization $R^{opt}(\Psi_{pz})$: $(A^{opt}(\Psi_{pz}), B^{opt}(\Psi_{pz}), C^{opt}(\Psi_{pz}))$ can be computed.

$\boxed{R^{opt}(\Psi_{pz}) :}$

$$A^{opt}(\Psi_{pz}) = \begin{pmatrix} 1.0233 & -0.0666 & -0.0759 & -0.0268 \\ 0.1917 & 0.8709 & -0.1423 & -0.0082 \\ -0.0045 & 0.1066 & 0.9308 & -0.1079 \\ 0.0417 & 0.0773 & 0.0655 & 0.7655 \end{pmatrix}$$

156   6. Pole and zero sensitivity minimization

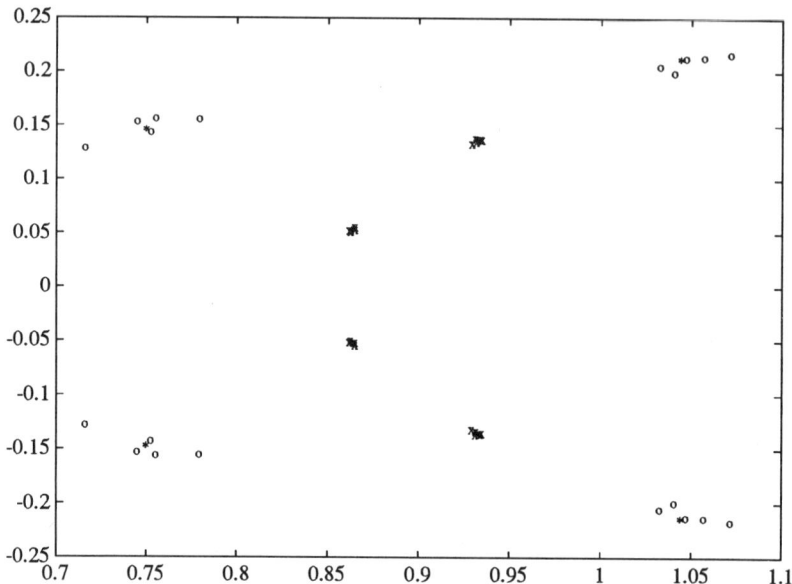

FIGURE 6.2. Poles and zeros for 5 different perturbations of $R^{opt}(\Psi_p)$.

$$B^{opt}(\Psi_{pz}) = \begin{pmatrix} 0.2010 \\ -0.5121 \\ 0.4433 \\ -0.1294 \end{pmatrix} \quad C^{opt}(\Psi_{pz}) = \begin{pmatrix} 0.0940 \\ 0.0933 \\ 0.0903 \\ 0.0846 \end{pmatrix}^T.$$

For this realization, the sensitivities of the first two poles are $\Psi_p(1) = \Psi_p(2) = 3.4723$, while the unweighted pole and zero sensitivities are, respectively, $\Psi_p = 17.6982$ and $\Psi_z = 24.0051$. A simulation identical to the previous ones has been performed by applying ten successive perturbations to this realization $R^{opt}(\Psi_{pz})$ with $\sigma = 5 \times 10^{-3}$. The results are displayed in Figure 6.4.

Comparing $R^{opt}(\Psi_{pz})$ with $R^{opt}(\Psi_p)$ and $R^{opt}(\Psi_z)$, one observes that this optimal realization is a very good compromise between pole and zero sensitivity measure, and the two poles closest to the unit circle have an acceptable sensitivity.

6.9 Conclusions

Most results on sensitivity minimizing realizations, including our new results of Chapters 4 and 5 dealt with transfer function sensitivities. It

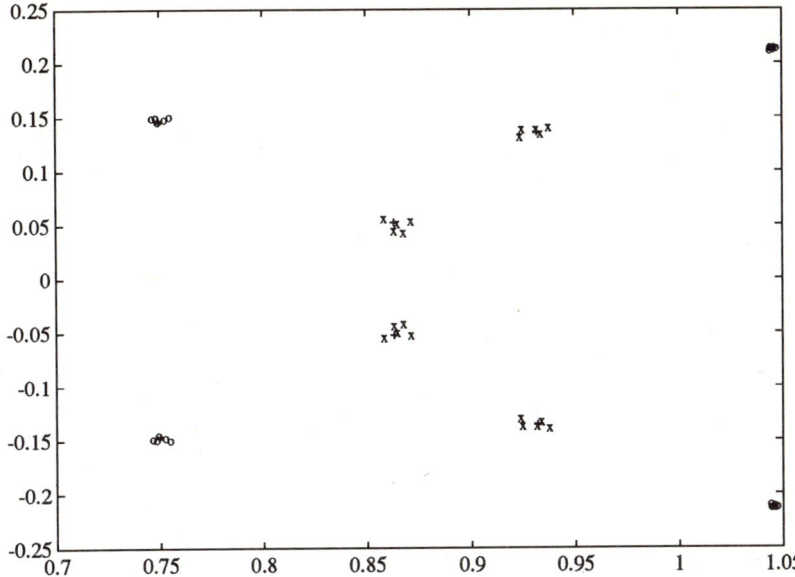

FIGURE 6.3. Poles and zeros for 5 different perturbations of $R^{opt}(\Psi_z)$.

turns out that in some applications it is more important to safeguard the exact location of certain poles and zeros of a system against FWL errors on the coefficients of its realization than it is to minimize the overall degradation of its transfer function. Examples of such applications have been mentioned in the introduction of this chapter.

We have therefore devoted this chapter to the computation of pole and zero sensitivities, and to the formulation of several pole, zero and pole-zero sensitivity measures. Even though formulas can be written for the errors on pole or zero locations induced by coefficient errors in the case of multiple poles or zeros, the sensitivities are properly defined only for simple - as opposed to multiple - poles or zeros. Since our optimal realization designs are based on minimization of sensitivity measures with respect to the set of all possible similarity transformations, we have therefore limited our analysis and our optimal design procedures to this case of simple poles and simple zeros. For this case we have given a complete solution to the optimal state space realization problem for different possible choices of sensitivity measures.

Even when limited to the case of simple poles and simple zeros, simulations have shown that our results can provide rather dramatic improvements in the effect of coefficient errors on pole and zero locations when optimal realizations are compared to the classical direct forms. A

158   6. Pole and zero sensitivity minimization

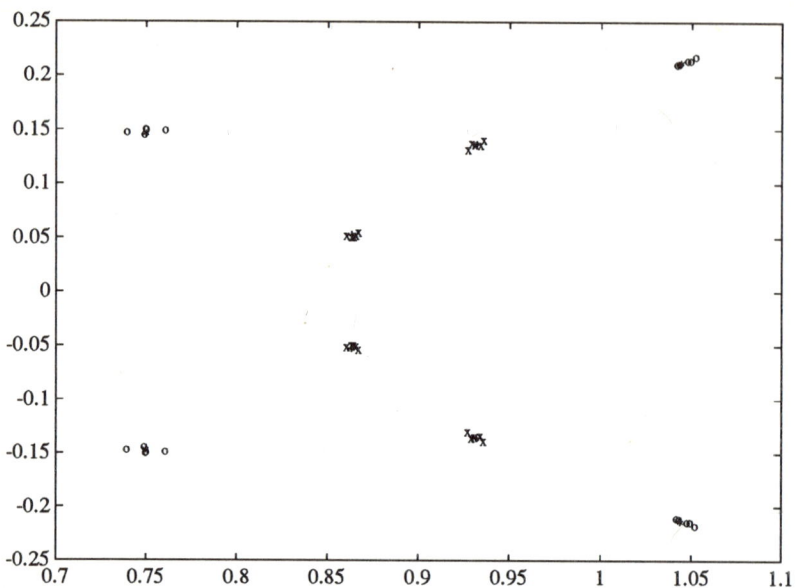

FIGURE 6.4. Poles and zeros for 5 different perturbations of $R^{opt}(\Psi_{pz})$.

major benefit of our method is that it allows for the minimization of a measure in which each pole and each zero sensitivity can be individually weighed. This allows for flexible designs tailored to each particular application.

With this chapter, we conclude our string of new results on sensitivity minimizing procedures. In the next chapter we return to the problem of designing realizations that minimize the roundoff noise gain. In fact we shall embed the analysis of coefficient errors into one of roundoff noise gain analysis, and show that both these FWL effects can be treated in a combined and coherent way.

7
A synthetic sensitivity - roundoff design

7.1 Introduction

In classical analysis, the two Finite Word Length effects due, respectively, to coefficient truncation and to arithmetic roundoff are investigated separately. We have seen in Chapter 3 that, when the L_1/L_2 measure is adopted as a transfer function sensitivity measure, the realizations that minimize this L_1/L_2 measure are not the same as those that minimize the roundoff noise gain, G. However, under an l_2 constraint on the states, the realizations that minimize G also minimize the upper bound of the L_1/L_2 measure. It is not clear what the connection might be - if any - between realizations that minimize the roundoff noise gain G (under l_2 constraints) and those that minimize the more realistic L_2 sensitivity measure.

Since parameter perturbation and signal roundoff exist simultaneously in actual implementations, and since both of them degrade the performance of a filter, it is obviously of interest to understand better the connections between the effects of coefficient errors and the effects of arithmetic roundoff errors. It would be even more appealing to find a global measure that takes account of both effects in a coherent way and that lends itself to a synthetic state space realization design that simultaneously and coherently optimizes the realization with respect to these two effects. In addition to a conceptual simplification, this would eliminate the drawback that the separate minimization of the sensitivity measure of a transfer function to coefficient perturbation, and of the roundoff noise gain, lead to two different optimal solution sets. To our knowledge, no such unified measure has been given in optimal FWL state space filter design. It is the object of the present chapter to formulate one and to solve the corresponding optimal realization problem.

The relationships between sensitivity to coefficient errors and roundoff noise have already received some attention in the literature. Fettweis [Fet72] showed that rounding can be interpreted as coefficient fluctuation, and Jackson [Jac76] showed that sensitivities can be bounded by roundoff noise. Using a statistical approach, Kawamata et al [KH85] showed that coefficient perturbation can be interpreted as roundoff noise, and that roundoff-noise-optimal realizations therefore typically also have low sensitivity. We shall adopt a similar statistical approach in formulating our global FWL error measure.

We propose a measure, called *Total Noise Gain* (G_T, in short), that is defined as the variance of the output degradation due to both coefficient errors (treated as random) and arithmetic roundoff errors. We show that this measure is a linear combination of the L_2 sensitivity measure studied in Chapter 5 and the classical roundoff noise gain. This will give yet another a posteriori justification to our separate and thorough study of this L_2 sensitivity measure in Chapter 5. By minimizing this unified measure under a dynamical range constraint, the optimal synthetic FWL state space design problem can be formulated. We shall solve it using an optimization technique that is essentially the same as that used for the minimization of the L_2 sensitivity measure, save for the additional difficulty of the dynamic range constraint. The unified measure studied in this chapter was first adopted by the authors for the FWL optimal design of a general two-degree-of-freedom compensator [LG91].

An outline of this chapter is as follows. In Section 7.2, we define and justify what we call the *Total Noise Gain* (G_T) which takes into account both parameter truncation errors and signal roundoff errors. We shall see that G_T is a linear combination of the L_2 sensitivity measure studied in Chapter 5 and of the classical roundoff noise gain. This will give further justification and a physical interpretation to this L_2 sensitivity measure. The optimal FWL realization problem is formulated in Section 7.3 as a minimization, over all equivalent realizations, of the Total Noise Gain G_T under the classical dynamic range constraint. The optimal solution is described and an algorithm for its construction is given. Some examples are given in Section 7.4 in order to illustrate the optimal design procedures. Some concluding remarks are given in Section 7.5.

7.2 A synthetic FWL noise gain

In this section, we first separately analyse the FWL effects of coefficient truncation and of arithmetic roundoff on the plant output. We then define a synthetic measure, called Total Noise Gain that takes account of both effects. We have seen in Chapter 3 that the minimization of a roundoff noise gain with respect to all equivalent realizations is ill-posed unless a dynamical range constraint is imposed on the states of the realization. The optimal realization problem will thus be formulated in terms of minimizing this synthetic measure under a constraint on the dynamic range of the states.

Let the ideal infinite precision state space model of a system $H(z)$ be given by

$$\begin{aligned} x(t+1) &= Ax(t) + Bu(t) \\ y(t) &= Cx(t) + du(t). \end{aligned} \quad (7.1)$$

We now consider a FWL implementation of the state equations (7.1) in which the parameters in (A, B, C, d) are truncated to B_c bits and in which the signals are rounded off to B_s bits. We first examine separately the effect on the output of coefficient truncation and of signal roundoff,[1] and we then lump these two errors together in order to define our synthetic error measure.

With (A, B, C, d) replaced by their FWL version (A^*, B^*, C^*, d^*) but without accounting for signal roundoff, (7.1) becomes

$$\begin{aligned} x^*(t+1) &= A^* x^*(t) + B^* u(t) \\ y^*(t) &= C^* x^*(t) + d^* u(t). \end{aligned} \quad (7.2)$$

When, additionally, the roundoff in arithmetic operations is also considered, the above model has to be modified. We consider a fixed point implementation with roundoff before multiplication: see Section 2.4. In addition, we assume, as in Section 3.5, that the signal $u(t)$ has already been rounded off to B_s bits. This leads to the actual computational model of the system:

$$\begin{aligned} x'(t+1) &= A^* Q[x'(t)] + B^* u(t) \\ y'(t) &= C^* Q[x'(t)] + d^* u(t) \end{aligned} \quad (7.3)$$

where $y'(t)$ is the actual output of the realization and $Q[.]$ denotes the quantization operation. The quantizer $Q[.]$ makes $Q[x]$ have a B_s bit expression where x has $(B_s + B_c)$ or more than $(B_s + B_c)$ bits.

---

[1] Following common practice, we use the term 'truncation' for the coefficients and 'roundoff' for the signals. However, as explained in Section 2.2, we assume throughout this book that roundoff is used to reduce an infinite wordlength number to a finite wordlength one, whether that number represents a coefficient or a signal.

7. A synthetic sensitivity - roundoff design

We define the roundoff noise

$$e_x(t) \triangleq x'(t) - Q[x'(t)]. \tag{7.4}$$

As we have seen in Section 2.4, such a roundoff noise process is modelled as independent white noise having zero mean and a variance $\sigma_n^2 = (1/12)2^{-2Bs}$. From (7.3) and (7.4), it follows that

$$\begin{aligned} x'(t+1) &= A^*x'(t) + B^*u(t) - A^*e_x(t) \\ y'(t) &= C^*x'(t) + d^*u(t) - C^*e_x(t). \end{aligned} \tag{7.5}$$

The degradation of the output of the realization due to a FWL implementation of the parameters in (A, B, C, d) and to a quantization of the states can be measured by the difference between the desired output of the plant, $y(t)$, and the actual output of the realization, $y'(t)$:

$$\begin{aligned} \Delta y(t) &= y(t) - y'(t) = [y(t) - y^*(t)] + [y^*(t) - y'(t)] \\ &\triangleq \Delta y^*(t) + \Delta y'(t). \end{aligned} \tag{7.6}$$

In (7.6) we have conceptually separated the overall output error $\Delta y(t)$ into two errors $\Delta y^*(t) \triangleq y(t) - y^*(t)$ and $\Delta y'(t) \triangleq y^*(t) - y'(t)$, which account for the FWL effects on the coefficients and on the arithmetic operations, respectively. Of course these two effects cannot be separately measured but this conceptual separation will prove to be helpful for the analysis. We first consider the arithmetic roundoff error term $\Delta y'(t)$.

It follows from (7.2) and (7.5) that

$$\begin{aligned} E_x(t+1) &= A^*E_x(t) + A^*e_x(t) \\ \Delta y'(t) &= C^*E_x(t) + C^*e_x(t). \end{aligned} \tag{7.7}$$

where

$$E_x(t) \triangleq x^*(t) - x'(t). \tag{7.8}$$

So, $\Delta y'(t)$ can be interpreted as the output of a Multiple Output Single Input (MISO) system excited by the roundoff noise $e_x(t)$ of the state $x'(t)$ and it can be computed with the state space equation (7.7).

Now, we consider the coefficient error term $\Delta y^*(t)$. Let $\{h_i\}$ and $\{h_i^*\}$ be the impulse responses of the system $H(z)$ with infinite and finite coefficient implementation, respectively. Then

$$\Delta y^*(t) = y(t) - y^*(t) = \sum_{i=0}^{\infty}[h_i - h_i^*]u(t-i). \tag{7.9}$$

Denoting $\Delta h_i \triangleq h_i - h_i^*$, one can see that $\Delta y_i^*(t)$ is in fact the output of a system whose impulse response is $\{\Delta h_i\}$, excited by $u(t)$. Since $e_x(t)$ and $u(t)$ are independent, so are $\Delta y'(t)$ and $\Delta y_i^*(t)$. The steady-state variance of the plant output error $\Delta y(t)$ is therefore the sum of the variances of these two independent error signals:

$$\sigma^2 \triangleq E[\Delta y(t)^2] = \sigma_1^2 + \sigma_2^2 \qquad (7.10)$$

where $\sigma_1^2 \triangleq E[\Delta y^*(t)^2]$ and $\sigma_2^2 \triangleq E[\Delta y'(t)^2]$.

We first compute σ_1^2, the contribution to the output variance of coefficient errors. Let $\{p_i\}$ denote the theoretical parameters of the realization of the system $H(z)$ and assume that this realization has N parameters. Their FWL version is $\{p_i^*\}$ with $p_i = p_i^* + \Delta p_i$ for $i = 1, \ldots, N$. Clearly,

$$\Delta h_i = \frac{1}{2\pi j} \oint_{|z|=1} \{H(z) - H^*(z)\} z^{i-1} dz. \qquad (7.11)$$

With a first order approximation, one has

$$\Delta H(z) = H(z) - H^*(z) = \sum_{i=1}^{N} \frac{\partial H(z)}{\partial p_i} (\Delta p_i). \qquad (7.12)$$

In an actual implementation, the coefficient perturbations $\{\Delta p_i\}$ can be computed if the ideal parameters and the number of bits are given. However, in a design stage the explicit values of the system matrices - and sometimes also the wordlength - are not specified yet. For example, in a FWL optimal state space design, the realization is known only up to similarity transformations. In such case, the errors of coefficient quantization cannot be a priori determined, but they are known to be in the range $[-2^{-B_c}, 2^{-B_c}]$. Thus, for design purposes it is useful to adopt a *statistical approach* to coefficient quantization errors.

The statistical approach was first proposed by Knowles and Olcayto [KO68] for the analysis of coefficient quantization errors in direct and parallel forms. It was later used by Avenhaus [Ave72] and Crochiere [Cro74]-[Cro75] for the estimation of the coefficient wordlength required to achieve a prespecified FWL performance for a filter implementation, and subsequently extended by Moroney et al. [MWH80] for the same purpose, but in the FWL controller design problem. Kawamata et al. [KH85] and Iwatsuki et al. [IKH90] later used this statistical approach for the optimal FWL state space realization problem, thereby connecting the classical L_1/L_2 sensitivity measure with the roundoff noise gain

7. A synthetic sensitivity - roundoff design

G. They also experimentally validated this statistical model for the coefficient errors.

The basic idea of the statistical approach is to consider each state space implementation of the filter as a realization of an 'ensemble' of structures, over which the coefficient errors $\{\Delta p_i\}$ are considered as zero mean uniformly distributed independent random variables, that are also independent of the input signal process $u(t)$. Thus, $E[\Delta p_i] = 0$, $E[\Delta p_i \Delta p_j] = \sigma_c^2 \delta_{ij}$ with $\sigma_c^2 = (1/12)2^{-2B_c}$.

With the above assumptions, with a unit variance white noise input signal $u(t)$, and using the first order approximation (7.12) for $\Delta H(z)$, one has

$$\sigma_1^2 \triangleq E\{[\Delta y^*(t)]^2\}$$
$$= \sum_{i=1}^{N}\left\{\frac{1}{2\pi j}\oint_{|z|=1}\left(\frac{\partial H(z)}{\partial p_i}\right)\left(\frac{\partial H(z)}{\partial p_i}\right)^H z^{-1}dz\right\}\sigma_c^2. \tag{7.13}$$

To compute (7.13), one needs the sensitivity functions $\{\partial H/\partial p_i\}$ of the transfer function with respect to the parameters of its representation. If $H(z)$ is represented in the form of a state space realization (A, B, C, d), then the $\{p_i\}$ are the coefficients of these matrices. Assuming again that these matrices are fully parametrized and using the analysis of Section 5.2 (see, in particular, expression (5.3)), then σ_1^2 in (7.13) specializes to

$$\sigma_1^2 = \left\{\|\frac{\partial H(z)}{\partial A}\|_2^2 + \|\frac{\partial H(z)}{\partial B}\|_2^2 + \|\frac{\partial H(z)}{\partial C^T}\|_2^2 + \|\frac{\partial H(z)}{\partial d}\|_2^2\right\}\sigma_c^2$$
$$= \{tr[W_A + W_o + W_c] + 1\}\sigma_c^2, \tag{7.14}$$

where

$$W_A = \frac{1}{2\pi j}\oint_{|z|=1}\left(\frac{\partial H(z)}{\partial A}\right)\left(\frac{\partial H(z)}{\partial A}\right)^H z^{-1}dz$$
$$= \sum_{i=0}^{\infty} h(i)h^T(i), \tag{7.15}$$

with

$$h(i) \triangleq \sum_{k+j=i} g(k)f^T(j)$$
$$= \sum_{k+j=i}[(CA^k)^T(A^jB)^T], \tag{7.16}$$

$g(k)$ and $f(j)$ being the impulse responses of $G(z)$ and $F(z)$, respectively: see Section 5.2.

One immediately recognizes that, in this analysis based on random coefficient perturbations, the variance of the output due to coefficient errors in the matrices A, B, C, d naturally comes out to be the L_2 sensitivity measure M_{L_2} studied in Chapter 5 (see, for example, (5.11)), at least up to an additive constant term σ_c^2 that is coordinate independent anyway.[2]

We now compute σ_2^2, the contribution to the output variance of the arithmetic roundoff errors. From (7.7), one can get the following state error:

$$E_x(t) = \sum_{i=0}^{\infty}(A^*)^i A^* e_x(t-1-i).$$

Therefore

$$\Delta y'(t) = C^* \sum_{i=0}^{\infty}(A^*)^i A^* e_x(t-1-i) + C^* e_x(t).$$

Noting that $E\{e_x(i)e_x^T(j)\} = \sigma_n^2 \delta_{ij} I$, then the steady-state error variance can be computed as

$$\sigma_2^2 = \{tr[(A^*)^T W_o^* A^*] + C^*(C^*)^T\}\sigma_n^2, \tag{7.17}$$

where

$$W_o^* \triangleq \sum_{i=0}^{\infty}(A^{*T})^i (C^*)^T C^* (A^*)^i \tag{7.18}$$

is the observability Gramian of the system, satisfying the Lyapunov equation

$$W_o^* = (A^*)^T W_o^* A^* + (C^*)^T C^*. \tag{7.19}$$

So,

$$\sigma_2^2 = tr(W_o^*)\sigma_n^2. \tag{7.20}$$

Now, with (7.10), (7.14) and (7.20) one can compute the overall output error variance σ^2 of the system:

$$\sigma^2 = tr(W_o^*)\sigma_n^2 + [tr(W_A + W_o + W_c) + 1]\sigma_c^2, \tag{7.21}$$

---

[2] As the astute reader will have observed, by this we mean once again that the last term is independent of the particular realization (A, B, C, d) since d is the same in all these realizations.

where σ_n^2 and σ_c^2 are the variances of the roundoff noise at the states and of the coefficient errors, treated as random noise. These two variances are due to the FWL effects and depend on the wordlengths B_s of the signals and B_c of the coefficients, respectively.

For a given realization (A, B, C, d) of $H(z)$, expression (7.21) allows one to evaluate the degradation of the system performance due to the FWL implementation of the realization, but the most interesting problem is to minimize this degradation by choosing optimal realizations.

Before we embark on this exercise, we note that (7.21) contains both $tr(W_o^*)$ and $tr(W_o)$. In the first term the observability Gramian is computed from the Lyapunov equation (7.19) using FWL versions of A, B, C, while in the second term the Gramian is computed using the same Lyapunov equation but with the infinite precision versions of A, B, C. In our derivations we shall neglect the difference between these two expressions for the trace of the observability Gramian.

Deleting the constant (realization independent) term in (7.21), and dividing by the variance of the coefficient errors, σ_c^2, one can then define a global measure G_T, called *Total Noise Gain*, as follows:

$$G_T \triangleq tr(W_A + W_o + W_c) + \frac{\sigma_n^2}{\sigma_c^2} tr(W_o)$$
$$= M_{L_2} + \rho^2 G. \qquad (7.22)$$

Comments

1. The *Total Noise Gain* G_T takes into account both the roundoff noise of the states and the parameter truncation errors. It is therefore a good candidate as an optimization criterion for a synthetic "sensitivity plus roundoff noise" design.

2. G_T is a linear combination of the classical roundoff noise gain G used as an optimization criterion by Mullis, Roberts and Hwang (see [MR76a] and [Hwa77]), and of the new L_2 sensitivity measure M_{L_2} that we have introduced in Chapter 5. It gives a nice physical interpretation to this L_2 measure which could not easily be obtained from the more classical L_1/L_2 measure.

3. Assuming no roundoff noise, Iwatsuki et al. used a statistical approach on the quantization errors to derive an expression for a statistical sensitivity measure [IKH90]. Under some simplifying

assumptions, they showed that their sensitivity measure is approximately equal to the upper bound $\bar{M}_{L_{12}}$ of the 'classical' L_1/L_2 sensitivity measure. Our *Total Noise Gain* G_T not only unifies the FWL effects of parameters and signal roundoff, but it is also derived under weaker assumptions than in [KH85] and [IKH90].

4. The factor ρ^2 that determines the respective weights of the coefficient error term and the roundoff noise term in G_T is entirely determined by the hardware, that is by the wordlengths B_s of the states and B_c of the parameters:

$$\rho^2 = \sigma_n^2/\sigma_c^2 = 2^{2(B_c-B_s)}. \qquad (7.23)$$

In classical FWL studies, the sensitivity analysis is performed assuming no roundoff noise, that is $B_s = \infty$. In such case we have $G_T = M_{L_2}$. On the other hand, B_c is taken as ∞ in the classical roundoff noise study. In practical implementations, B_s and B_c are of course both finite.

5. If $B_c > B_s$, this means that in the hardware design choices greater care has been taken for coefficient errors than for signal roundoff errors. The minimization of the *Total Noise Gain* (7.22) with respect to all equivalent realizations will then have the logical effect that the realization design will bring the roundoff noise gain down more than the coefficient error effects, since $\rho^2 > 1$.

6. On the other hand, if $B_c < B_s$, it means that the hardware choices have emphasized small signal roundoff errors. To improve the performance of the actual implementation, it it is then better to minimize the effects of coefficient errors, that is to pay more attention to the minimization of the sensitivity measure. This is precisely what the unified measure G_T does since in this case $\rho^2 < 1$.

7. In [Fet72], Fettweis studied the connections between coefficient Word Length and roundoff noise. Consider a B-bit product of a coefficient c and a signal $x(t)$ whose bit numbers are B_c and B_s respectively, with $B = B_c + B_s$. Fettweis argued that any design procedure that enables one to reduce the wordlength of the coefficients without impairment to the response characteristic offers the possibility to increase the number of digits for the signals.

168   7. A synthetic sensitivity - roundoff design

Now, if one wants to reduce the implementation complexity by decreasing the coefficient bit number B_c while keeping B_s unchanged, then to maintain the same performance one needs to reduce the sensitivity. This is exactly what the unified measure G_T does by automatically adjusting the weighting factor ρ^2 to the respective bit lengths.

We have thus formulated a synthetic measure that we call *Total Noise Gain*. We have discussed some of its properties and have argued that this measure makes a lot of intuitive sense as a global measure that takes account of all the FWL effects. In the next section, we use this measure as an optimization criterion and solve the corresponding optimal realization problem.

7.3 Optimizing the Total Noise Gain

We have seen in Chapter 3 that the minimization of a noise gain without proper scaling of the states leads to absurd solutions. An l_2-norm scaling is usually introduced in order to maintain the amplitudes of the states within an acceptable range determined by the hardware, thereby reducing the probability of overflow. We shall do so here too: we therefore impose the following constraint on the realizations (or on the coordinate choice for the system representation):

$$(W_c)_{i,i} = 1 \quad \forall i, \qquad (7.24)$$

where W_c is the controllability Gramian of the realization as defined before.

So, the optimal FWL system design can be formulated as the following constrained minimization problem:

$$\min_{(A,B,C) \in S_H} G_T = M_{L_2} + \rho^2 G \quad \text{subject to} \quad (W_c)_{i,i} = 1 \quad \forall i. \qquad (7.25)$$

In this section, we present a solution to this optimal design problem. We first examine how the finite wordlength noise gain, G_T, depends on a similarity transformation by a matrix T. We recall that G_T can, equivalently, be written as:

$$G_T = tr(W_A) + tr(W_c) + (1 + \rho^2) tr(W_o). \qquad (7.26)$$

Let (A_0, B_0, C_0, d) be some initial realization in S_H and let $\{W_c^0, W_o^0, W_A^0\}$ be the corresponding triplet appearing in the definition (7.22) of G_T, where $W_A^0 = tr\left(\sum_{i=0}^{\infty} h_0(i) h_0^T(i)\right)$ (see (7.15)). Then a similarity transformation matrix T transforms this triplet into $\{T^{-1} W_c^0 T^{-T}, T^T W_o^0 T, W_A(T)\}$ in the new coordinate space. The expression for $W_A(T)$ has already been computed in (5.10) but is recalled here for reader comfort:

$$W_A = \sum_{i=0}^{\infty} T^T h_0(i) T^{-T} T^{-1} h_0^T(i) T. \tag{7.27}$$

The *Total Noise Gain* G_T can thus be expressed as the following function of the initial realization and of T:

$$\begin{aligned} G_T(T) &= tr(W_A(T)) + tr[T^{-1} W_c^0 T^{-T}] + (1+\rho^2) tr[T^T W_o^0 T] \\ &= tr[\sum_{i=0}^{\infty} T^T h_0(i) T^{-T} T^{-1} h_0^T(i) T] \\ &\quad + tr[T^{-1} W_c^0 T^{-T}] + (1+\rho^2) tr[T^T W_o^0 T]\}. \end{aligned} \tag{7.28}$$

The problem of identifying those realizations that are the solution of problem (7.25) is equivalent to that of finding those similarity transformations T that minimize $G_T(T)$ subject to the constraints (7.24) since, with the initial realization (A_0, B_0, C_0, d) and those T, the optimal realizations can easily be obtained. The optimal design problem can thus be reformulated as follows:

$$\min_{T: \det T \neq 0} G_T(T) \text{ subject to } (T^{-1} W_c^0 T^{-T})_{ii} = 1 \ \forall i. \tag{7.29}$$

To solve this problem we note that $G_T(T)$ can be recast as follows:

$$\begin{aligned} G_T(T) &= tr\left[\sum_{i=0}^{\infty} P h_0(i) P^{-1} h_0^T(i)\right] + tr(P^{-1} W_c^0) + tr(P M_o^0) \\ &\triangleq R(P) \end{aligned} \tag{7.30}$$

where $P = TT^T$ and $M_o^0 \triangleq (1+\rho^2) W_o^0$. It then follows from the above that

$$\min_{T: \det T \neq 0} G_T(T) \iff \min_{P: P > 0} R(P) \tag{7.31}$$

170   7. A synthetic sensitivity - roundoff design

both with the constraint
$$(T^{-1}W_c^0 T^{-T})_{ii} = 1 \quad \forall i. \tag{7.32}$$

The expression (7.30) of $R(P)$ is identical to the expression (5.12) of the L_2 sensitivity minimization problem of Chapter 5. This is not so surprising given that, except for the replacement of the term W_o by the term $(1+\rho^2)W_o$, the Total Noise Gain G_T is identical to the L_2 sensitivity measure. The minimization problem (7.31)-(7.32) would then be completely solved by the methods of Chapter 5 if it were not for the fact that here we need to solve a constrained optimization problem, that is to optimize $R(P)$ under the constraint (7.32). As we shall see, the solution method follows essentially the same line as that developed for the earlier unconstrained problem.

To solve the constrained minimization problem (7.31)-(7.32), our first step will be to replace the constraint (7.32) on T by an equivalent constraint on P, so that we can solve a constrained minimization problem involving P only. To do this, we need the following technical lemma.

Lemma 7.1 : *Let $M \in \mathbb{R}^{n \times n}$ be a symmetric and nonnegative definite matrix. There exists an orthogonal matrix U such that all the diagonal elements of $U^T M U$ are equal to 1 if and only if $tr[M] = n$.*

Proof: By SVD, there exists an orthogonal matrix U_m and a diagonal matrix D such that $M = U_m^T D^2 U_m$. Let $U = U_m^T U_0$, where U_0 is some orthogonal matrix, then $U^T M U = U_0^T D^2 U_0$. According to Lemma 3.2, there exists an U_0 such that $U_0^T D^2 U_0$ has all its diagonal elements equal to one if and only if $tr[D^2] = n$. The lemma then follows from the fact $tr[M] = tr[D^2]$. ∎

Note that $G_T(TU) = G_T(T)$ for any orthogonal matrix U. Therefore, the constraint $(T^{-1}W_c^0 T^{-T})_{ii} = 1, i = 1, 2, \ldots, n$ in (7.32) is equivalent to $tr[T^{-1}W_c^0 T^{-T}] = n$ since, according to Lemma 7.1, one can find a U such that $(U^T T^{-1}W_c^0 T^{-T} U)_{ii} = 1, i = 1, 2, \ldots, n$, if and only if $tr[T^{-1}W_c^0 T^{-T}] = n$. Thus condition (7.32) can be replaced by a condition on P, namely
$$tr(W_c^0 P^{-1}) = n. \tag{7.33}$$

The minimization of G_T can once again be reformulated as follows:

$$\min_{P:P>0} R(P) \text{ subject to } tr(W_c^0 P^{-1}) = n. \tag{7.34}$$

7. A synthetic sensitivity - roundoff design

Our first result shows that the minimum of (7.34) exists.

Lemma 7.2 : *For an asymptotically stable system, the minimum of $R(P)$ under the constraint (7.33) exists and is achieved by a nonsingular P.*

Proof: The proof follows directly from Lemma 7.A of Appendix 7. ∎

To compute the minimum and to show that it is unique, we define the Lagrangian

$$L(P,\lambda) = R(P) + \lambda[tr(W_c^0 P^{-1}) - n]. \qquad (7.35)$$

Using Lagrange's method, we obtain the following equations:

$$\begin{aligned}\frac{\partial L}{\partial P} &= \sum_{i=0}^{\infty}\{h_0(i)P^{-1}h_0^T(i) - P^{-1}h_0^T(i)Ph_0(i)P^{-1}\} + \\ &\quad + M_o^0 - P^{-1}W_c^0 P^{-1} - \lambda P^{-1}W_c^0 P^{-1} = 0, \qquad (7.36)\\ \frac{\partial L}{\partial \lambda} &= tr[W_c^0 P^{-1}] - n = 0. \qquad (7.37)\end{aligned}$$

These are necessary conditions that must be satisfied by any solution of the optimization problem (7.31)-(7.32). Since the solution set is restricted to nonsingular P, (7.36) is equivalent with

$$P\left[M_o^0 + \sum_{i=0}^{\infty} h_0(i)P^{-1}h_0^T(i)\right]P = (1+\lambda)W_c^0 + \sum_{i=0}^{\infty} h_0^T(i)Ph_0(i) \quad (7.38)$$

The following theorem shows that the equations (7.36), (7.37) or, equivalently, (7.37), (7.38) have a unique solution.

Theorem 7.1 : *With W_c^0, M_o^0 and $\{h_0(i)\}$ defined above, the set of equations (7.36)-(7.37) have a unique solution.*

Proof: Let (P_0, λ_0) and (P, λ) be two solutions of (7.36)-(7.37) with $\lambda_0 \leq \lambda$. Using a pre-transformation as in the proof of Theorem 5.1, it can be shown that, without loss of generality, $P_0 = I$ is a solution of (7.36)-(7.37). Such pre-transformation leaves λ_0 unchanged.

By SVD, P can be decomposed as $P = VX^2V^T$, where V is some orthogonal matrix and $X = diag(x_1, x_2, \ldots, x_n), x_i \geq x_{i+1} > 0$. So,

7. A synthetic sensitivity - roundoff design

corresponding to the solution (P, λ), (7.38) can be rewritten as

$$X^2 \left[\bar{M}_o^0 + \sum_{i=0}^{\infty} \bar{h}_0(i) X^{-2} \bar{h}_0^T(i) \right] X^2 = (1+\lambda) \bar{W}_c^0$$

$$+ \sum_{i=0}^{\infty} \bar{h}_0^T(i) X^2 \bar{h}_0(i)$$

(7.39)

where $\bar{M}_o^0 \triangleq V^T M_o^0 V$, $\bar{W}_c^0 = V^T W_c^0 V$ and with $tr(\bar{W}_c^0 X^2) = n$ and $\bar{h}_0(i) = V^T h_0(i) V \; \forall i$.

Similarly, for the solution (I, λ_0), (7.38) can be rewritten as

$$I \left[\bar{M}_o^0 + \sum_{i=0}^{\infty} \bar{h}_0(i) I \bar{h}_0^T(i) \right] = (1+\lambda_0) \bar{W}_c^0 + \sum_{i=0}^{\infty} \bar{h}_0^T(i) I \bar{h}_0(i).$$

(7.40)

The (k,k)-th equation of (7.39) is then given by

$$x_k^4 \left[\bar{M}_o^0(k,k) + \sum_{i=0}^{\infty} \left(\sum_{j=1}^{n} h_{kj}^2(i) x_j^{-2} \right) \right] = (1+\lambda) \bar{W}_c^0(k,k)$$

$$+ \sum_{i=0}^{\infty} \left(\sum_{j=1}^{n} h_{jk}^2(i) x_j^2 \right), \text{ or}$$

$$x_k^4 \left[\bar{M}_o^0(k,k) + \sum_{j=1}^{n} \left(\sum_{i=0}^{\infty} h_{kj}^2(i) \right) x_j^{-2} \right] = (1+\lambda) \bar{W}_c^0(k,k)$$

$$+ \sum_{j=1}^{n} \left(\sum_{i=0}^{\infty} h_{jk}^2(i) \right) x_j^2.$$

(7.41)

Similarly, for (7.40) we get

$$\bar{M}_o^0(k,k) + \sum_{j=1}^{n} \left(\sum_{i=0}^{\infty} h_{kj}^2(i) \right) = (1+\lambda_0) \bar{W}_c^0(k,k) + \sum_{j=1}^{n} \left(\sum_{i=0}^{\infty} h_{jk}^2(i) \right) \forall k$$

(7.42)

where $h_{jk}(i)$ is the (j,k)-th element of $\bar{h}_0(i)$. From (7.41), one obtains for $k = n$:

$$x_n^4 \left[\bar{M}_o^0(n,n) + \sum_{j=1}^{n} \left(\sum_{i=0}^{\infty} h_{nj}^2(i) \right) x_j^{-2} \right] = (1+\lambda) \bar{W}_c^0(n,n)$$

$$+ \sum_{j=1}^{n} \left(\sum_{i=0}^{\infty} h_{jn}^2(i) \right) x_j^2.$$
(7.43)

By noting that $x_n \leq x_i$ for all $i = 1, 2, \ldots, n-1$ and $\lambda_0 \leq \lambda$, this leads to the following inequality :

$$x_n^4 \left[\bar{M}_o^0(n,n) + \sum_{j=1}^{n} \left(\sum_{i=0}^{\infty} h_{nj}^2(i) \right) x_n^{-2} \right] \geq (1+\lambda)\bar{W}_c^0(n,n)$$
$$+ \sum_{j=1}^{n} \left(\sum_{i=0}^{\infty} h_{jn}^2(i) \right) x_n^2$$
$$\geq (1+\lambda_0)\bar{W}_c^0(n,n)$$
$$+ \sum_{j=1}^{n} \left(\sum_{i=0}^{\infty} h_{jn}^2(i) \right) x_n^2$$

or

$$\bar{M}_o^0(n,n)x_n^2 + \sum_{j=1}^{n} \left(\sum_{i=0}^{\infty} h_{nj}^2(i) \right) \geq (1+\lambda_0)\bar{W}_c^0(n,n)x_n^{-2}$$
$$+ \sum_{j=1}^{n} \left(\sum_{i=0}^{\infty} h_{jn}^2(i) \right). \quad (7.44)$$

Clearly, equality holds in (7.44) if and only if $x_i = x_n$ for all $i = 1, 2, \ldots, n-1$ and $\lambda = \lambda_0$.

Since $tr(W_c^0 P^{-1}) = n$, that is $tr(W_c^0 X^{-2}) = n$, and also $tr(W_c^0) = n$, we have $x_n \leq 1$. For such an x_n, we have

LHS[3] of (7.44) $\leq \bar{M}_o^0(n,n) + \sum_{j=1}^{n} \left(\sum_{i=0}^{\infty} h_{nj}^2(i) \right) \triangleq L$

and

RHS of (7.44) $\geq (1+\lambda_0)\bar{W}_c^0(n,n) + \sum_{j=1}^{n} \left(\sum_{i=0}^{\infty} h_{jn}^2(i) \right) \triangleq R.$

However, setting $k = n$ in equation (7.42) shows that $L = R$. This leads to the fact that (7.44) is true only for $x_n = 1$. Therefore, $x_i = 1$ for $i = 1, 2, \ldots, n$ and $\lambda = \lambda_0$. By this, the theorem is proved. ∎

---

[3] We use LHS to denote the left hand side and RHS to denote the right hand side.

7. A synthetic sensitivity - roundoff design

By combining the results of Lemma 7.2 and Theorem 7.1 we have the required result.

Theorem 7.2 : *The optimization problem (7.31)-(7.32) has a unique solution. It is obtained as the unique solution of (7.36)-(7.37).*

Proof: The proof is an immediate consequence of the previous results.
∎

It appears difficult to solve (7.36)-(7.37) analytically in order to get the optimal P^{opt}. However the solution can be obtained iteratively by the following algorithm:

$$P(k+1) = P(k) - \mu_1 \frac{\partial L(P,\lambda)}{\partial P}\Big|_{P=P(k),\lambda=\lambda(k)},$$

$$\lambda(k+1) = \lambda(k) - \mu_2 \frac{\partial L(P,\lambda)}{\partial \lambda}\Big|_{P=P(k),\lambda=\lambda(k)}, \quad (7.45)$$

where μ_1 and μ_2 are positive step sizes. Clearly, when these step sizes are chosen to be small enough, the algorithm (7.45) converges to the unique solution of (7.36)-(7.37).

Assume that the optimal P, P^{opt}, has been obtained as the steady state solution of (7.45). Unlike in previous chapters, the optimal similarity transformation, T^{opt}, can not be taken as any arbitrary square root of P^{opt}, because it needs to satisfy the constraint

$$[(T^{opt})^{-1} W_c^0 (T^{opt})^{-T}]_{ii} = 1$$

for all i. Let $P^{opt} = U\Sigma^2 U^T$ be an SVD of P^{opt} with $\Sigma = diag(\sigma_1, \sigma_2, \ldots, \sigma_n)$. Now consider $\Sigma^{-1} U^T W_c^0 U \Sigma^{-1}$ and let

$$V_0^T \Pi^2 V_0 = \Sigma^{-1} U W_c^0 U^T \Sigma^{-1}$$

be its SVD, where Π^2 is diagonal and can be shown to satisfy $tr(\Pi^2) = n$. By Lemma 7.1 there exists an orthogonal matrix V_1 that satisfies $(V_1 \Pi^2 V_1^T)_{ii} = 1$ for all i. The algorithm given by Hwang to construct minimum roundoff noise realizations is a procedure for constructing such V_1. With $V = V_1 V_0$ and $T^{opt} = U\Sigma V^T$ we now have $T^{opt} T^{opt^T} = P^{opt}$ and, simultaneously, $[(T^{opt})^{-1} W_c^0 (T^{opt})^{-T}]_{ii} = 1$ for all i, that is we have obtained an optimal transformation that transforms the initial realization (A_0, B_0, C_0, d) to one that optimizes the Total Noise Gain G_T subject to the l_2 scaling constraint.

7. A synthetic sensitivity - roundoff design

Comment:

We have already noted the identity, up to a scaling factor for one of the terms, between the expression (7.30) of $R(P)$ for the total noise gain and the expression (5.12) of $R(P)$ for the L_2 sensitivity measure. We now observe that (7.30) is also identical to the corresponding expression (6.52) for the pole-zero sensitivity measure, M_{pz}, except for the replacement of the infinite sum by a finite sum in the first term. Thus, the constrained minimization technique developed in this chapter for the design of realizations minimizing G_T under an l_2 constraint can also be used for the design of optimal realizations minimizing M_{pz} under a similar l_2 constraint on the dynamic range. This remark is in line with our final comment of Section 6.7.

7.4 A numerical example

In this section, we present an example to illustrate our theoretical results. We consider a 4-th order filter, given in direct form.

$\boxed{R_c :}$

$$A_c = \begin{pmatrix} 3.5897 & -4.8513 & 2.9241 & -0.6630 \\ 1.0000 & 0 & 0 & 0 \\ 0 & 1.0000 & 0 & 0 \\ 0 & 0 & 1.0000 & 0 \end{pmatrix} \quad B_c = \begin{pmatrix} 1 \\ 0 \\ 0 \\ 0 \end{pmatrix}$$

$$C_c = \begin{pmatrix} 0.2371 & 0.0359 & 0.2163 & 0.0105 \end{pmatrix} \times 10^{-3}$$

$$d = 3.1239 \times 10^{-5}.$$

We first compute an l_2-scaled realization that minimizes the L_2 sensitivity measure, that is we solve the problem (7.25) with $\rho = 0$ using the iterative algorithm (7.45). The optimal transformation from R_c is given by:

$$T^{opt}(M_{L_2}^{sc}) = \begin{pmatrix} -296.1541 & 216.3613 & 27.8686 & 143.0086 \\ -331.0805 & 258.7708 & 16.2682 & 96.5274 \\ -366.2937 & 304.9637 & 15.0423 & 37.1790 \\ -399.7398 & 356.7817 & 27.9302 & -35.8941 \end{pmatrix}$$

and the corresponding realization $R^{opt}(M_{L_2}^{sc})$ is given by:

7. A synthetic sensitivity - roundoff design

$\boxed{R^{opt}(M_{L_2}^{sc}):}$

$$A^{opt}(M_{L_2}^{sc}) = \begin{pmatrix} 0.8922 & 0.0206 & 0.0162 & -0.0887 \\ -0.0160 & 0.8838 & 0.0074 & 0.1009 \\ -0.0787 & -0.1222 & 0.8984 & -0.0506 \\ 0.0483 & -0.0364 & 0.1730 & 0.9154 \end{pmatrix}$$

$$B^{opt}(M_{L_2}^{sc}) = \begin{pmatrix} -0.3440 \\ -0.4042 \\ 0.0948 \\ -0.1124 \end{pmatrix} \quad C^{opt}(M_{L_2}^{sc}) = \begin{pmatrix} -0.1655 \\ 0.1303 \\ 0.0107 \\ 0.0450 \end{pmatrix}^T.$$

The optimal L_2 sensitivity measure for this l_2-scaled realization and the corresponding classical roundoff noise gain are:

$$M_{L_2}(R^{opt}(M_{L_2}^{sc})) = 28.6418 \text{ and } G(R^{opt}(M_{L_2}^{sc})) = 0.5623.$$

We have also computed the minimum of M_{L_2} without l_2 scaling, which is 27.6952.

We now compute an l_2-scaled realization that minimizes the classical roundoff noise gain G, namely an MRH structure that satisfies (3.57). This corresponds to solving (7.25) with $\rho = \infty$. The optimal transformation from R_c is given by:

$$T^{opt}(G) = \begin{pmatrix} -238.1998 & -113.2005 & -193.8319 & -166.9584 \\ -290.5165 & -146.0922 & -162.8762 & -156.0722 \\ -342.1780 & -189.5528 & -118.2495 & -149.1466 \\ -392.2533 & -243.8683 & -56.9049 & -149.7712 \end{pmatrix}$$

and the corresponding optimal realization is:

$\boxed{R^{opt}(G):}$

$$A^{opt}(G) = \begin{pmatrix} 0.9157 & -0.1052 & 0.1052 & 0.0732 \\ 0.0117 & 0.8932 & 0.0769 & -0.0834 \\ -0.0853 & -0.0026 & 0.8820 & 0.1080 \\ -0.1002 & 0.0878 & 0.0538 & 0.8989 \end{pmatrix}$$

$$B^{opt}(G) = \begin{pmatrix} 0.2023 \\ -0.4044 \\ -0.1909 \\ 0.2012 \end{pmatrix} \quad C^{opt}(G) = \begin{pmatrix} -0.1450 \\ -0.0756 \\ -0.0780 \\ -0.0790 \end{pmatrix}^T.$$

The L_2 sensitivity measure and the roundoff noise gain of this MRH structure are

$$M_{L_2}(R^{opt}(G)) = 29.2693 \text{ and } G(R^{opt}(G)) = G_{min} = 0.5555.$$

Comment:
We observe that the effect of the errors due to coefficient quantization is much more serious than that due to signal roundoff if the coefficients and the signals are implemented with the same number of bits (which would correspond to choosing $\rho = 1$ in our global measure G_T). These computations imply that more bits should be used for coefficient implementation than for the signals. This conclusion could not have been derived from the computation of the classical L_1/L_2 sensitivity measure since it has no obvious physical meaning and hence can not be compared with the roundoff noise gain G. The L_2 measure, on the other hand, is the variance of the effect of coefficient errors on the output and is thus comparable with the roundoff noise gain, which is the variance of the effect of the signal roundoff errors on the output.

It follows directly from the above analysis that $M_{L_2}(R^{opt}(M_{L_2}^{sc}))$ and $\rho^2 G(R^{opt}(G))$ will have the same order of magnitude if $\rho^2 = 64$. This corresponds to $B_c - B_s = 3$, that is to taking 3 more bits for coefficient implementation than for the encoding of the signals.

The minimization of G_T with $\rho^2 = 64$ using the algorithm (7.45) produces the following optimal transformation from R_c to $R^{opt}(G_T)$:

$$T^{opt}(G_T) = \begin{pmatrix} -194.2497 & -274.9319 & 155.3204 & 17.0603 \\ -218.4101 & -267.1176 & 188.2996 & 73.8997 \\ -245.5136 & -246.4952 & 234.2001 & 133.2234 \\ -278.1531 & -209.7700 & 294.9119 & 193.0983 \end{pmatrix}$$

and the corresponding optimal realization is:

$\boxed{R^{opt}(G_T) :}$

$$A^{opt}(G_T) = \begin{pmatrix} 0.8851 & 0.0554 & -0.0079 & 0.0867 \\ 0.0424 & 0.8980 & 0.0042 & 0.1365 \\ 0.0892 & -0.0876 & 0.8694 & 0.0169 \\ -0.0867 & -0.0873 & -0.1218 & 0.9372 \end{pmatrix}$$

$$B^{opt}(G_T) = \begin{pmatrix} 0.4345 \\ -0.1354 \\ 0.3094 \\ 0.0062 \end{pmatrix} \quad C^{opt}(G_T) = \begin{pmatrix} -0.1099 \\ -0.1303 \\ 0.0973 \\ 0.0375 \end{pmatrix}^T \quad (7.46)$$

The L_2 sensitivity measure and the roundoff noise gain for this globally optimal realization corresponding to $\rho^2 = 64$ are, respectively,

$$M_{L_2}(R^{opt}(G_T)) = 28.8830 \quad \text{and} \quad G(R^{opt}(G_T)) = 0.5572.$$

Finally, for comparison we have also considered the l_2-scaled control canonical form, R_c^{sc}, which is obtained from R_c by the similarity transformation $D = W_c^{1/2}(1,1)I$, where W_c is the controllability Gramian of R_c. The L_2 sensitivity measure and the roundoff noise gain for R_c^{sc} are, respectively,

$$M_{L_2} = 9.7742 \times 10^6 \quad \text{and} \quad G = 1.4162 \times 10^5.$$

Comments:

1. Once again this example shows that the FWL performance of the direct (control canonical) form is extraordinarily worse than that of any of the optimal forms.

2. For this example, all three optimal realizations yield comparable results: they all have good performances both with respect to roundoff noise gain and with respect to transfer function sensitivity. This confirms an often observed finding that realizations that minimize the roundoff noise gain also tend to have good sensitivity performance and vice-versa.

3. The realization $R^{opt}(G_T)$ is obviously a compromise between the other two optimal realizations, as should be.

4. The main lesson to be learned from this example, and as a matter of fact from this chapter, is that the global measure G_T gives us a handle on the optimal selection of the respective bit lengths, B_c and B_s, for the coefficients and for the signals. It is the comparison of the two terms of G_T that allows us to make this selection.

7.5 Conclusions

In this chapter we have synthesized the FWL effects of coefficient errors and of signal roundoff errors into a unique measure that we have called the *Total Noise Gain*. This has been achieved by considering the coefficient errors as random fluctuations and by computing the variance of the output errors due to these fluctuations. The Total Noise Gain G_T then turns out to be a linear combination of the classical roundoff noise gain $G = tr(W_o)$, with W_o the observability Gramian, and of the newly-defined L_2 sensitivity measure M_{L_2}: $G_T = M_{L_2} + \rho^2 G$, where ρ^2 is determined by the respective wordlengths used for the coefficients and for the signals.

By minimizing this unified measure under a dynamic range constraint on the states to minimize the probability of overflow, an optimal synthetic FWL state space design problem has been formulated. We have shown that the optimal realizations always exist and we have given an iterative algorithm to compute them.

In this and the previous chapters, we have studied the optimal FWL state space design problem in terms of optimizing a range of possible criteria. In these design criteria, sensitivity functions have been calculated as if all parameters are potentially encoded with error. In actual FWL implementations, such parameters as 0 and 1 suffer no FWL implementation error, and hence the sensitivity function with respect to these parameters should not be considered in the overall sensitivity measure. As a consequence, the theoretical measures discussed so far are in fact larger than the actual sensitivity measures taking into account error-free parameters, and, more importantly perhaps, the optimal realizations that result from our different design schemes are typically fully parametrized. In the next chapter, we shall aim at producing realizations, either optimal or suboptimal, that are sparse, in that they have as many integer coefficients as possible.

Appendix 7: Existence of a constrained minimum.

Lemma 7.A: *Consider three symmetric positive definite matrices M_1, M_2, and M_3. Let $P \in \mathbb{R}^{n \times n}$ belong to the set of nonnegative definite matrices and let $R(P) \triangleq f(P) + tr(PM_1) + tr(P^{-1}M_2)$, where $f(P)$ is a scalar positive differentiable function of P. Then the minimum of $R(P)$ under the constraint $tr(P^{-1}M_3) = n$ exists and is achieved by a positive definite P.*

Proof: Our proof is an adaptation of the proof of Lemma 4.A, where the adaptation is to take account of the constraint on the trace.

Let $P = U^T \Sigma^2 U$ with U an orthogonal matrix and

$$\Sigma = diag(\sigma_1, ..., \sigma_i, ..., \sigma_n), \sigma_i \geqslant 0 \ \forall i.$$

In the proof of Lemma 4.A we have defined a nonsingular set P_E for a given large enough positive number E:

$$P_E \triangleq \{P : |\theta_i| \leqslant \pi, C_-^2(E) \leqslant \sigma_i^2 \leqslant C_+^2(E), \forall i\}$$

with

$$C_-^2(E) = \lambda_{min}(M_1)/E \leqslant 1, C_+^2(E) = E/\lambda_{min}(M_2) \geqslant 1.$$

We have also shown that $R(P) > E$ outside P_E.

The basic idea in the following proof is to show that $tr(P^{-1}M_3) = n$ defines a closed set of $\{P\}$ that belongs to P_E. Therefore, $R(P)$ must have its minimum in this set since it is a continuous function of $P > 0$.

Clearly, for all P in P_E, the $\{\sigma_i^2\}$ satisfy

$$C_+^{-2}(E) \leqslant \sigma_i^{-2} \leqslant C_-^{-2}(E) \ \forall i$$

with $C_+^{-2}(E) < 1$ and $C_-^{-2}(E) > n$ for some large enough E (For example, $E > n \times max\{\lambda_{min}(M_1), \lambda_{min}(M_2)\}$).

Now, without loss of generality let us assume that $M_3 = I$. Therefore, $tr(P^{-1}M_3) = n$ becomes

$$\sum_{i=1}^{n} \sigma_i^{-2} = n. \tag{7.A.1}$$

So, when $C_+^{-2}(E) \leqslant \sigma_i^{-2} \ \forall i$, it follows from (7.A.1) that

$$\sigma_i^{-2} \leqslant n - (n-1)C_+^{-2}(E) \triangleq C_0 < n. \tag{7.A.2}$$

Clearly, $\{\sigma_i^{-2}\}$ outside $\Sigma_c \triangleq \{C_+^{-2}(E) \leqslant \sigma_i^{-2} \leqslant C_0, \forall i\}$ will either violate (7.A.1), or else $R(P) > E$.

The constraint (7.A.1) can be seen as a plane in a coordinate space with variables $\{\sigma_i^{-2}\}$. It is easy to see that with the constraint Σ_c on these variables the constrained plane is a closed set, which we denote by P_c. Outside this set P_c, either $R(P) > E$ or (7.A.1) is violated, while any point P inside this set is positive definite and satisfies (7.A.1). Therefore, under the constraint $tr(P^{-1}M_3) = n$, $R(P)$ has its minimum in this closed set since it is continuous. ∎

8

Sparse optimal and suboptimal realizations

8.1 Introduction

A linear system of order n can be modelled by its transfer function using $2n + 1$ parameters. In state space representations, the so-called direct forms, also called canonical forms, such as the controllable and observable forms are also parametrized with $2n + 1$ parameters. These forms have the advantage of a simple implementation structure and hence of high computational speed. However, we have seen in our previous chapters, both theoretically and through a range of simulations, that these realizations usually yield poor performance when implemented with FWL coefficients and arithmetics. To overcome this performance degradation, optimal FWL design problems have been formulated and solved. This has been thoroughly studied in the previous chapters.

We have observed that optimal realizations that minimize a sensitivity measure, a roundoff noise gain or some combined sensitivity/roundoff noise measure yield a much better performance than the canonical forms. However, the derivation of the optimal design schemes of the previous chapters can be criticized on two grounds.

- In our various definitions of sensitivity and in our roundoff noise gain analyses, every parameter in a realization is considered to have a FWL implementation error and each arithmetic operation is assumed to produce a roundoff noise. These assumptions do not express what exactly happens in an actual implementation where some parameters, such as 0 and ±1, can be implemented without FWL error. Hence these particular parameters - which we shall call *trivial parameters* in this chapter - do not contribute any error to the transfer function, and any arithmetic operation with these parameters and FWL signals produces no roundoff noise at all. The assumption that the coefficient vectors and matrices are fully parametrized rather than structured were made because they make the analysis so much easier[1]. By not taking into account the fact that these trivial parameters are encoded without error, our theoretical analysis heavily penalizes against canonical forms, and one would be justified in questioning the practical relevance of all the optimal state space realization design schemes discussed so far. Fortunately for the authors of this book, for all the other authors in this field, and also for you (the reader), the simulations vindicate the computation of optimal realizations: they show that these have a much better performance against FWL errors than the canonical forms, even though the *actual* gap is not as large as the *theoretical* one.

- The other criticism one could (and should) level at the FWL optimal realizations is that they are generally fully parametrized. This means that they increase the complexity of implementation and slow down the arithmetic processing. For an n-th order single input single output filter, this implies that there are $(n+1)^2$ multiplies for the computation of each output sample. This is the price to be paid to have a better FWL performance. It is questionable whether this large number of multiplies can be justified in most applications. On the basis of computational speed, it is therefore desirable that the filter have a nice performance as well as a small number of coefficients. Clearly, in many applications some form of speed/performance tradeoff needs to be considered.

Given that a coordinate optimization over the set of sparse realizations is presently out of reach mathematically or, stated otherwise, that no optimization can presently be performed using *actual* sensitivity measures (that take account of trivial parameters), the next best thing is to investigate whether the class of realizations that optimize some of the design criteria studied so far do not contain sparse realizations, or whether perhaps sparse realizations can be obtained that are sufficiently close to the optimal ones and whose sensitivity or roundoff noise measures are close to optimal.

The traditional method of reducing the number of parameters to be implemented in a digital filter is to decompose an n-th order filter into cascade or parallel connections of second (and possibly first) order

---

[1] In fact the whole theory of perturbations of structured matrices is only in its infancy according to our numerical analysis friends.

sections. This idea was originally aimed at reducing the pole-zero sensitivity of a transfer function with respect to the coefficients of its two polynomials (see [Kai66] and Chapter 6). The idea of decomposing a high order filter into a series of smaller order subfilters and of optimizing the realization of each subfilter was then developed, and the name *block optimal structures* was coined: see e.g. [RM87].

Here we follow a rather more systematic approach. First we examine the subsets of optimal realizations with respect to a number of the design criteria studied in earlier chapters, in particular the various sensitivity measures. We show that these subsets in fact contain sparse realizations. Since the optimal realization algorithms of the previous chapters do not naturally yield sparse structures, we show how they can be obtained by an orthogonal similarity transformation starting from an arbitrary realization in this optimal subset. The key point is that the optimal subset is invariant under orthogonal similarity transformations, that is by applying an orthogonal similarity transformation to a realization $(A^{opt}, B^{opt}, C^{opt})$ in the optimal subset, a new realization is obtained that is still in the optimal subset. In Section 8.2, we apply this idea to construct FWL-optimal Hessenberg and Schur realizations. The Hessenberg and Schur forms both possess $\frac{1}{2}n(n-1)$ trivial zero parameters. A third family of sparse optimal realizations based on the symmetric property of balanced forms is also presented in Section 8.2. These three families of sparse realizations belong to the optimal realization subset: they are therefore called *sparse optimal realizations*.

In Section 8.3 we discuss in some detail the gap between *actual* and *theoretical* sensitivity measures. We explain why, for a sparse optimal realization, the actual sensitivity is smaller than the theoretical optimal one that is based on the assumption that errors are made on all coefficients, and we illustrate this with a simulation. In addition we show that the optimal Schur realizations have the added property that their actual pole sensitivity is very small, and even as small as that of normal forms (which are optimal with respect to pole sensitivity) in the case of distinct poles.

In Section 8.4 we discuss the sparseness issue in the case of optimal realization problems under l_2 scaling constraints, such as the optimal roundoff noise problem. The optimal realization sets for these problems are again defined up to some orthogonal transformation, but these orthogonal transformations are constrained by the l_2 scaling requirement.

We show that sparse quasi-optimal realizations can be obtained using a remarkable algorithm that was successively developed by Chan [Cha79], Bomar and Hung [BH84] and Amit and Shaked [AS88]. The algorithm was conceived for the minimum roundoff noise problem, but we show that it can be adapted to a range of other constrained problems.

Finally, in Section 8.5 we briefly review some other more classical methods for the synthesis of sparse realizations such as parallel or cascade connections, or suboptimal triangular realizations. We conclude with a discussion in Section 8.6.

8.2 Sparse optimal realizations

It is well known that many commonly used measures are invariant under orthogonal similarity transformations and that a matrix can be made sparse with such a transformation. The sensitivity minimizing schemes of Chapters 3, 4, 5 and 6 are all based on measures that remain unchanged under an orthogonal similarity transformation. Therefore, it would be interesting to transform the fully parametrized optimal realizations computed in these chapters into sparse realizations using orthogonal similarity transformations. That this is at all possible - and we shall show that it is - means that, for each of the minimal sensitivity design schemes studied so far, the optimal realization set contains a subset of sparse realizations (A^{opt}, B^{opt}, C^{opt}, d), and that these can be obtained from an arbitrary fully parametrized optimal realization by orthogonal similarity transformations.

The basis for our analysis is that the optimal realizations that minimize the (frequency weighted or not) sensitivity measures are not unique and that, in each case, the optimal realizations are defined only up to an arbitrary[2] orthogonal matrix. This freedom can be used to generate sparse realizations which belong to the optimal realization set. In addition, since an orthogonal matrix of order n has $\frac{1}{2}n(n-1)$ degrees of freedom, it is to be expected that the sparsest realizations have at most that many trivial parameters. In the next three subsections we present some sparse realizations obtained from an arbitrary optimal realization using this freedom of orthogonal similarity transformations. We shall see that they have precisely that number of zero parameters.

---

[2] At least in the case of unconstrained problems.

The results of this section apply to any one of the (weighted or unweighted) L_1/L_2, L_2 or pole-zero sensitivity minimizations, since all of these measures have been shown to be invariant under orthogonal transformations. We therefore denote by S^{opt} the optimal realization set corresponding to any one of these measures, and we now show that sparse optimal realizations can be constructed by starting from an arbitrary initial realization in S^{opt}, and then transforming it to sparse form by an appropriate orthogonal similarity transformation.

Before we get down to the nitty gritty of constructing such sparse optimal realizations, we make the final comment that orthogonal transformations have the important advantage of being numerically stable [DV85].

8.2.1 HESSENBERG OPTIMAL REALIZATIONS

Hessenberg matrices have found many applications such as in eigenvalue and frequency response computations [LL86]. In this subsection we first provide some general background on Hessenberg forms, then show how to obtain the optimal system Hessenberg realization from an optimal realization that minimizes a sensitivity measure.

Definition 8.1 : *A matrix $A = \{a_{ij}\} \in \mathbb{R}^{n \times n}$ is said to be in upper Hessenberg form if $a_{ij} = 0$ for $i - j \geqslant 2, i, j \in \{1, 2, ..., n\}$.*

A 4×4 upper Hessenberg matrix would have the form

$$\begin{pmatrix} x & x & x & x \\ x & x & x & x \\ 0 & x & x & x \\ 0 & 0 & x & x \end{pmatrix} \tag{8.1}$$

Lower Hessenberg matrices are defined similarly ($a_{ij} = 0$ for $j - i \geqslant 2$), but we shall confine our attention in the sequel to upper Hessenberg matrices, and when we say that a matrix is in Hessenberg form, we mean that it is in upper Hessenberg form.

It is well known that for any matrix A, there exists a nonsingular $T \in \mathbb{R}^{n \times n}$ such that $T^{-1}AT$ is in Hessenberg form. This T is not unique; in particular, T can be taken to be an orthogonal matrix (see [GL83]). In system theory, a realization (A, B, C, d) is called a system Hessenberg form if the matrix A is of Hessenberg form and all the elements of either B or C are zero except the first one [LL86]. Lutz and Hakimi [LH88] have used orthogonal similarity transformations to transform an optimal realization that minimizes the sensitivity measure into a corresponding system Hessenberg form, called Hessenberg optimal realization.

To explain their scheme, we first introduce a special class of orthogonal matrices known as Householder transformations. Householder transformations are matrices of the form

$$T(u) = I_n - 2\frac{uu^T}{u^T u}, \tag{8.2}$$

where I_n is the unit matrix of order n and u is an arbitrary $n \times 1$ column vector. Householder transformations are symmetric ($T(u) = T^T(u)$), orthogonal ($T(u)T^T(u) = I$), and involutory ($T^2(u) = I$). Thus a Householder transformation $T(u)$ has $T(u) = T^T(u) = T^{-1}(u)$. Householder transformations have the following important properties:

- P_1: $T(u)v = v$ if u and v are orthogonal;

- P_2: one can use a Householder transformation to zero selected components of a non-zero vector x by proper choice of u.

Suppose we are given a non-zero $x \in \mathbb{R}^{n \times 1}$ and we want $T(u)x$ to be a multiple of a unit vector e. It can be shown [GL83] that when

$$u = x + \rho e, \rho = \pm ||x||_2, \tag{8.3}$$

then

$$T(u)x = -\rho e. \tag{8.4}$$

In particular, when e is the i-th column vector $e_i(n)$ of the unit matrix I_n of order n, $T(u)$ determined by (8.2) with (8.3) transforms x into a vector whose elements are zero except for the i-th one. A numerical algorithm for computing this u is given in [GL83]. The next result shows that any realization (A, B, C, d) can be transformed to system Hessenberg form using a succession of Householder transformations.

Theorem 8.1 : *An arbitrary state space realization (A, B, C, d) can be transformed to a system Hessenberg form (A_H, B_H, C_H, d) where A_H is in Hessenberg form and*

$$B_H = (\mp ||B||_2, 0, 0, ..., 0)^T. \tag{8.5}$$

by a succession of $n-1$ Householder transformations.

8. Sparse optimal and suboptimal realizations

Proof: First, one can see that $T(u_1)$, determined by (8.2) and (8.3) with $x = B, e = e_1(n)$, transforms B into B_H. Denote

$$A_i \triangleq T(u_i)...T(u_2)T(u_1)AT(u_1)T(u_2)...T(u_i)$$

and $A_i(j:k,l)$ the vector that is obtained from the l-th column vector of A_i by deleting all its components except the j-th to the k-th. With $x = A_1(2:n,1)$ and $e = e_1(n-1)$, (8.3) produces a vector u_{21} of order $(n-1)$, and hence one can form u_2 in such a way that $u_2 \triangleq (0 \ u_{21}^T)^T$. It is easy to see that all the components of $A_2(1:n,1)$ are zero except the first two. Since B_H and u_2 are orthogonal, $T(u_2)$ does not change B_H (see property P_1). Now, with $x = A_2(3:n,2)$ and $e = e_1(n-2)$, (8.3) will yield a vector u_{31} of order $(n-2)$, and hence one can form u_3 in such a way that $u_3 \triangleq (0 \ 0 \ u_{31}^T)^T$. The transformation $T(u_3)$ produces A_3 such that all components of $A_3(1:n,2)$ are zero except the first three. With the property P_1, one can see that $T(u_3)$ does not change B_H and that $A_3(3:n,1)$ remains a zero vector just as $A_2(3:n,1)$. Continuing in this way, we can find $T(u_4), ..., T(u_{n-1})$ such that $A_H = T_H^T A T_H$ is in upper Hessenberg form with

$$T_H = T(u_{n-1})...T(u_2)T(u_1). \tag{8.6}$$

The transformed B vector is $T_H^T B = B_H$ as shown in (8.5). ∎

Since this transformation to Hessenberg form applies to any realization, it applies in particular to a realization that belongs to some optimal set, S^{opt}. Since the similarity transformation T_H is orthogonal, the transformed realization still belongs to S_H. We have thus proved the following corollary.

Corollary 8.1 : Let (A, B, C, d) be an arbitrary optimal realization that belongs to some optimal realization set S^{opt}. Then (A, B, C, d) can be transformed to system Hessenberg form with B_H as given by (8.5). This realization has at least $\frac{1}{2}(n-1)n$ zero elements.

Proof: Follows directly from the theorem above. ∎

With its $\frac{1}{2}(n-1)n$ zero elements, the sensitivity-optimal Hessenberg realizations are as sparse as can be: see our discussion above. Compared with fully parametrized optimal realizations, this represents a considerable reduction in implementation complexity, which will speed up the data processing.

8.2.2 SCHUR OPTIMAL REALIZATIONS

We now investigate another sparse realization, called Schur realization, which can also be obtained from an arbitrary realization by an orthogonal similarity transformation. We first define what we call a Schur realization.

Definition 8.2 : *A realization* (A_S, B_S, C_S, d_S) *is called Schur realization if the matrix* A_S *is in the following real Schur form:*

$$A_S = \begin{pmatrix} A_{11} & x & . & . & x & . & x \\ 0 & A_{22} & . & . & x & . & x \\ . & . & . & . & . & . & . \\ . & . & . & . & A_{ii} & . & . \\ . & . & . & . & . & . & . \\ 0 & 0 & . & . & 0 & . & A_{mm} \end{pmatrix} \qquad (8.7)$$

where each A_{ii} *is either a real number or a real* 2×2 *matrix having complex conjugate eigenvalues.*

We note that what makes a realization (A, B, C, d) a Schur realization is a property of the A matrix only. As far as our sensitivity minimization problems are concerned, the Schur realizations derive their interest from the following classical result.

Theorem 8.2 : *Any matrix* $A \in \mathbb{R}^{n \times n}$ *can be transformed to the real Schur form* A_S *of (8.7) by an orthogonal similarity transformation matrix* U.

Proof: see e.g. [GL83]. ■

With B_S and C_S arbitrary and A_S in the form (8.7), a Schur realization (A_S, B_S, C_S) has at least $\frac{1}{2}n(n-1) - p$ zero elements where n is the dimension of A_S and p is the number of diagonal block-matrices of dimension 2×2 in A_S. This is clearly a sparse realization, but not as sparse as might be hoped for given our discussion above. We show in the next theorem that this number of zero elements can be increased to $\frac{1}{2}n(n-1)$ - just as for Hessenberg forms - by further orthogonal similarity transformations.

Theorem 8.3 : *Any realization* (A, B, C, d) *can be transformed to a Schur realization* (A_S, B_S, C_S, d) *that has at least* $\frac{1}{2}n(n-1)$ *zero parameters by an orthogonal similarity transformation.*

8. Sparse optimal and suboptimal realizations

Proof: First, by Theorem 8.2 above, (A, B, C) can be transformed to a Schur realization $(\bar{A}_S, \bar{B}_S, \bar{C}_S)$ by an orthogonal similarity transformation, with $\bar{A}_S$ as in (8.7) and $\bar{B}_S, \bar{C}_S$ arbitrary. This form has at least $\frac{1}{2}n(n-1) - p$ zero elements, with n and p defined above. Denote

$$U = diag(U_{11}, U_{22}, ..., U_{mm}), \tag{8.8}$$

where U_{ii} is either a 1×1 orthogonal matrix (i.e. ± 1) or a 2×2 orthogonal matrix, having the same dimension as $\bar{A}_{ii}$ in $\bar{A}_S$. Notice that a similarity transformation by the matrix U will change neither the optimality nor the Schur structure of the realization, that is if $(\bar{A}_S, \bar{B}_S, \bar{C}_S)$ is an optimal Schur realization, so is $(U^T \bar{A}_S U, U^T \bar{B}_S, \bar{C}_S U)$. So this realization in the new coordinate system has at least $\frac{1}{2}(n-1)n - p$ zeros in $U^T \bar{A}_S U$. Note that

$$U^T \bar{B}_S = [\bar{B}_1^T U_{11}, \bar{B}_2^T U_{22}, ..., \bar{B}_i^T U_{ii}, ..., \bar{B}_m^T U_{mm}]^T. \tag{8.9}$$

Let $\bar{A}_{ii}$ be a block of dimension 2×2 in $\bar{A}_S$. Then we denote

$$U_{ii} = \begin{pmatrix} cos\delta_i & sin\delta_i \\ -sin\delta_i & cos\delta_i \end{pmatrix} \tag{8.10}$$

and $\bar{B}_i = [b_{1i}, b_{2i}]^T$. One can see that with the following choice of δ_i:

$$\delta_i = tan^{-1}(\frac{b_{1i}}{b_{2i}}),$$

the first element of $U_{ii}^T \bar{B}_i$ will be zero. Therefore, with a series of p such 2×2 orthogonal matrices U_{ii}, the new vector B_S has the following form

$$B_S = U^T \bar{B}_S = [0, x, 0, x, ..., 0, x]^T, \tag{8.11}$$

which means that the realization in the new coordinate system has at least $\frac{1}{2}n(n-1)$ zero parameters. ∎

Since any realization can be transformed to this particularly sparse Schur form via an orthogonal transformation, we have thus proved the following corollary.

Corollary 8.2 : Let (A, B, C, d) be an arbitrary optimal realization that belongs to some optimal realization set S^{opt}. Then (A, B, C, d) can be transformed to an optimal Schur realization (A_S, B_S, C_S, d), with A_S in the form (8.7) and B_S in the form (8.11). This realization has at least $\frac{1}{2}(n-1)n$ zero elements.

Proof: Follows directly from the theorem above. ∎

In the same way, we can alternatively make the vector C_S, instead of B_S, have the same sparse form as above.

We have shown how to use the $\frac{1}{2}n(n-1)$ degrees of freedom of orthogonal transformations to compute Hessenberg or Schur realizations within an optimal realization set, S^{opt}. These sparse forms can thus be used for any of the sensitivity measures mentioned at the beginning of this section, since they are all invariant under orthogonality transformations. The optimal Hessenberg and Schur realizations have $\frac{1}{2}n(n-1)$ zero elements, leading to significantly simpler implementations and an improvement in computational speed.

8.2.3 SPARSE BLOCK-BALANCED REALIZATIONS

In the particular case where the design criterion is the unweighted L_1/L_2 sensitivity measure, the optimal realization set consists of those realizations whose controllability and observability Gramians are identical, as we have shown in Corollary 3.1. We have also seen in Theorem 3.3 that the internally balanced realizations belong to this particular optimal set. Using a particular symmetry property of balanced realizations, Iwatsuki et al [IKH90] have obtained a sparse 'block-balanced' realization by applying an appropriate orthogonal similarity transformation to this symmetric balanced realization. Their procedure is as follows.

As already noted in Chapter 3, it has been shown (see [Kun78] and [WK83]) that there exists a balanced realization (A_b, B_b, C_b) with the following symmetry property:

$$A_b = Q A_b^T Q, B_b = Q C_b^T, \qquad (8.12)$$

where Q is a signature matrix and the number of $+1$ and -1 in this diagonal matrix are determined by the Cauchy index of the system [FN83]. Let p and q be the number of $+1$ and -1 in this signature matrix, respectively, then $p + q = n$, where n is the order of the system.

By a series of elementary row and column permutations, one can produce a Q matrix whose first p diagonal elements are $+1$, and whose last q diagonal elements are -1. The corresponding balanced realization has the structure:

$$A_b = \begin{pmatrix} A_{11} & A_{12} \\ -A_{12}^T & A_{22} \end{pmatrix}, B_b = \begin{pmatrix} B_1 \\ B_2 \end{pmatrix}, C_b^T = \begin{pmatrix} B_1 \\ -B_2 \end{pmatrix}. \qquad (8.13)$$

The block matrices A_{11} and A_{22} in the above matrix A_b are $p \times p$ and $q \times q$ symmetric matrices, respectively. Since there exist two orthogonal matrices U_{11} and U_{22} of appropriate dimension such that $U_1^T A_{11} U_1$ and $U_2^T A_{22} U_2$ are both diagonal, one can use a block orthogonal similarity transformation matrix $U = diag(U_1, U_2)$ to transform the balanced realization into (A_0, B_0, C_0), where A_0 possesses $N_{pq} = p(p-1) + (n-p)(n-p-1)$ zero parameters. This number depends on the Cauchy index. For some cases, N_{pq} is smaller than $\frac{1}{2}n(n-1)$, the number of zeros in a Hessenberg or Schur realization, for other cases it is larger. The observability and controllability Gramians of this sparse form, (A_0, B_0, C_0), are identical and block-diagonal; hence the name block-balanced realization.

We stress again that these block-balanced structures are of interest only in the case where the balanced realization is optimal, that is when the filter design criterion is the unweighted L_1/L_2 sensitivity measure. In that sense optimal Hessenberg and Schur realizations are more general than these structures.

8.3 Theoretical versus actual sensitivity measure

In the introduction of this chapter, we levelled a two-pronged attack on our optimal sensitivity machinations of the previous chapters.

- The sensitivity measures adopted for the computations of optimal realizations consider all parameters in a realization (A, B, C) to be liable to error whereas any sensible person knows that a coefficient 0, a 1, or even any integer multiple of the quantization step can be implemented without error. Our way out of this criticism was to say that numerical analysts do not have much to say about norms of sparse matrices.

- The optimal realizations computed by the algorithms of the previous chapters typically lead to fully parametrized matrices, which makes for complex[3] implementations and slow data processing.

---

[3]The word complex is of course dangerous because our more mathematically inclined readers might immediately think about $\mathbb{C}$, the set of complex numbers in which they feel so comfortable. Our more alert readers have of course understood that this book is written for shop floor engineers who will have had no difficulty realizing that complex just means complicated here.

We have seen in the previous section that the optimal realization sets contain sparse matrices, and we have shown how to compute them, thereby fending off the second criticism. We shall now argue that the sparse optimal matrices thus computed have the added benefit that their *actual sensitivity measures* - that is, taking account of error-free 0 and 1 coefficients - are in fact smaller than the *theoretical measures* that they minimize. This means that sparse realizations can not only reduce the complexity of the implementation and hence speed up the processing, but also improve the actual performance. Even though this does not fully answer the first criticism (an optimization over a set of sparse matrices might yield an even better realization), it certainly gives an added incentive for the use of these sparse optimal realizations.

Recall first that a fully parametrized optimal realization and a corresponding optimal sparse realization obtained by an orthogonal similarity transformation have the same *theoretical* sensitivity measure. This is because in the sensitivity definitions used in the previous chapters, it has been assumed that every parameter in a realization (A, B, C, d) is potentially implemented with error and the sensitivity with respect to every one of these parameters is taken into account in the over-all sensitivity measure. The *actual* sensitivity measure of a sparse optimal realization is of course smaller than that of a fully parametrized optimal one, since the trivial parameters are implemented exactly. This will be illustrated by an example at the end of this section, but first we show that the Schur realizations presented in subsection 8.2.2 have the additional nice feature that their actual pole sensitivity measure is near optimal and can be made optimal by a clever orthogonal transformation.

8.3.1 Pole sensitivity of Schur realizations

We have shown in Sections 6.3 and 6.4 that the pole sensitivity measure depends strongly on the chosen realization, that the partial sensitivity measure Ψ_{pk} of a particular pole $\lambda_k \in \{\lambda(A)\}$ is larger than or equal to one and that, for a filter of order n, the minimal value, n, of the global pole sensitivity measure $\Psi_p = \sum_{k=1}^{n} \Psi_{pk}$ is achieved if and only if the matrix A is normal.

An outstanding property of a Schur realization (A_S, B_S, C_S) is that its poles are determined only by the main block diagonal elements of A_S. This property allows one to analyse its pole sensitivity behaviour easily. By applying an orthogonal similarity transformation of the form

(8.8)-(8.10) to any Schur realization, the actual pole sensitivity performance can be improved even though the theoretical pole sensitivity measure remains unchanged. To show this, we first study the pole sensitivity of a Schur realization.

For a real pole λ_k corresponding to a 1×1 block matrix in A_S, it is easy to see that that actual sensitivity measure is[4]

$$\|\frac{\partial \lambda_k}{\partial A_S}\|_F^2 = 1.$$

Indeed, the zero elements under the diagonal of A_S have no implementation error at all, and hence this real pole is unaffected by elements other than itself. This means that the actual pole sensitivity measure of a real pole in a Schur realization reaches the minimal possible value. Hence if A_S has all real eigenvalues, the actual pole sensitivity measure, Ψ_p, is equal to n even if A_S is not normal.

We now show that, even if a Schur matrix A_S has complex poles, its actual sensitivity measure can be made very small by orthogonal transformations. Consider a pair of complex conjugate poles corresponding to a 2×2 block matrix in A_S. Denote by $M = \{m_{ij}\}$ this 2×2 real diagonal block matrix of A_S having complex conjugate eigenvalues λ_1 and λ_2. Then the following condition is satisfied

$$\Delta^2 = -4m_{12}m_{21} - (m_{11} - m_{22})^2 > 0. \tag{8.14}$$

By direct computation, we have the following sensitivity measures for $k = 1, 2$:

$$|\frac{\partial \lambda_k}{\partial m_{11}}|^2 = |\frac{\partial \lambda_k}{\partial m_{22}}|^2 = \frac{1}{4}[1 + \frac{(m_{11} - m_{22})^2}{\Delta^2}],$$

$$|\frac{\partial \lambda_k}{\partial m_{12}}|^2 = \frac{m_{21}^2}{\Delta^2}, \quad |\frac{\partial \lambda_k}{\partial m_{21}}|^2 = \frac{m_{12}^2}{\Delta^2}. \tag{8.15}$$

For this pair of complex poles the actual sensitivity measure with respect to M is given by

$$\Psi_p = \sum_{k=1}^{2} \|\frac{\partial \lambda_k}{\partial M}\|_F^2 = \frac{(m_{12} - m_{21})^2}{\Delta^2}. \tag{8.16}$$

---

[4]In this computation, errors are only considered on the nontrivial parameters in A_S.

For $m_{11} = m_{22}, m_{12} = -m_{21}$, the 2×2 matrix M is normal and the pole sensitivity measure Ψ_p reaches its minimal value, 2. Denote

$$U \triangleq \begin{pmatrix} cos\theta & sin\theta \\ -sin\theta & cos\theta \end{pmatrix} \text{ with } \theta \in [0, \pi]. \qquad (8.17)$$

and

$$R \triangleq U^T M U \triangleq \begin{pmatrix} r_{11} & r_{12} \\ r_{21} & r_{22} \end{pmatrix}.$$

Then,

$$r_{11} = m_{11}cos^2\theta - (m_{12} + m_{21})sin\theta cos\theta + m_{22}sin^2\theta$$
$$r_{12} = m_{12}cos^2\theta + (m_{11} - m_{22})sin\theta cos\theta - m_{21}sin^2\theta$$
$$r_{21} = m_{21}cos^2\theta + (m_{11} - m_{22})sin\theta cos\theta - m_{12}sin^2\theta$$
$$r_{22} = m_{22}cos^2\theta + (m_{12} + m_{21})sin\theta cos\theta + m_{11}sin^2\theta \qquad (8.18)$$

From (8.18) one can see that with a proper choice of θ, it is always possible to change one of the four r_{ij} to some desired value lying between bounds determined by the m_{ij}. Observing (8.15), we can see that the closer the two diagonal elements are, the more insensitive the measure is with respect to these elements because Δ^2 is unchanged by an orthogonal transformation. So one can choose θ such that $r_{11} = r_{22}$, which makes $\frac{\partial \lambda_k}{\partial r_{jj}} = 1/4$ for $k, j = 1, 2$ (see (8.15)). This can be achieved by letting

$$\theta = \frac{1}{2} tan^{-1} [\frac{m_{11} - m_{22}}{m_{12} + m_{21}}] \qquad (8.19)$$

Since the matrices M and R are both fully parametrized, the actual pole sensitivity measure is identical to the theoretical one. Hence, this reduction of the pole sensitivity measure with respect to the diagonal elements will correspondingly increase the measure with respect to the two off-diagonal elements, r_{12} and r_{21}, since the orthogonal transformation U does not change the theoretical pole sensitivity measure. So, what have we gained? The idea is that for a given bit number, B_c, for coefficient implementation, one can choose a FWL number v^* that is as close as possible to the off-diagonal element, say r_{21}, to which the pole sensitivity is highest. (This implies $r_{12}^2 > r_{21}^2$, see (8.15)). Using (8.18) one can then find a θ^* (and hence a U via (8.17)) such that this off-diagonal element is exactly equal to v^*. As a result, this parameter r_{21} having the highest sensitivity is implemented exactly with finite wordlength, while the two diagonal elements r_{11} and r_{22} have near-minimal sensitivity. Performing a similar orthogonal transformation for each diagonal block will improve the actual pole sensitivity behaviour.

The pole sensitivity properties that we have exhibited in this subsection apply to any matrix A_S in real Schur form. They show that realizations (A_S, B_S, C_S) with A_S in real Schur form have an actual pole sensitivity measure, Ψ_p, that is optimal (and equal to n) in the case of real poles and near-optimal when the system contains complex conjugate poles. Since the transformations required to bring a matrix in real Schur form and the further transformations proposed to reduce the pole sensitivity of Schur matrices having complex eigenvalues are orthogonal similarity transformations, these considerations apply to the case where the realizations belong to some sensitivity-optimal realization set, S^{opt}. Thus, let (A, B, C) be a realization that minimizes any one of the sensitivity design criteria mentioned at the beginning of this chapter, and let S^{opt} be the corresponding optimal realization set to which (A, B, C) belongs. Then (A, B, C) can be transformed to a Schur realization (A_S, B_S, C_S) within S^{opt}, with a double advantage:

- A_S is sparse, hence a higher computational speed;
- the actual pole sensitivity is reduced.

Consider in particular the case where the design criterion is precisely the pole sensitivity measure. Then

- the optimal realizations in the case of simple poles are the set of normal matrices, as we have seen in Chapter 6. Since this set allows for orthogonal similarity transformations (see Theorem 6.2), it contains a Schur normal form that has the advantage of being sparse. In either case, the theoretical and actual pole sensitivity measures are equal to n.

- if the poles are all real and distinct, then any Schur realization, whether normal or not, will do as well as a normal form in terms of the actual pole sensitivity measure, even though the theoretical sensitivity measures are different.

8.3.2 A NUMERICAL EXAMPLE

In this subsection, we return to Example 3.2 of Chapter 3 to illustrate the effectiveness of the sparse Hessenberg and Schur optimal re-

198   8. Sparse optimal and suboptimal realizations

alizations. In that example, we have computed the optimal realization $R^{opt}(M_{L_{12}}) \triangleq (A^{opt}(M_{L_{12}}), B^{opt}(M_{L_{12}}), C^{opt}(M_{L_{12}}))$ that minimizes the L_1/L_2 sensitivity measure for a sixth-order narrow band low pass filter.

$\boxed{R^{opt}(M_{L_{12}}):}$

$$A^{opt}(M_{L_{12}}) =$$
$$\begin{pmatrix} 0.9878 & 0.1123 & -0.0063 & 0.0275 & 0.0088 & -0.0057 \\ -0.1123 & 0.9709 & 0.1354 & -0.0195 & -0.0221 & 0.0090 \\ -0.0063 & -0.1354 & 0.9530 & 0.1163 & 0.0282 & -0.0191 \\ -0.0275 & -0.0195 & -0.1163 & 0.9264 & -0.1142 & 0.0336 \\ 0.0088 & 0.0221 & -0.0282 & 0.1142 & 0.9095 & 0.1212 \\ 0.0057 & 0.0090 & 0.0191 & 0.0336 & -0.1212 & 0.9041 \end{pmatrix}$$

$$B^{opt}(M_{L_{12}}) =$$
$$\left(\begin{pmatrix} 0.1121 & 0.1543 & 0.1901 & 0.1783 & -0.1072 & -0.0553 \end{pmatrix} \right)^T$$

$$C^{opt}(M_{L_{12}}) =$$
$$\begin{pmatrix} 0.1121 & -0.1543 & 0.1901 & -0.1783 & -0.1072 & -0.0553 \end{pmatrix}.$$

$R^{opt}(M_{L_{12}})$ is a fully parametrized balanced realization. This means that 49 multiplications and 49 additions are needed for each output sample computation.

We use $R^{opt}(M_{L_{12}})$ as the starting realization from which we first compute a sparse Hessenberg form using the procedure described in subsection 8.2.1. We first use the orthogonal matrix $T(u_1)$ given by (8.2), where u_1 is obtained by (8.3) with $x = B^{opt}(M_{L_{12}})$, $e = e_1(6)$ and ρ is taken positive, to transform $B^{opt}(M_{L_{12}})$ into a vector whose elements are all zero except the first one. Using the command 'HESS' in MATLAB, we then get the corresponding optimal Hessenberg realization $R_H = (A_H, B_H, C_H)$:

8. Sparse optimal and suboptimal realizations

$\boxed{R_H:}$

$$A_H =$$
$$\begin{pmatrix} 0.9183 & -0.0735 & 0.0168 & -0.0004 & 0.0090 & 0.0058 \\ 0.1509 & 0.9455 & 0.0781 & 0.0123 & -0.0030 & -0.0017 \\ 0.0000 & -0.1333 & 0.9537 & -0.0904 & 0.0157 & 0.0062 \\ 0.0000 & 0.0000 & 0.1436 & 0.9296 & 0.0173 & 0.0352 \\ 0.0000 & 0.0000 & 0.0000 & -0.0873 & 0.9521 & -0.1698 \\ 0.0000 & 0.0000 & 0.0000 & 0.0000 & 0.1943 & 0.9534 \end{pmatrix}$$

$$B_H =$$
$$\begin{pmatrix} -0.3448 & 0.0000 & 0.0000 & 0.0000 & 0.0000 & 0.0000 \end{pmatrix}^T$$

$$C_H =$$
$$\begin{pmatrix} -0.0044 & -0.0498 & 0.0071 & 0.2440 & 0.0288 & -0.2365 \end{pmatrix}.$$

Similarly, starting from $R^{opt}(M_{L_{12}})$ we first use the command 'SCHUR' in MATLAB to transform $A^{opt}(M_{L_{12}})$ into a Schur form, and we then transform this optimal Schur realization into the following sparser optimal Schur realization by using the algorithm given in the proof of Theorem 8.3:

$\boxed{R_S:}$

$$A_S =$$
$$\begin{pmatrix} 0.9202 & -0.0378 & -0.0187 & -0.0084 & -0.0245 & -0.0646 \\ 0.1111 & 0.9101 & -0.0123 & 0.0221 & -0.0208 & -0.0390 \\ 0.0000 & 0.0000 & 0.9718 & 0.1981 & 0.0090 & -0.0017 \\ 0.0000 & 0.0000 & -0.1997 & 0.9727 & -0.0012 & 0.0225 \\ 0.0000 & 0.0000 & 0.0000 & 0.0000 & 0.9371 & 0.1461 \\ 0.0000 & 0.0000 & 0.0000 & 0.0000 & -0.1804 & 0.9406 \end{pmatrix}$$

$$B_S =$$
$$\begin{pmatrix} -0.2277 & 0.0000 & -0.0719 & 0.0000 & -0.2487 & 0.0000 \end{pmatrix}^T$$

$$C_S =$$
$$\begin{pmatrix} -0.0397 & -0.3226 & -0.0698 & -0.0370 & 0.0505 & -0.0665 \end{pmatrix}.$$

Both R_H and R_S belong to the optimal realization set that minimizes the L_1/L_2 sensitivity measure, but note that they both contain 15 zero elements. This represents a saving of about 30% computations

200   8. Sparse optimal and suboptimal realizations

compared with the number performed using the fully parametrized $R^{opt}(M_{L_{12}})$. In addition, the actual sensitivity measure of these sparse forms must be smaller than the theoretical one for the reasons explained above. Since there is no expression to compute the actual sensitivity measure, we have verified this assertion by simulations.

Using the same procedure as described in Section 3.8, we have computed the actual frequency responses of the FWL realizations $R^{opt}(M_{L_{12}})$, R_H and R_S with $B_c = 8$ bits. The results are shown in Figure 8.1 for $R^{opt}(M_{L_{12}})$ and R_H, and in Figure 8.2 for $R^{opt}(M_{L_{12}})$ and R_S. In both cases, the ideal frequency response of the filter is also presented for comparison purposes.

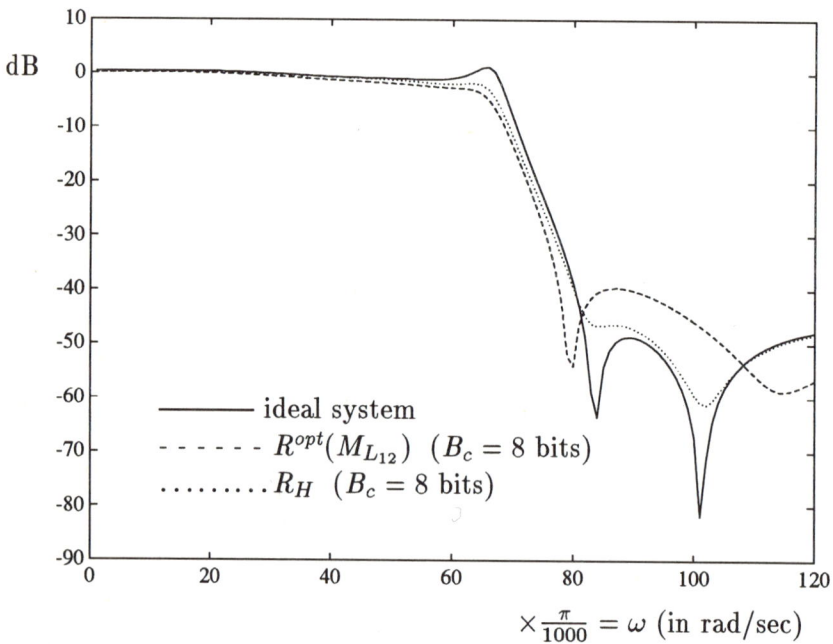

FIGURE 8.1. Magnitude frequency responses of ideal system, of $R^{opt}(M_{L_{12}})$ and of R_H

The figures clearly show that the sparse optimal realizations have a better performance than that of the fully parametrized balanced form. This confirms our arguments that the actual performance of an arbitrary sensitivity-optimal realization can be improved by using an orthogonally equivalent sparse realization.

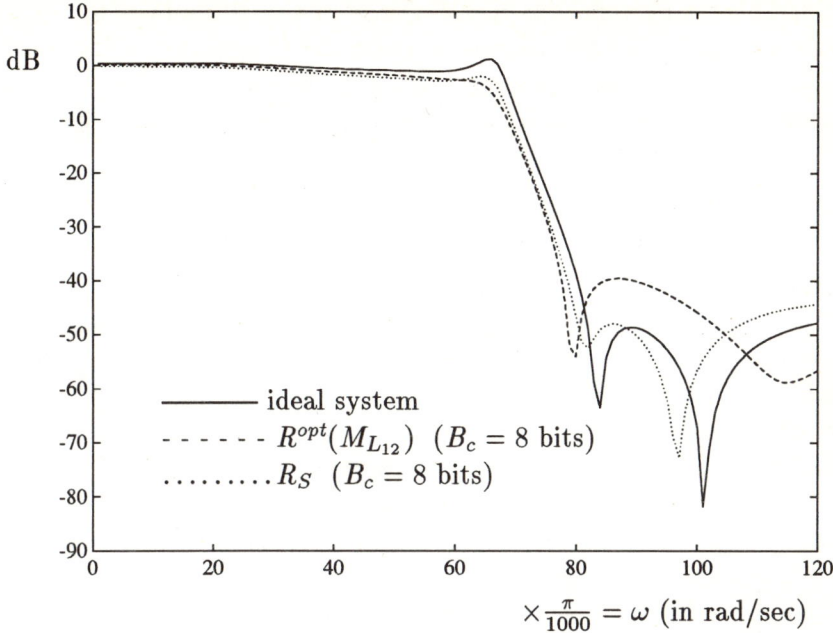

FIGURE 8.2. Magnitude frequency responses of ideal system, of $R^{opt}(M_{L_{12}})$ and of R_S

8.4 Sparse quasi-optimal realizations

We have shown in the last two sections that the degrees of freedom available in a sensitivity-optimal realization set can be used to find sparse optimal realizations that still belong to the optimal set. On the one hand, since the freedom is characterized by an orthogonal matrix that has only $\frac{1}{2}n(n-1)$ free parameters, the number of trivial parameters in a sparse optimal realization is at most $\frac{1}{2}n(n-1)$. On the other hand, we have also argued that the definitions and analyses are not quite consistent with what happens in an actual implementation, and we have revealed the fact that sparse optimal structures have an actual FWL performance that is better than the theoretical one.

Given the speed/performance tradeoff mentioned in the introduction, and given the inability of the sensitivity measure to capture the actual sensitivity performance of sparse realizations, and the absence of mathematical algorithms to optimize these measures under sparseness constraints, we now examine procedures and algorithms for computing sparse suboptimal realizations. The intuitive idea is to allow for a slight departure from optimality, and then to search for realizations that have considerable sparseness. The resultant realizations have lost the theoretical optimality, but the sacrifice of theoretical optimality is the price paid for sparseness, which brings with it higher computational speed and simpler realization structures. In this section, we present several approaches towards the implementation of this intuitive idea.

8.4.1 CONSTRAINED SIMILARITY TRANSFORMATIONS

In a sensitivity-optimal realization set the freedom is given by the set of all arbitrary orthogonal matrices, and we have shown in the previous section how this freedom can be used to construct sparse optimal realizations such as optimal Hessenberg and Schur realizations. On the other hand, in any roundoff noise optimal realization set, the freedom is a constrained orthogonal matrix. This constraint is due to the dynamic range limitation, that is the l_2 scaling. It seems difficult to find sparse optimal realizations that minimize the roundoff noise gain using this constrained orthogonal matrix set.

One approach to the construction of sparse near-minimum roundoff noise gain realizations is a procedure initially proposed by Chan [Cha79] for minimizing roundoff noise in a structure subject to the constraint that certain structure coefficients remain unaltered. Recall first that a fully parametrized (A, B, C, d) has $n^2 + 2n + 1$ parameters, while a canonical (or direct) form has $2n + 1$ parameters. By starting with a canonical form, Chan showed that up to $(n^2 - n - 1)$ of the trivial coefficients can be left unchanged while maintaining the constraints. However, a serious problem with this approach is that depending on the coefficients that are held fixed, the constrained minimum roundoff noise gain can vary widely and is often far from the unconstrained minimum. This is easy to understand because the starting realization, which is in canonical form, is usually far from an optimal MRH[5] structure and the trajectory to this optimal realization from the initial one is very complicated.

These problems were noted by Bomar and Hung [BH84] who proposed an approach that is in some sense dual to that of Chan. Instead of starting with a canonical structure, a roundoff noise optimal MRH structure is taken as the initial structure. The method is then based on applying a succession of $(n^2 - n - 1)$ similarity transformations

---

[5]We repeat that MRH stands for a roundoff-noise-optimal Mullis, Roberts and Hwang structure.

to this initial realization. Before each similarity transformation, the coefficient that is nearest to an exact power-of-two[6] is identified and the transformation then transforms that coefficient to an exact power-of-two without significantly increasing the roundoff noise. Eventually one gets a realization that has $(n^2 - n - 1)$ power-of-two coefficients. One significant feature of this kind of structures is that they typically have better *actual* sensitivity performance than the MRH structure, since all $(n^2 - n - 1)$ power-of-two coefficients are implemented without FWL errors, but in addition they have a much higher computational speed. Thus, structures are synthesized that possess essentially minimum roundoff noise, have a better sensitivity than the MRH structures and yet require only $(3n + 2)$ nontrivial multipliers.

Evidently, a roundoff noise source does occur at any multiplication involving a power-of-two parameter, except if it is 0 or ± 1. Noting this fact, Amit and Shaked [AS88] adapted the Bomar and Hung algorithm by replacing the target power-of-two parameters with 0 or ± 1 elements. In addition, they introduced a roundoff noise gain criterion that accounts for roundoff noise only at multiplications by coefficients other than 0 or ± 1. Thus their criterion is a reflection of the *actual* rather than the theoretical roundoff noise gain.

In this subsection, we present the algorithm of Bomar and Hung [BH84], whose objective is the minimization of the roundoff noise gain with sparseness constraints. We will subsequently generalize it to more general optimal parametrization problems with structure constraints.

Let (A_0, B_0, C_0, d) be a given realization in some realization set $S_H = \{(A, B, C, d): H(z) = C(zI - A)^{-1}B + d\}$, and let $T(t)$ be a nonsingular matrix which is a function of the continuous variable t. When $T(t)$ is used as a similarity transformation, the resultant realization belongs to the realization set S_H and is a function of t:

$$\begin{aligned} A(t) &= T^{-1}(t)A_0 T(t) \\ B(t) &= T^{-1}(t)B_0 \\ C(t) &= C_0 T(t). \end{aligned} \qquad (8.20)$$

The new realization has Gramians that are also functions of t:

$$W_c(t) = T^{-1}(t)W_c^0 T^{-T}(t), W_o(t) = T^T(t)W_o^0 T(t), \qquad (8.21)$$

---

[6]What is meant by a power-of-two coefficient is really a negative power of two, or equivalently any number of the form 2^{-k} with $k \in \{1, B_c\}$. Such a coefficient can be implemented without error, and more importantly the multiplication by such coefficient is very fast, since it is just a shifting operation.

where W_c^0 and W_o^0 are the Gramian pair corresponding to the initial realization (A_0, B_0, C_0, d).

Fundamental to the algorithm of Bomar and Hung is the numerical generation of the matrix $T(t)$ such that an initial minimum roundoff noise structure, say an MRH structure, evolves in some desired fashion as t increases. The evolution of a specific parameter of $(A(t), B(t), C(t))$ can be steered towards a fixed (say, power-of-two) number by controlling the derivative of that parameter in the realization (8.20), while maintaining other quantities constant with respect to time t, namely:

- the scaling constraint:
$$(W_c(t))_{ii} = 1, \quad i = 1, 2, ..., n \tag{8.22}$$

- the noise gain
$$G(t) = tr W_o(t), \tag{8.23}$$

- and the parameters that have already been moved to fixed desirable values, say
$$\theta_i(t) = \theta_i, \quad i = 1, \ldots, m. \tag{8.24}$$

Noting that
$$\frac{d[T(t)T^{-1}(t)]}{dt} = 0$$
leads to
$$\frac{dT^{-1}(t)}{dt} = -T^{-1}(t)\frac{dT(t)}{dt}T^{-1}(t).$$

It follows from (8.20), (8.21) and (8.23) that
$$\begin{aligned} \dot{A}(t) &= A(t)\Phi(t) - \Phi(t)A(t) \\ \dot{B}(t) &= -\Phi(t)B(t) \\ \dot{C}(t) &= C(t)\Phi(t) \end{aligned} \tag{8.25}$$

and
$$\begin{aligned} \dot{W}_c(t) &= -[\Phi(t)W_c(t) + W_c(t)\Phi^T(t)] \\ \dot{G}(t) &= tr[\Phi^T(t)W_o(t) + W_o(t)\Phi(t)], \end{aligned} \tag{8.26}$$

where $\dot{x}$ denotes $\frac{dx}{dt}$ and $\Phi(t) \triangleq T^{-1}(t)\dot{T}(t)$, or equivalently,
$$\dot{T}(t) = T(t)\Phi(t). \tag{8.27}$$

8. Sparse optimal and suboptimal realizations

In terms of the individual elements $a_{ij}(t)$ and $\phi_{ij}(t)$ of $A(t)$ and $\Phi(t)$, respectively, the first equation of (8.25) becomes

$$\begin{aligned}
\dot{a}_{ij}(t) &= \sum_{k=1}^{n}[a_{ik}(t)\phi_{kj}(t) - \phi_{ik}a_{kj}(t)] \\
&= \sum_{l,k=1}^{n}[\delta_{jk}a_{il}(t) - \delta_{il}a_{kj}(t)]\phi_{lk}(t) \\
&= \sum_{l,k=1}^{n} S_{lk}^{A}(t,i,j)\phi_{lk}(t) \qquad (8.28)
\end{aligned}$$

where

$$S_{lk}^{A}(t,i,j) \triangleq \delta_{jk}a_{il}(t) - \delta_{il}a_{kj}(t)$$

with δ_{ij} the Kronecker delta.

Similarly, one has

$$\begin{aligned}
\dot{b}_{i}(t) &= \sum_{l,k=1}^{n} S_{lk}^{B}(t,i)\phi_{lk} \\
\dot{c}_{i}(t) &= \sum_{l,k=1}^{n} S_{lk}^{C}(t,i)\phi_{lk} \\
\dot{W}_{c}(t)_{ii} &= \sum_{l,k=1}^{n} S_{lk}^{W_c}(t,i)\phi_{lk} \\
\dot{G}(t) &= \sum_{l,k=1}^{n} S_{lk}^{G}(t)\phi_{lk} \qquad (8.29)
\end{aligned}$$

where

$$\begin{aligned}
S_{lk}^{B}(t,i) &= -\delta_{il}b_{k}(t) \\
S_{lk}^{C}(t,i) &= \delta_{ik}c_{l}(t) \qquad (8.30) \\
S_{lk}^{W_c}(t,i) &= -2\delta_{il}(W_{c}(t))_{kl} \\
S_{lk}^{G}(t) &= 2(W_{o}(t))_{lk}. \qquad (8.31)
\end{aligned}$$

Clearly, if we consider the elements of $\Phi(t)$ as variables, then every derivative in (8.28) and (8.29) is a linear function of these variables. If $M = (m_1 \; m_2 \; \ldots \; m_k)$ is any matrix in $\mathbb{R}^{n \times k}$, then we denote

$$vec M \triangleq (m_1^T \; m_2^T \; \ldots \; m_k^T)^T. \qquad (8.32)$$

It then follows from (8.28) and (8.29) that

$$\begin{aligned}
vec\dot{A}(t) &= S^A(t)vec\Phi(t) \\
vec\dot{B}(t) &= S^B(t)vec\Phi(t) \\
vec\dot{C}(t) &= S^C(t)vec\Phi(t) \\
\dot{W}_c(t)_{ii} &= S^{W_c}(t,i)vec\Phi(t), \quad i=1,2,...,n \\
\dot{G}(t) &= S^G(t)vec\Phi(t),
\end{aligned} \qquad (8.33)$$

where $S^A(t) \in \mathbb{R}^{n^2 \times n^2}$, $S^B(t)$ and $S^c(t) \in \mathbb{R}^{n \times n^2}$, $S^{W_c}(t,i) \in \mathbb{R}^{1 \times n^2}$ and $S^G(t) \in \mathbb{R}^{1 \times n^2}$ are made up, respectively, of the $\{S^A_{lk}\}$, $\{S^B_{lk}\}$, $\{S^c_{lk}\}$, $\{S^{W_c}_{lk}(t,i)\}$ and $\{S^G_{lk}(t)\}$ as defined above.

We now define the parameter vector $\Theta(t)$ as

$$\Theta(t) \triangleq \begin{pmatrix} vecA(t) \\ vecB(t) \\ vecC(t) \end{pmatrix}. \qquad (8.34)$$

The parameter update equations are then

$$\begin{aligned}
\dot{\Theta}(t) &= S^\Theta(t)vec\Phi(t) \\
\dot{T}(t) &= T(t)\Phi(t),
\end{aligned} \qquad (8.35)$$

where

$$S^\Theta(t) = \begin{pmatrix} S^A(t) \\ S^B(t) \\ S^C(t) \end{pmatrix}. \qquad (8.36)$$

To any given $\Phi(t)$, there corresponds a $T(t)$ and a $\dot{\Theta}(t)$ which supplies the change of direction and speed of every parameter. However, $\Phi(t)$ is not arbitrary because of the constraints and we now explain how these can be taken into account. The algorithm starts from an MRH structure and moves one parameter at a time to a desired value (power-of-two, or 0 or ± 1, say) while keeping the constraints (8.22) satisfied and keeping $G(t)$ at its minimum value. Now, suppose that at the m-th stage of the algorithm m parameters have already been set to desired values. Without loss of generality, that is after a possible reordering of the vector Θ, we denote these fixed values $\theta_1, \ldots, \theta_m$ and the corresponding rows of S^Θ are denoted $S_1^\Theta, \ldots, S_m^\Theta$. We then define:

$$L_m(t) \triangleq \begin{pmatrix} S^{W_c}(t,1) \\ S^{W_c}(t,2) \\ \vdots \\ S^{W_c}(t,n) \\ S^G(t) \\ S_1^\Theta(t) \\ \vdots \\ S_m^\Theta(t) \end{pmatrix} \in \mathbb{R}^{(m+n+1)\times n^2}. \quad (8.37)$$

The three types of constraints (8.22)-(8.23)-(8.24) at the $(m+1)$-st iteration can then be expressed globally as

$$0 = L_m(t) vec\Phi(t). \quad (8.38)$$

The constraints (8.38) mean that $vec\Phi(t)$ must belong to the null space $\mathcal{N}[L_m(t)]$ of $L_m(t)$. We examine how to construct a $\Phi(t)$ that satisfies this constraint.

Suppose that $n + 1 + m \leq n^2$. Then by SVD, one has

$$\begin{aligned} L_m(t) &= U(t) \begin{pmatrix} \Lambda(t) & 0 \\ 0 & 0 \end{pmatrix} V^T(t), \text{ or} \\ L_m(t)V(t) &= U(t) \begin{pmatrix} \Lambda(t) & 0 \\ 0 & 0 \end{pmatrix}, \end{aligned} \quad (8.39)$$

where $U(t) \in \mathbb{R}^{(n+1+m)\times(n+1+m)}$ and $V(t) \in \mathbb{R}^{n^2 \times n^2}$ are orthonormal and

$$\Lambda(t) = diag\{\lambda_1(t), \lambda_2(t), \ldots, \lambda_{n+1+m}(t)\}$$

with $\lambda_i(t) \geqslant 0$. Clearly, if

$$V(t) = [v_1(t) \ v_2(t) \ \ldots \ v_{n+1+m}(t) \ v_{n+2+m}(t) \ \ldots \ v_{n^2}(t)],$$

then (8.39) implies

$$L_m(t)v_i(t) = 0, i = n+2+m, \ldots, n^2. \quad (8.40)$$

So, the null space of $L_m(t)$ can be expressed as

$$\mathcal{N}(L_m(t)) = span\{v_{n+2+m}, \ldots, v_{n^2}\} \quad (8.41)$$

which is not empty as long as

$$n^2 > n+1+m \quad \text{or} \quad m < n^2 - n - 1. \tag{8.42}$$

Therefore, a general $vec\Phi(t)$ satisfying (8.38) can be chosen as

$$vec\Phi(t) = \sum_{k=n+2+m}^{n^2} \alpha_k v_k(t) \tag{8.43}$$

for any real $\{\alpha_k, \; k = n+2+m, ..., n^2\}$. In addition to satisfying the constraints, $\Phi(t)$ should be chosen such that a new parameter, say $\theta_{m+1}(t)$, converges to its desired value (say, a power-of-two) as fast as possible, that is by making $|\dot{\theta}_{m+1}(t)|$ as large as possible. This requirement can be satisfied by choosing $vec\Phi(t)$ as the normalized orthogonal projection of $S^{\Theta}_{m+1}(t)$ onto $\mathcal{N}(L_m(t))$, as follows. Let $p(t)$ be the projection of $S^{\Theta}_{m+1}(t)$ on $\mathcal{N}(L_m(t))$:

$$\begin{aligned} p(t) &= \sum_{k=n+2+m}^{n^2} \alpha_k v_k(t) \quad \text{with} \\ \alpha_k(t) &= v_k^T(t) S^{\Theta}_{m+1}(t), k = n+2+m, ..., n^2. \end{aligned} \tag{8.44}$$

Then choose $vec\Phi(t)$ as the normalized version of $p(t)$:

$$vec\Phi(t) = \pm \frac{p(t)}{p^T(t)p(t)}. \tag{8.45}$$

As to the sign in (8.45), if θ_{m+1} is larger than the nearest desirable value, then the minus sign is taken, otherwise the plus sign is used.

The update procedure can be summarized as follows. Given the realization (A_m, B_m, C_m) at the m-th stage, one can compute $S^{\Theta}(t)$ in (8.36) and $L_m(t)$ in (8.37). The matrix $\Phi(t)$ can then be obtained from (8.39), (8.44) and (8.45), while $T(t+\Delta)$ and $\theta_{m+1}(t+\Delta)$ are updated by numerical integration algorithms using (8.35). Here Δ is the integration step size. The selection of this step size and the numerical aspects of this algorithm are discussed in detail in [BH84], where it is also observed that the more parameters one wants to move to exact desirable numbers, the larger the departure one has to accept with respect to the minimum noise gain. Hence sparseness is typically achieved at the price of a certain departure from optimality.

Theoretically, this procedure can be repeated until the dimension of the null space of $L_m(t)$ is zero, that is until:

$$n + 1 + m = n^2. \tag{8.46}$$

Thus, $n^2 - n - 1$ parameters can be transformed to desirable values. A fully parametrized (A, B, C) structure has $n^2 + 2n$ parameters. Thus, starting from a fully parametrized MRH structure, this algorithm yields a structure that has only $3n + 1$ nontrivial parameters, which is only $n + 1$ more nontrivial parameters than in a canonical form.

8.4.2 EXTENSIONS OF THE BOMAR AND HUNG ALGORITHM

By setting the target values as 0 and ±1, Amit et Shaked [AS88] have obtained what they call '0 − 1 realizations' with this algorithm. In these realizations, the parameters that are 0 or ±1 do not produce any roundoff error at multiplications. By using a roundoff noise gain criterion that accurately reflects this fact - as opposed to just using the trace of the observability Gramian - they show by simulations that the '0 − 1 realizations' typically yield a better *actual* roundoff noise gain than the fully parametrized MRH realizations that minimize the *theoretical* gain.

The algorithm of Bomar and Hung can be further generalized by noting that one can change not only the target values (as in the Amit-Shaked variant) but also the measure with respect to which the realizations are optimized. In the Bomar and Hung algorithm, the roundoff noise gain measure and the l_2 scaling constraint form the initial constraint matrix $L_0(t)$. In fact, any measure and constraint equations that can be written into linear functions of $vec\Phi(t)$ can form the initial constraint matrix. As an example, we show how the algorithm can be applied to the pole-zero minimization problem that has been investigated in Chapter 6.

It follows from the results of Chapter 6 that the pole-zero sensitivity measure defined in (6.41) can be written as

$$M_{pz}(T) = \sum_{k=1}^{2n} \{tr(T^T H_1^0(k)T) tr(T^{-1} H_2^0(k) T^{-T})\} + tr(T^T H_3^0 T) + tr(T^{-1} H_4^0 T^{-T}), \tag{8.47}$$

with appropriate definitions of $H_1^0(k)$, $H_2^0(k)$, H_3^0 and H_4^0. Define $H_1(k) \triangleq T^T H_1^0(k) T$, $H_2(k) \triangleq T^{-1} H_2^0(k) T^{-T}$, $H_3 \triangleq T^T H_3^0 T$

210   8. Sparse optimal and suboptimal realizations

and $H_4 \triangleq T^{-1}H_2^o T^{-T}$. One then has

$$M_{pz}(T) = \sum_{k=1}^{2n} tr(H_1(k))tr(H_2(k)) + tr(H_3) + tr(H_4). \qquad (8.48)$$

We introduce the fictitious variable t and recall that $\dot{T}(t) = T(t)\Phi(t)$, as defined by (8.27). It then follows that

$$\begin{aligned}
\dot{M}_{pz}(T(t)) &= \sum_{k=1}^{2n} \{\frac{dtr(H_1(k))}{dt}tr(H_2(k)) + tr(H_1(k))\frac{dtr(H_2(k))}{dt}\} + \\
&\quad \frac{dtr(H_3)}{dt} + \frac{dtr(H_4)}{dt} \\
&= tr\{\Phi^T(t)\sum_{k=1}^{2n}tr(H_2(k))H_1(k) + \sum_{k=1}^{2n}tr(H_2(k))H_1(k)\Phi(t)\} \\
&\quad - tr\{\Phi(t)\sum_{k=1}^{2n}tr(H_1(k))H_2(k) + \sum_{k=1}^{2n}tr(H_1(k))H_2(k)\Phi^T(t)\} \\
&\quad + tr\{\Phi^T(t)H_3 + H_3\Phi(t)\} - tr\{\Phi(t)H_4 + H_4\Phi^T(t)\}, \\
&= tr\{\Phi^T(t)W(t) + W(t)\Phi(t)\} \qquad (8.49)
\end{aligned}$$

where

$$W(t) \triangleq \sum_{k=1}^{2n}\{tr(H_2(k))H_1(k) - tr(H_1(k))H_2(k)\} + H_3 - H_4. \qquad (8.50)$$

We note that (8.49) is identical to (8.26) with $\dot{M}_{pz}(T(t))$ replacing $\dot{G}(t)$ and $W(t)$ replacing $W_o(t)$. Thus one could, for example, use the Bomar-Hung algorithm to compute sparse l_2-scaled realizations that are close to minimizing the pole-zero sensitivity measure while maximizing the number of 0 and ±1 elements.

It is easy to see that with some manipulations this algorithm can be similarly adapted to construct sparse realizations that are close to minimizing the L_2-sensitivity measure or the global measure G_T of Chapter 7.

8.5 Sparse suboptimal realizations

The algorithm presented in the previous section has the advantage of being very general. It can transform an optimal realization into a near-optimal sparse[7] one while maintaining some constraints. However, this algorithm has several drawbacks: it takes a long time to compute the final structure, and it cannot predict a priori the positions of the non-trivial parameters in that final sparse structure. Therefore, other suboptimal sparse realizations have been proposed that are easier to compute, and we now briefly present some of them.

The most common method of achieving both low roundoff noise gain and sparseness is to decompose an n-th order transfer function into a parallel or cascade connection of first- and second-order subfilters. This allows one to separate closely spaced poles and zeros into different subfilters. If the second-order subfilters, say, are realized in direct form, one gets $(2.5n+1)$ multiplies per output sample. A further improvement in the roundoff noise gain (at the expense of a slight increase in the computational cost) can be achieved if each subfilter is itself realized as a minimum roundoff noise realization.

In a *parallel realization* the transfer function $H(z)$ is expressed as a partial fraction expansion:

$$H(z) = h_0 + \sum_{i=1}^{\kappa} H_i(z) \tag{8.51}$$

where

$$H_i(z) = \frac{b_{1i}z + b_{i2}}{z^2 + a_{i1}z + a_{i2}} \tag{8.52}$$

in the case of complex conjugate poles, or

$$H_i(z) = \frac{b_{1i}}{z + a_{i1}} \tag{8.53}$$

in the case of a simple pole. Such a parallel decomposition of $H(z)$ is sketched in Figure 8.3.

The parallel decomposition of $H(z)$ into the individual subfilters $H_i(z)$ is essentially unique, but the realization of these subfilters can

---

[7] Here the traditional notion of sparseness - meaning many zero coefficients - is extended to include other desirable coefficients as well, such as ±1 or powers-of-two.

212   8. Sparse optimal and suboptimal realizations

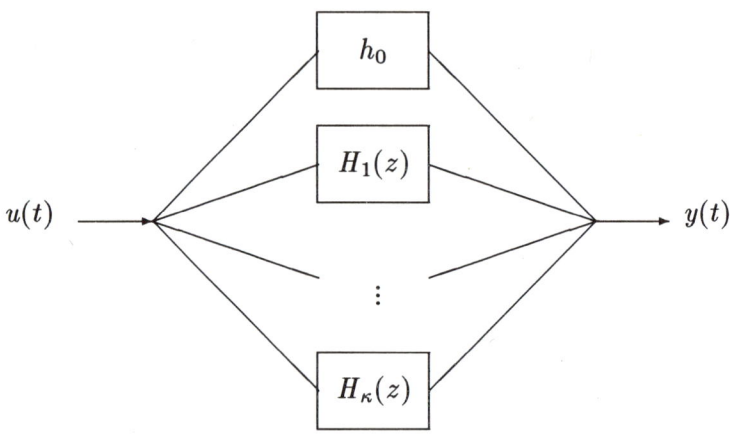

FIGURE 8.3. A parallel decomposition of $H(z)$

be performed in different ways, for example a direct form or an MRH structure.

An alternative decomposition of $H(z)$ is as a cascade of first- or second-order subfilters:

$$H(z) = K \prod_{i=1}^{n} \frac{(z - v_i)}{(z - \lambda_i)} = K \prod_{i=1}^{m} H_i(z) \qquad (8.54)$$

where $\{v_i\}$ and $\{\lambda_i\}$ are, respectively, the zeros and the poles of $H(z)$, K is a real number, and the $H_i(z)$ are first or second order transfer functions. That is

$$H_i(z) = \frac{z^2 + b_{i1}z + b_{i2}}{z^2 + a_{i1}z + a_{i2}} \qquad (8.55)$$

or

$$H_i(z) = \frac{z + b_{1i}}{z + a_{i1}}. \qquad (8.56)$$

A cascade decomposition of $H(z)$ is illustrated in Figure 8.4.

The cascade decomposition (8.54) is essentially non-unique. This is due to the different ways of pairing the poles and zeros in the subfilters: if one exchanges the zero polynomials of any two subsystems of the same order, another cascade decomposition is obtained. These two different decompositions may have different FWL behaviour, and the pairing of poles and zeros within the individual sections is therefore an important aspect in FWL implementations. In addition, the ordering of

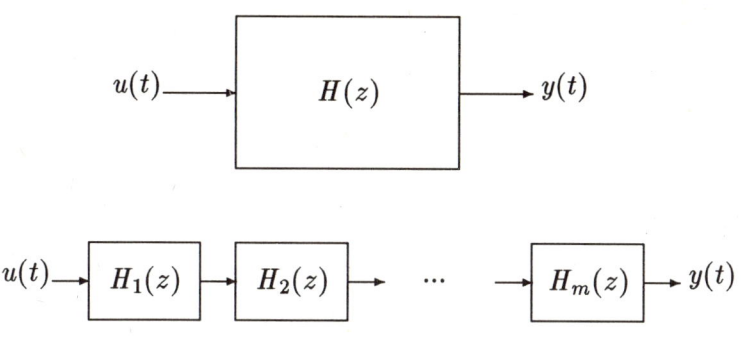

FIGURE 8.4. A cascade decomposition of $H(z)$

the individual sections also affects the FWL performance of the filter. Determining the optimal pole-zero pairings and the optimal position of each pair in the cascade connection in order to minimize some FWL performance criterion is by no means an easy problem. There does not seem to be a satisfactory theory to deal with this pairing problem.

In both cascade and parallel forms, each subsystem can be realized in an infinite number of ways. But whereas the realization of $H(z)$ obtained by connecting the sub-realizations of a parallel form belongs to the global realization set S_H, this is not the case for a cascade connection. More precisely, if one writes the realization (A, B, C) of $H(z)$ obtained as the parallel connection of first- and second-order subsystems, then the parameters in A, B and C are all parameters that appear in the realizations of the subsystems, whereas the parameters of the matrices A, B and C obtained from a cascade connection contain products of the parameters of the subsystem realizations.

The state transition matrix A obtained from a parallel connection is block diagonal, with blocks of dimensions 1×1 or 2×2, while the matrix A obtained from a cascade connection is block triangular with diagonal blocks of the same size again. Thus, both of these sparse forms lead to realizations in Schur form. However, this Schur form does typically not belong to the optimal realization subset corresponding to some sensitivity measure, and it is not clear how an optimal realization in Schur form could be decomposed into a particular cascade realization.

The most common way of achieving sparse low roundoff noise gain or low sensitivity realizations is to use parallel or cascade decompositions and to optimize the first and second order subsystems with respect to the performance criterion. If each subfilter is optimized in isolation, then the cascade connection of these subfilters will typically not satisfy the global l_2 scaling constraint even though each subfilter does. This is because each local l_2 scaling is performed on the assumption that the input of the subfilter is white noise, which is no longer true when these subfilters are cascaded.

The design procedure in which each first- or second-order subsystem is optimized in isolation is sometimes called *sectional optimal*. A design in which the subsystems are optimized in a way that preserves the global constraints is called *block optimal*. Both the sectional optimal and the block optimal designs of second order subfilters are explained in great detail by Roberts and Mullis [RM87], who also present several design examples.

Finally, another method of obtaining a sparse suboptimal realization is an algorithm due to Smith and Bomar [SB88]. Starting from a controllable (or observable) form, one can use a series of similarity transformations, each of which is an identity matrix for all elements except for one diagonal and one off-diagonal element on the same row. Thus, such a matrix is a very sparse triangular matrix. Since the inverse of this similarity transformation matrix has the same structure, the roundoff noise gain can be optimized with respect to the two parameters that are changed at each iteration while at the same time maintaining the l_2 scaling. This special transformation will retain $\frac{1}{2}n(n-1)$ zeros of the initial direct form, thus producing a saving of $\frac{1}{2}n(n-1)$ elements compared to an optimal form. One advantage of this algorithm is that the positions of the trivial zero elements can be known a priori due to the special structure of the transformations. We refer to [SB88] for a detailed discussion.

8.6 Conclusion

It is most often the case that in order to describe a physical phenomenon a simple and concise mathematical model is first used. Once a theory has been elaborated based on this simple model, it is then re-evaluated on the basis of additional physical considerations. The initial model is then often found to be too simplistic and a modified theory is developed that is more consistent with the physical phenomenon.

This process has taken place with the development of an optimal FWL state space design theory. The initial sensitivity and roundoff noise measures have been based on the assumptions that all parameters were encoded with error, including parameters such as 0 and ±1. Not only did this not accurately reflect reality, but it automatically led to fully parametrized optimal realizations. By pointing out some weak points in the optimal FWL design theories established in the previous chapters, we have in this chapter described an array of methods that lead to sparse realizations.

Our first approach has been to investigate the optimal realization sets described in the previous chapters, where the optimality is with respect to the very measures that we have just criticized on the grounds of their not being realistic. Noting that there is some degree of freedom in the optimal realization sets, we have used this fact to find sparse realizations in or near these sets. Since the optimal sets corresponding to the various sensitivity measures are all characterized up to an arbitrary orthogonal matrix, we have shown how to use this orthogonal matrix to obtain optimal Hessenberg and Schur realizations, both of which possess $\frac{1}{2}n(n-1)$ trivial parameters.

Noting that the degrees of freedom in roundoff noise gain minimization problems is constrained by l_2 scaling, we have presented a numerical algorithm due to Bomar and Hung that searches for sparse realizations in the neigbourhood of the optimal realization set while maintaining these constraints. We have shown that this algorithm can be extended to other problems.

Some other commonly used techniques have also been briefly described in this chapter, such as the decomposition of a high order filter into a series of second or first order subfilters and the optimization of the subfilter realizations.

9

Parametrizations in control problems

9.1 Introduction

The results described in the previous chapters all have a signal processing flavour, in that the objective has been to minimize the performance degradation of filters due to errors caused either by the FWL implementation of their coefficients, or by the FWL computations, or both. The performance degradation has been encapsulated by a variety of measures, and the solutions have been in the form of optimal implementations in the class of all similar realizations of a given transfer function.

By comparison, the effects of FWL errors in digital controllers on the performance of the controlled system have received much less attention. In the most common situation where the system to be controlled is a continuous time system controlled by a sampled data controller, such errors arise from four different sources:

- errors on the input of the controller arising from the analog to digital (A/D) conversion of the output of the system;

- errors on the output of the controller (that is the control input of the system) arising from the finite wordlength of the digital to analog (D/A) converter ;

- errors due to the finite wordlength implementation of the controller coefficients;

- errors due to the roundoff of the states in the controller.

The earliest work in the area of controller design was mainly concerned with an analysis of the performance degradation of the closed loop system due to the FWL implementation of the controller coefficients [MWH80]. The main aim of this early work, focused on Linear Quadratic Gaussian (LQG) controllers, was not to synthesize optimal implementations of FWL controllers, but to develop analysis tools for the estimation of the tradeoff between bit length in the controller and performance degradation of the closed loop system. The objective was to estimate the coefficient bit length that was required for the implementation of the controller coefficients in order to maintain the performance degradation of the LQG cost within an acceptable level.

It was not until the late eighties that the question of implementation of digital controllers and its impact on closed loop performance came under scrutiny [WK89], [LG90a,b], [LS90], [LG91], [LSG92]. The design of a discrete time controller through the use of either optimal control theory or model reference design has been well established, but these design procedures produce an infinite precision ideal controller. This means that this controller achieves the performance for which it is designed only when implemented in infinite precision or, more realistically, on expensive, floating point computers. The advent of cheap microprocessors, and the requirement in some applications (such as in robotics) for high computational speed and hence low bit lengths, has led to an interest for the numerical accuracy aspects of controller implementations.

One reason that renders FWL error minimization sometimes even more crucial for controller design than for filter design is that digital controllers either contain an integrator or have poles that are close to $z = 1$. We have seen in previous chapters that, by just about any sensitivity or roundoff noise gain measure, this is the situation that leads to the most severe FWL effects. The reason for this is that digital controllers are either obtained through discretization of controllers designed in continuous time, or they are synthesized directly in discrete time on the basis of discrete time models of continuous time systems. Performance requirements often command that reasonably fast sampling be used. In [MG90] Middleton and Goodwin argue that sampling frequencies between 5 and 10 times the desired closed loop bandwidth are reasonable. With such sampling rates, the poles and zeros of the discretized system and those of the controller will typically cluster around $z = 1$: see e.g. [MG90], and also [ÅSH84]. Thus, direct form implementations of digital controllers will typically be prone to FWL errors, just as digital filters are.

The task of the implementation design step in the global control design procedure is to make sure that the actual digitally controlled closed loop system performs as close as possible to the ideal closed loop system, subject to expense and speed requirement constraints. It is clear that there exist some parallels between filter and controller implementation, so that the concepts existing in digital signal processing could be borrowed for digital controller implementation problems. However, because of the presence of a feedback loop, many of these concepts do not apply for the control problem directly. *One of the main differences between the optimal digital implementation of an open loop system and that of a controller is that the optimal controller implementation must be designed not for achieving the best performance of the controller itself, but for achieving the best performance of the closed loop system, where only the controller is implemented with Finite Word Length.*

Depending on the control application at hand, various performance criteria for the closed loop system under FWL digital control and various controller implementation strategies can be adopted, going from the simplest to the most sophisticated. Without attempting to be exhaustive, let us just enumerate some reasonable performance criteria and some optimization strategies of increasing level of complexity.

1. In a regulation problem, where it is desired to maintain the output of a system at a desired setpoint, a classical measure of performance (without accounting for FWL errors) is the variance of the deviation between the output and this setpoint, or a combination of this variance and of the control variance as in Linear Quadratic Gaussian (LQG) control. The FWL errors in the controller will typically add some additional contribution to this variance, since the FWL controller is suboptimal with respect to the ideal controller and since the roundoff on the signals introduces an additional noise. A measure of performance with which two FWL controllers can be compared could then be the total output variance (or the total LQG cost) including the term due to the FWL errors.

2. In a tracking problem, or in a model reference design scheme, it is the closed loop transfer function that is the object of interest. A reasonable FWL control design criterion could then be some norm of the error between the ideal closed loop transfer function and the closed loop transfer function of the system with the FWL controller. Alternatively, some norm of the difference in closed loop step responses could be used as the performance criterion to be minimized.

3. The regulation criteria mentioned above will more naturally focus

on roundoff noise minimization since roundoff errors contribute an additive noise term to the output variance or to the LQG cost. On the other hand, the tracking criteria will focus naturally on coefficient wordlength effects, and hence on sensitivity issues, since deviations between ideal and FWL closed loop transfer functions reflect coefficient errors only.

4. The computational speed is often also an important factor by which different controllers or controller implementations can be judged and compared. The computational speed is mainly a function of controller complexity. In this respect, it is interesting to observe that this complexity is influenced by

   - the sparseness of the controller implementation,
   - the order of the controller transfer function,
   - the bitlengths with which the controller coefficients and signals are implemented.

5. This last point made Williamson and Kadiman observe that 'reducing the wordlength of controllers is another form of controller reduction' [WK89]. This judicious observation, often overlooked in the presently popular literature on controller reduction, was adopted by Liu, Skelton and Grigoriadis who added a penalty on the signal bitlengths of the individual controller states in their LQG optimal design [LSG92].

Let us now turn our attention to the various ways of minimizing the closed loop performance degradation criteria mentioned above. In the first three strategies we assume that, if the system to be controlled is a continuous time system, then a discrete time model of that system has been obtained first.

(i). The most commonly used, but naive, method is to compute an ideal discrete time controller using the designer's favourite method and then to apply a relevant FWL filter design criterion, among those presented in Chapters 3 to 8, to the discrete time controller considered as a filter.

(ii). At the next level of sophistication are the FWL closed loop design methods of the late eighties, in which an ideal discrete time controller is again designed first, and in which the implementation of this controller is then optimized in such a way that the behaviour of the *closed loop system* with the ideal and with the FWL controller are 'as close as possible' to one another. This distance can be measured by any one of the performance criteria discussed above.

(iii). A third level of sophistication is where the FWL digital controller is optimized directly over all FWL controller realizations with given bitlength. That is, the controller is computed by minimizing, over all FWL controllers, some performance criterion that incorporates the effects of FWL errors. The difference with the previous approach is that it is not just the *implementation* of the controller transfer function that is optimized, but this transfer function itself. For example, in an LQG design, this leads to controller and observer gains that are different from the ideal - infinite precision - gains obtained from the LQG Riccati equations, as we shall explain in Section 9.3.

(iv). Finally, the *nec plus ultra* of FWL optimal controller design would apparently be achieved if the design of the FWL controller were performed by minimizing not the performance degradation of the discrete time closed loop system, but that of the continuous-discrete closed loop system. Indeed, the actual closed loop system contains a continuous time system and a discrete time FWL controller. Thus, the true quality test of a FWL controller is not obtained by comparing its performance with that of an infinite wordlength controller in closed loop with the discrete time approximation of the true system, but by comparing their performances in closed loop with the continuous time system itself.

Given that the first FWL controller design results are only a few years old, it is not surprising that only a few of the performance criteria and of the minimization strategies have been addressed so far. The problem that has probably received most attention is the FWL LQG design problem. This is because the criterion lends itself naturally to the incorporation of additive noise error terms. In [WK89], Williamson and Kadiman gave a solution to the LQG design problem using strategy (ii) above, and they gave some partial results for the solution of problem (iii). A complete solution to (iii) was given by Liu et al. in [LS90] and [LSG92]. In these solutions, only the signal roundoff error effects have been addressed, since these are naturally incorporated as an additive noise term in an LQG criterion. The key results on FWL LQG design will be presented in Section 9.3.

The model reference or pole placement FWL controller design problem has received less attention, except by the authors of the present book. In [LG90b], the minimization of the performance degradation of a closed loop system with a FWL observer-pole placement controller has been solved separately for a sensitivity performance criterion and for a roundoff noise criterion using strategy (ii) above. These results will be presented in Section 9.2. A design that optimizes some closed loop performance measure jointly and coherently with respect to both coefficient errors and roundoff errors is of course always to be preferred. One example of such a synthetic design, which parallels the synthetic filter design of Chapter 7, will be presented in Chapter 10. This approach is presently limited to strategy (ii): an ideal controller is computed first using any control design criterion (LQG, model reference, H_∞, etc) and the realization of this ideal controller is subsequently optimized with respect to a closed loop performance criterion that incorporates both coefficient errors and signal roundoff errors in a synthetic way. Finally, some preliminary results exist on strategy (iv) [AGG92], but they will not be covered in this book. The idea is to extend to the controller sensitivity minimization problem the recent sampled data control design scheme of Keller and Anderson [KA92], in which a discrete time controller is computed that minimizes some hybrid operator, which is a function of the sampled continuous time system in feedback with the discrete time controller. The main novelty of the Keller-Anderson scheme[1] is that the sampled data control design scheme incorporates the effect of the controller not on the discrete time approximation of the continuous system as in all classical design schemes, but on the actual sampled continuous time system.

An outline of this chapter is as follows. In Section 9.2, we study the state-estimate feedback design problem, where the state variable observer and the feedback gains are implemented in finite precision and where arithmetic operations are performed with roundoff. One of our contributions here is to derive the expressions of the sensitivity measure and of the roundoff noise gain for *closed loop systems*. Another is to give procedures for computing the optimal realization sets that minimize either the sensitivity measure, or the roundoff noise gain under a

---

[1]The same idea, but using different operators, has also been intensively pursued by Chen and Francis [CF91a,b].

dynamic range constraint on the states of the controller. In Section 9.3 we review the most relevant results on the minimization of roundoff error effects in FWL LQG compensators, and we show the similarity between the roundoff noise analysis of LQG and pole placement regulators. To end this chapter, some conclusions are given in Section 9.4.

Throughout this and the next chapter we shall assume that, if the system to be controlled lives in continuous time, then the signals in the A/D converter, the D/A converter and the FWL controller are all implemented in fixed point with the same bitlength, B_s, for the fractional part and hence the same roundoff error variance. An analysis of the roundoff noise effects taking into account different wordlengths in these different components of a sampled data control system can be performed: see e.g. [LSG92].

9.2 Implementation of a pole placement controller

In this section, we investigate the optimal FWL implementation of a pole placement controller using state estimate feedback. This is a special case of model reference control and is one of the fundamental methods of control design. The optimal design that we shall pose and solve is to find a realization of the controller that minimizes some performance degradation criterion, which is a measure of the difference between the behaviour of the closed loop system with an ideal controller and that of the closed loop system with a FWL controller. The performance measures that we study are, successively, a sensitivity measure of the closed loop transfer function with respect to the controller coefficients, and a roundoff noise gain of the closed loop system with respect to roundoff in the controller.

9.2.1 THE IDEAL POLE PLACEMENT CONTROLLER

The purpose of pole placement is to design a state estimate feedback controller so that the poles of the closed loop system have prescribed values. The infinite precision design procedures can be found in many textbooks, for example in [ÅW84] and [MG90]. Whether or not the actual system is continuous or discrete, here we consider that the system is described by a linear time invariant strictly proper discrete time state space model:

9. Parametrizations in control problems

$$\begin{aligned} x(t+1) &= A_0 x(t) + B_0 u(t) \\ y(t) &= C_0 x(t) \end{aligned} \qquad (9.1)$$

with $A_0 \in \mathbb{R}^{n \times n}, B_0 \in \mathbb{R}^n, C_0^T \in \mathbb{R}^n$. We also assume that this system is minimal, that is the pair $[A_0, B_0]$ is completely reachable and the pair $[C_0, A_0]$ is completely observable. The transfer function can be expressed in terms of the state space matrices as

$$H_0(z) = C_0(zI - A_0)^{-1} B_0. \qquad (9.2)$$

When the state variables $x(t)$ are measurable, the poles can be assigned at any desired set of locations by a linear state feedback control law:

$$u(t) = -K_0 x(t) + r(t). \qquad (9.3)$$

Here $r(t)$ is an external reference signal, and $K_0 \in \mathbb{R}^n$ is the vector of feedback gains determined such that $\{\lambda_i(A_0 - B_0 K_0)\}$ gives the set of desired closed loop poles (see e.g. [Kai80]).

Since usually only the output process $y(t)$ is available, the state feedback law (9.3) is not implementable. The way around this difficulty is to replace the unknown state $x(t)$ by a state estimate obtained by an "observer", that takes the following form:

$$\hat{x}(t+1) = A_0 \hat{x}(t) + B_0 u(t) + L_0[y(t) - C_0 \hat{x}(t)]. \qquad (9.4)$$

It is a dynamical system that generates $\hat{x}(t)$, an estimate of $x(t)$. The vector L_0 is called the observer gain. The observer equation can be rewritten as:

$$\hat{x}(t+1) = [A_0 - L_0 C_0]\hat{x}(t) + B_0 u(t) + L_0 y(t). \qquad (9.5)$$

The control law is then obtained by replacing the state $x(t)$ in the feedback law (9.3) by the state $\hat{x}(t)$ of the observer (9.5), yielding the state estimate feedback controller:

$$u(t) = -K_0 \hat{x}(t) + r(t). \qquad (9.6)$$

Comments:

1. It follows from (9.5) that the poles of the observer are the eigenvalues of $(A_0 - L_0 C_0)$. If the pair $[C_0, A_0]$ is completely observable, these poles can be freely assigned by a proper choice of L_0.

2. When the system is without noise disturbances, the gain L_0 of the observer is chosen such that these poles are at preassigned values. In particular,

$$|\lambda_i(A_0 - L_0 C_0)| < 1 \quad \forall i = 1, \ldots, n. \tag{9.7}$$

This guarantees the asymptotic convergence of the estimated state to the actual state. The speed of convergence can be selected by a proper choice of L_0. When the system is disturbed by noises, L_0 is often chosen as the optimal gain of a Kalman filtering problem.

3. The observer shown above is a dynamical output feedback device, and hence so is the controller.

4. We have assumed that the system is strictly proper. This makes the following derivations slightly easier and clearer. In the case of a proper but not strictly proper system, the procedures for FWL optimal design are exactly the same, but they involve a few more manipulations.

9.2.2 FINITE PRECISION ASPECTS IN A CLOSED LOOP COMPENSATOR: PROBLEM FORMULATION

In this subsection we study the effects of finite wordlength implementation and finite precision arithmetic on the performance (sensitivity and roundoff noise gain) of the closed loop system.

First we observe that the observer-controller (9.6) can be rewritten as a *two-input one-output* dynamical system:

$$\begin{aligned} \hat{x}(t+1) &= \Phi_0 \hat{x}(t) + B_0 r(t) + L_0 y(t) \\ u(t) &= -K_0 \hat{x}(t) + r(t) \end{aligned} \tag{9.8}$$

where

$$\Phi_0 \triangleq A_0 - B_0 K_0 - L_0 C_0. \tag{9.9}$$

The corresponding input/output relation is

$$\begin{aligned} u(t) &= [I - K_0(zI - \Phi_0)^{-1} B_0] r(t) - [K_0(zI - \Phi_0)^{-1} L_0] y(t) \\ &= C_1(z) r(t) - C_2(z) y(t). \end{aligned} \tag{9.10}$$

The compensator is now parametrized by (Φ_0, B_0, K_0, L_0) which, in general, has fewer parameters than $(A_0, B_0, C_0, K_0, L_0)$ in (9.6). In the sequel, we study the compensator with the parametrization (9.8). Just as for any dynamical system, the input-output relation (9.10) admits an infinite number of realizations which, in infinite precision, are all equivalent. They are all related by similarity transformations; these transform (Φ_0, B_0, K_0, L_0) into a set $\{(T^{-1}\Phi_0 T, T^{-1}B_0, K_0 T, T^{-1}L_0)\}$ of similar compensator realizations. For any one of these realizations, the ideal (i.e. infinite precision) closed loop transfer function $H_c^0(z)$ from $r(t)$ to $y(t)$ is the same:

$$H_c^0(z) = F[H_0(z), (\Phi_0, B_0, K_0, L_0)] \qquad (9.11)$$

$$= \frac{C_1(z)H_0(z)}{1 + C_2(z)H_0(z)} \qquad (9.12)$$

$$= C_0[zI - (A_0 - B_0 K_0)]^{-1} B_0. \qquad (9.13)$$

We note that the ideal closed loop transfer function is independent of the observer gain L_0, that is the observer dynamics cancel in the input-output relation of the closed loop system. All these similar compensator realizations form what we call the compensator realization set S_c:

$$S_c \triangleq \{(\Phi_0, B_0, K_0, L_0) : H_c^0(z) = F[H_0(z), (\Phi_0, B_0, K_0, L_0)]\} \qquad (9.14)$$

where $F(.)$ is the function defined by (9.12) and $H_c^0(z)$ is defined by (9.13). When the ideal compensator realization (Φ_0, B_0, K_0, L_0) is implemented with finite wordlength, the actual closed loop transfer function $H_c(z)$ is obtained by replacing (Φ_0, B_0, K_0, L_0) by its FWL version (Φ, B, K, L) in (9.11)-(9.12),

$$H_c(z) = F[H_0(z), (\Phi, B, K, L)] \neq H_c^0(z). \qquad (9.15)$$

Under a similarity transformation, (Φ_0, B_0, K_0, L_0) is transformed into $(T^{-1}\Phi_0 T, T^{-1}B_0, K_0 T, T^{-1}L_0)$ yielding a different FWL implementation, say, (Φ^*, B^*, K^*, L^*). Unlike the infinite precision implementation, the two FWL realizations (Φ, B, K, L) and (Φ^*, B^*, K^*, L^*) will yield different closed loop transfer functions determined by (9.15). In addition, unlike the ideal $H_c^0(z)$, the actual $H_c(z)$ is a function of all four quantities Φ, B, K, L. So a reasonable question is how to find a compensator realization such that the actual closed loop system with this FWL compensator performs as much as possible like the ideal closed loop system. One way to compare performances is to compare the actual closed loop transfer $H_c(z)$ with the ideal one, $H_c^0(z)$. This leads to sensitivity minimization. Another way is to minimize the effect of the compensator roundoff noise on the closed loop response. In each case, appropriate measures need to be defined.

Our task in the remainder of this section is thus as follows: for a given discrete time plant model $H_0(z)$, an ideal state feedback control gain K_0, and an ideal observer gain L_0, find

1. a computable measure of the sensitivity of the closed loop transfer function $H_c(z)$ with respect to the parameters of the realization (Φ, B, K, L);

2. an expression for the roundoff noise gain of the closed loop realization;

3. the realization set (i.e. the set of realizations (Φ, B, K, L)) that minimizes the sensitivity measure, the roundoff noise gain, or both, subject to a dynamic range constraint on the states of the observer-controller.

9.2.3 Sensitivity analysis and optimal structures

In this subsection we derive formulæ for the sensitivity of the closed loop transfer function with respect to the parameters of the compensator. We assume that a given coordinate space has been chosen for the state space representation of $H_0(z)$. We call (A_0, B_0, C_0) the exact (infinite precision) implementation of $H_0(z)$ in that coordinate space (see (9.1)), (Φ_0, B_0, K_0, L_0) the infinite precision desired compensator (9.8) in that same coordinate system and (Φ, B, K, L) the corresponding FWL implementation of this compensator.

The block diagram of the plant $H_0(z)$ with its FWL state estimate feedback controller is given in Figure 9.1.

The state equations of the closed loop system with FWL compensator coefficients are then

$$\begin{pmatrix} x(t+1) \\ \hat{x}(t+1) \end{pmatrix} = \begin{pmatrix} A_0 & -B_0 K \\ LC_0 & \Phi \end{pmatrix} \begin{pmatrix} x(t) \\ \hat{x}(t) \end{pmatrix} + \begin{pmatrix} B_0 \\ B \end{pmatrix} r(t)$$

$$y(t) = \begin{pmatrix} C_0 & 0 \end{pmatrix} \begin{pmatrix} x(t) \\ \hat{x}(t) \end{pmatrix}. \tag{9.16}$$

We denote:

$$\bar{A} = \begin{pmatrix} A_0 & -B_0 K \\ LC_0 & \Phi \end{pmatrix}, \bar{B} = \begin{pmatrix} B_0 \\ B \end{pmatrix}, \bar{C} = \begin{pmatrix} C_0 & 0 \end{pmatrix}. \tag{9.17}$$

9. Parametrizations in control problems

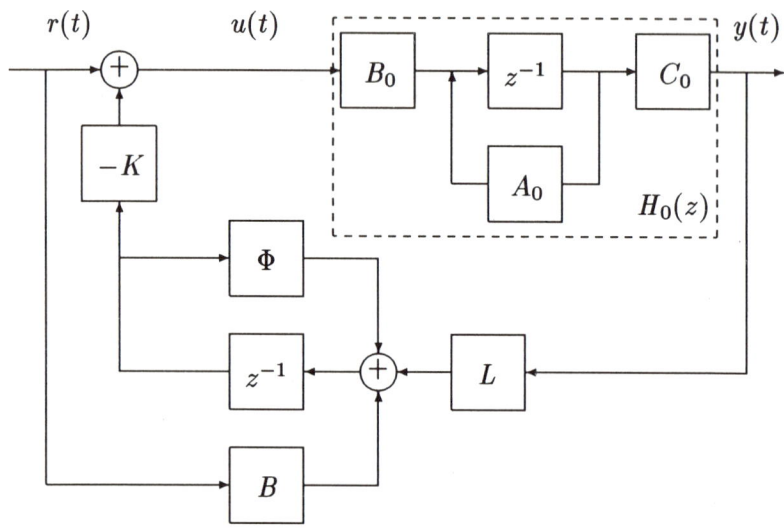

FIGURE 9.1. Closed loop system with state estimate feedback controller

The closed loop transfer function with FWL compensator coefficients is

$$H_c(z) = \bar{C}(zI - \bar{A})^{-1}\bar{B}. \tag{9.18}$$

Because of the FWL effects (i.e. $(\Phi,B,K,L) \neq (\Phi_0,B_0,K_0,L_0)$), the observer dynamics do not cancel in the closed loop transfer function, that is

$$H_c(z) \neq H_c^0(z) \triangleq C_0[zI - (A_0 - B_0K_0)]^{-1}B_0. \tag{9.19}$$

Thus the 'separation principle', which is at the heart of much of the linear state estimate feedback control theory, whether it be by pole placement or by LQG control, breaks down in practice at the implementation phase of the control law. This is a key observation that should haunt the nights of many linear or linear quadratic control theoreticians.

Comment:

As stated in the introduction, one way to alleviate the effects of FWL coefficient errors in the compensator is to apply one of the optimal design methods of Chapters 3 to 7 to the design of an optimal FWL realization of the compensator $(C_1(z)\ C_2(z))$. This would lead, for example, to the implementation of this compensator in MRH structure if the 'classical' L_1/L_2 sensitivity measure were used. This design philosophy would be an application of strategy (i) described in the introduction, and it would clearly be far from optimal. *A much better, but more difficult, approach is to minimize a sensitivity measure of the closed loop transfer function with respect to the parameters of the compensator since we are interested in the performance of the closed loop system, not of the compensator as such.*

We thus compute the sensitivities of $H_c(z)$ with respect to the controller coefficient matrices Φ, B, K, L evaluated at the exact (i.e. infinite precision) values Φ_0, B_0, K_0 and L_0[2]:

$$\frac{\partial H_c(z)}{\partial \Phi} = -H_c^0(z)G_o(z)F_{c2}^T(z) \triangleq G_{o1}(z)F_{c2}^T(z) \qquad (9.20)$$

$$\frac{\partial H_c(z)}{\partial B} = -H_c^0(z)G_o(z) \triangleq G_{o1}(z) \qquad (9.21)$$

$$\frac{\partial H_c(z)}{\partial L} = -(H_c^0(z))^2 G_o(z) \triangleq G_{o3} \qquad (9.22)$$

$$\frac{\partial H_c(z)}{\partial K^T} = -H_c^0(z)(1 + H_{c0}(z))F_{c2}(z) \triangleq F_{c4} \qquad (9.23)$$

where $H_c^0(z)$, as given by (9.19), is the desired (or infinite precision) closed loop transfer function and

$$G_o(z) \triangleq [zI - (A_0 - L_0C_0)^T]^{-1}K_0^T \qquad (9.24)$$

$$F_{c2}(z) \triangleq [zI - (A_0 - B_0K_0)]^{-1}B_0 \qquad (9.25)$$

$$H_{c0}(z) \triangleq K_0[zI - (A_0 - L_0C_0)]^{-1}B_0. \qquad (9.26)$$

These four sensitivities are functions of frequency. To keep life simple, we now adopt an L_1/L_2 sensitivity measure for the closed loop transfer function $H_c(z)$ with respect to the controller parameters Φ, B, K, L:

$$M_{cl,L_{12}} = \parallel \frac{\partial H_c(z)}{\partial \Phi} \parallel_1^2 + \parallel \frac{\partial H_c(z)}{\partial B} \parallel_2^2 + \parallel \frac{\partial H_c(z)}{\partial L} \parallel_2^2 + \parallel \frac{\partial H_c(z)}{\partial K^T} \parallel_2^2. \qquad (9.27)$$

The remainder of this subsection could almost be guessed by our readers by now. We use a Cauchy-Schwartz inequality to bound the first term, just as was done in Chapter 3. An upper bound for this

---

[2] The formulæ for the derivation of these expressions can be found in Appendix 9.

closed loop sensitivity measure is thus given by

$$\bar{M}_{cl,L_{12}} = \| G_{o1}(z) \|_2^2 \| F_{c2}(z) \|_2^2 + \| G_{o1}(z) \|_2^2 + \| G_{o3}(z) \|_2^2$$
$$+ \| F_{c4}(z) \|_2^2, \qquad (9.28)$$
$$\triangleq tr(K_{o1})tr(K_{c2}) + tr(K_{o1}) + tr(K_{o3}) + tr(K_{c4}) \qquad (9.29)$$

with obvious definitions for K_{o1}, K_{c2}, K_{o3}, K_{c4}. These are all weighted Gramians given by the following general form:

$$\| X(z) \|_2^2 = tr(K)$$

where

$$(9.30)$$

$$K = \frac{1}{2\pi j} \oint_{|z|=1} X(z) X^T(z^{-1}) z^{-1} dz. \qquad (9.31)$$

It is easy to compute the effect of similarity transformations on the Gramians K_{o1}, K_{c2}, K_{o3} and K_{c4} appearing in (9.29). The optimal sensitivity realization problem can then be stated as follows : given an arbitrary initial realization $(\Phi_0^0, B_0^0, K_0^0, L_0^0)$ and the corresponding weighted Gramians K_{o1}^0, K_{c2}^0, K_{o3}^0, K_{c4}^0, find the (set of) nonsingular transformations T such that

$$\bar{M}_{cl,L_{12}} = tr(T^T K_{o1}^0 T) tr(T^{-1} K_{c2}^0 T^{-T})$$
$$+ tr(T^T (K_{o1}^0 + K_{o3}^0) T) + tr(T^{-1} K_{c4}^0 T^{-T}) \qquad (9.32)$$

is minimized:

$$\min_{T: \det T \neq 0} \bar{M}_{cl,L_{12}}. \qquad (9.33)$$

The same minimization problem has been considered in Chapter 4 for the frequency weighted sensitivity minimization problem. The corresponding algorithms have been developed there. Here, we summarize the optimal design procedure of the compensator $(C_1(z) \; C_2(z))$ as follows. Starting with an initial compensator realization, say $(\Phi_0^0, B_0^0, K_0^0, L_0^0)$, compute first the four weighted Gramians $K_{o1}^0, K_{c2}^0, K_{o3}^0$ and K_{c4}^0 corresponding to this realization. An algorithm for computing weighted Gramians has been proposed in Appendix 4.B. An optimal similarity transformation matrix T^{opt} minimizing the sensitivity upper bound $\bar{M}_{cl,L_{12}}$ can then be computed using the algorithms given in Section 4.3.

A FWL optimal controller realization $(\Phi^{opt}, B^{opt}, K^{opt}, L^{opt})$ is then given by

$$\begin{aligned}\Phi^{opt} &= (T^{opt})^{-1}\Phi_0^0 T^{opt}, & B^{opt} &= (T^{opt})^{-1}B_0^0, \\ K^{opt} &= K_0^0 T^{opt}, & L^{opt} &= (T^{opt})^{-1}L_0^0.\end{aligned}$$

9.2.4 ROUNDOFF NOISE STUDY AND OPTIMAL REALIZATIONS

We now consider the case of exact coefficients and we perform a roundoff noise analysis in a closed loop system, where roundoff errors occur only in the controller. More precisely, we assume that each coefficient in (Φ_0, B_0, K_0, L_0) has an exact FWL expression, that is to say its FWL version (with B_c bits) is equal to its ideal value. The actual compensator (9.8) implemented with *Roundoff Before Multiplication* (RBM) can then be modelled as follows (see Section 3.5):

$$\begin{aligned}\hat{x}^*(t+1) &= \Phi_0 \hat{x}^*(t) + B_0 r(t) + L_0 y^*(t) - M^s(t) \\ u^*(t) &= -K_0 \hat{x}^*(t) + r(t) - n^s(t)\end{aligned} \quad (9.34)$$

where

$$\begin{aligned}M^s(t) &\triangleq \Phi_0 e_x(t) + B_0 e_r(t) + L_0 e_y(t) \\ n^s(t) &\triangleq -K_0 e_x(t) + e_r(t)\end{aligned} \quad (9.35)$$

with

$$\begin{aligned}e_x(t) &\triangleq \hat{x}^*(t) - Q[\hat{x}^*(t)] \\ e_r(t) &\triangleq r(t) - Q[r(t)] \\ e_y(t) &\triangleq y^*(t) - Q[y^*(t)].\end{aligned} \quad (9.36)$$

Here $Q[r(t)]$, say, is the rounded version of $r(t)$, rounded to B_s bits, and $y^*(t)$ is the output of the system driven by $Q[u^*(t)]$:

$$\begin{aligned}x^*(t+1) &= A_0 x^*(t) + B_0 Q[u^*(t)] \\ y^*(t) &= C_0 x^*(t).\end{aligned} \quad (9.37)$$

The state of the actual system with roundoff is different from the state of the ideal system (9.1), that is $x(t) \neq x^*(t)$, because the actual control signal $Q[u^*(t)]$ is different from the one computed with (9.6) where no rounding errors occur in the digital controller. The actual

232   9. Parametrizations in control problems

closed loop state space equations with roundoff before multiplication and exact coefficients are now given by

$$\begin{pmatrix} x^*(t+1) \\ \hat{x}^*(t+1) \end{pmatrix} = \bar{A}_0 \begin{pmatrix} x^*(t) \\ \hat{x}^*(t) \end{pmatrix} + \bar{B}_0 r(t) - N(t)$$

$$y^*(t) = \bar{C}_0 \begin{pmatrix} x^*(t) \\ \hat{x}^*(t) \end{pmatrix} \tag{9.38}$$

where

$$\bar{A}_0 \triangleq \begin{pmatrix} A_0 & -B_0 K_0 \\ L_0 C_0 & \Phi_0 \end{pmatrix}, \bar{B}_0 \triangleq \begin{pmatrix} B_0 \\ B_0 \end{pmatrix}, \bar{C}_0 \triangleq (C_0 \ 0) \tag{9.39}$$

and $N(t)$ is given by

$$N(t) \triangleq \begin{pmatrix} B_0(n^s(t) + e_u(t)) \\ M^s(t) \end{pmatrix}$$

$$= \begin{pmatrix} -B_0 K_0 & B_0 & 0 & B_0 \\ \Phi_0 & B_0 & L_0 & 0 \end{pmatrix} \begin{pmatrix} e_x(t) \\ e_r(t) \\ e_y(t) \\ e_u(t) \end{pmatrix}$$

$$\triangleq P_N \begin{pmatrix} e_x(t) \\ e_r(t) \\ e_y(t) \\ e_u(t) \end{pmatrix} \tag{9.40}$$

with $e_u(t) \triangleq u^*(t) - Q[u^*(t)]$. It is assumed that the four different roundoff errors are independent and, as explained in the introduction, it is also assumed for simplicity that the signals $y^*(t)$, $u^*(t)$, $\hat{x}^*(t)$ and $r(t)$ are all rounded off to the same bit length, and that the roundoff error signals therefore all have the same variance, denoted σ^2. So, if the state and output errors are denoted by

$$E(t) \triangleq \begin{pmatrix} x(t) - x^*(t) \\ \hat{x}(t) - \hat{x}^*(t) \end{pmatrix}, \quad \Delta y(t) \triangleq y(t) - y^*(t), \tag{9.41}$$

then one gets

$$E(t+1) = \bar{A}_0 E(t) + N(t)$$
$$\Delta y(t) = \bar{C}_0 E(t). \tag{9.42}$$

Hence, with $E(0) = 0$

$$E(t) = \sum_{i=0}^{t-1} \bar{A}_0^i N(t-1-i)$$

$$\triangle y(t) = \sum_{i=0}^{t-1} \bar{C}_0 \bar{A}_0^i N(t-1-i). \quad (9.43)$$

It then follows that the steady-state output error variance is given by

$$\sigma_{\triangle y}^2 \triangleq \lim_{t \to \infty} E[\{\triangle y(t)\}^2] = \sum_{i=0}^{\infty} \bar{C}_0 \bar{A}_0^i R_N (\bar{A}_0^i)^T \bar{C}_0^T$$

$$= tr\left[\sum_{i=0}^{\infty} (\bar{A}_0^i)^T \bar{C}_0^T \bar{C}_0 \bar{A}_0^i R_N\right] \quad (9.44)$$

with

$$R_N \triangleq E[N(t) N^T(t)] = P_N P_N^T \sigma^2 \quad (9.45)$$

where σ^2 is the unit roundoff error variance.

Clearly, the error variance can be rewritten as

$$\sigma_{\triangle y}^2 = tr[\bar{W}_o R_N] = tr(P_N^T \bar{W}_o P_N) \sigma^2 \quad (9.46)$$

where $\bar{W}_o \triangleq \sum_{i=0}^{\infty} (\bar{A}_0^i)^T \bar{C}_0^T \bar{C}_0 \bar{A}_0^i$ is the observability Gramian of the closed loop system. Denote

$$M_0 \triangleq P_N^T \bar{W}_o P_N \quad (9.47)$$

If T is a similarity transformation that transforms (A_0, B_0, C_0) into $(T^{-1} A_0 T, T^{-1} B_0, C_0 T)$, it then transforms the corresponding (Φ_0, B_0, K_0, L_0) into $(T^{-1} \Phi_0 T, T^{-1} B_0, K_0 T, T^{-1} L_0)$. In these new coordinates, P_N and $\bar{W}_o$ are given, respectively, by

$$P_N(T) = \begin{pmatrix} -T^{-1} B_0 K_0 T & T^{-1} B_0 & 0 & T^{-1} B_0 \\ T^{-1} \Phi_0 T & T^{-1} B_0 & T^{-1} L_0 & 0 \end{pmatrix}$$

$$= \begin{pmatrix} T^{-1} & 0 \\ 0 & T^{-1} \end{pmatrix} \begin{pmatrix} -B_0 K_0 T & B_0 & 0 & B_0 \\ \Phi_0 T & B_0 & L_0 & 0 \end{pmatrix} \quad (9.48)$$

and

$$\bar{W}_o(T) = \begin{pmatrix} T & 0 \\ 0 & T \end{pmatrix}^T \bar{W}_o \begin{pmatrix} T & 0 \\ 0 & T \end{pmatrix}.$$

Therefore, after transformation, M_0 becomes

$$M(T) = \begin{pmatrix} -B_0 K_0 T & B_0 & 0 & B_0 \\ \Phi_0 T & B_0 & L_0 & 0 \end{pmatrix}^T \bar{W}_o \begin{pmatrix} -B_0 K_0 T & B_0 & 0 & B_0 \\ \Phi_0 T & B_0 & L_0 & 0 \end{pmatrix}.$$

The expression of the error variance $\sigma^2_{\Delta y}(T)$ as a function of the coordinates is thus

$$\begin{aligned} \sigma^2_{\Delta y}(T) &= tr(M(T))\sigma^2 \\ &= [\eta^2 + G(T)]\sigma^2 \\ &= [\eta^2 + tr(T^T \Pi T)]\sigma^2, \end{aligned} \quad (9.49)$$

where

$$\Pi \triangleq \begin{pmatrix} -B_0 K_0 \\ \Phi_0 \end{pmatrix}^T \bar{W}_o \begin{pmatrix} -B_0 K_0 \\ \Phi_0 \end{pmatrix}] \quad (9.50)$$

and

$$\eta^2 \triangleq tr[\begin{pmatrix} B_0 \\ B_0 \end{pmatrix}^T \bar{W}_o \begin{pmatrix} B_0 \\ B_0 \end{pmatrix} + \begin{pmatrix} 0 \\ L_0 \end{pmatrix}^T \bar{W}_o \begin{pmatrix} 0 \\ L_0 \end{pmatrix}$$
$$+ \begin{pmatrix} B_0 \\ 0 \end{pmatrix}^T \bar{W}_o \begin{pmatrix} B_0 \\ 0 \end{pmatrix}]. \quad (9.51)$$

Since $\bar{W}_o$ is the observability Gramian of the closed loop system, it satisfies the Lyapunov equation:

$$\bar{W}_o = \bar{A}_0^T \bar{W}_o \bar{A}_0 + \bar{C}_0^T \bar{C}_0. \quad (9.52)$$

It then follows from (9.39) and (9.50) that $\Pi = \bar{W}_o(2,2)$.

The quantity $tr(M(T))$ is the ratio of the output error variance to the unit roundoff noise variance and can thus be considered as the *Closed Loop Roundoff Noise Gain*. It is the sum of a coordinate independent term, η^2, due to the roundoff errors e_r, e_y and e_u, and of a coordinate dependent term, $G(T)$, due to e_x. We note that the closed loop roundoff noise gain can be made arbitrarily large by a poor choice of controller coordinates.

The optimal design of the compensator in terms of minimizing the roundoff noise gain of the closed loop system is then to find the similarity transformations T^{opt} that solve the following problem:

$$\min_{T:\det T \neq 0} [G(T) = tr(T^T \Pi T) = tr(T^T \bar{W}_o(2,2) T)]. \quad (9.53)$$

Just as in the corresponding problem of filter design treated in Chapter 3, we note that the minimization does not make much sense since the roundoff noise gain can be made as small as possible by decreasing T (in the sense of making its norm smaller). However, if T is very small, the components of the state $\hat{x}^*(t)$ will be very large and, possibly, outside the acceptable range of the machine. To avoid overflow, the compensator states $\hat{x}^*(t)$ are scaled. Let $\bar{W}_c$ be the controllability Gramian of the closed loop system :

$$\bar{W}_c \triangleq \sum_{k=0}^{\infty} \bar{A}_0^k \bar{B}_0 \bar{B}_0^T (\bar{A}_0^T)^k \qquad (9.54)$$

$$= \begin{pmatrix} \bar{W}_c(1,1) & \bar{W}_c(1,2) \\ \bar{W}_c(2,1) & \bar{W}_c(2,2) \end{pmatrix}$$

where $\bar{A}_0, \bar{B}_0$ are defined by (9.39). Imposing an l_2 scaling on the compensator states corresponds with finding a coordinate transformation T applied to both $x(t)$ and $\hat{x}(t)$ such that

$$(T^{-1}\bar{W}_c(2,2)T^{-T})_{i,i} = 1 \text{ for i} = 1,2,\ldots,n. \qquad (9.55)$$

The minimization of the closed loop roundoff noise gain subject to an l_2 dynamic range constraint on the controller states can therefore be formulated as follows:

$$\min_{T:\,\det T \neq 0} G(T) = tr(T^T \bar{W}_o(2,2) T), \qquad (9.56)$$

where $\bar{W}_o$ satisfies (9.52), subject to

$$\qquad (9.57)$$

$$(T^{-1}\bar{W}_c(2,2)T^{-T})_{i,i} = 1 \text{ for } i = 1,2,\ldots,n. \qquad (9.58)$$

This problem can be solved using Theorems 3.6 and 3.7. Starting from an initial realization of the compensator, and using a solution T^{opt} provided by Theorem 3.7, one obtains an optimal compensator realization. It yields a closed loop system having minimal roundoff error propagation.

9.2.5 DESIGN EXAMPLE

We now consider an example that will both illustrate the design procedures described in this section, and demonstrate the sometimes important differences between the performances of a FWL compensator optimized independently of the closed loop system as if it were a filter, and those of a FWL compensator optimized in closed loop.

The discrete time or sampled data system to be controlled is given in its controllable realization (direct form) as:

$$A_0 = \begin{pmatrix} 3.7156 & -5.4143 & 3.6525 & -0.9642 \\ 1.0000 & 0 & 0 & 0 \\ 0 & 1.0000 & 0 & 0 \\ 0 & 0 & 1.0000 & 0 \end{pmatrix} \quad B_0 = \begin{pmatrix} 1 \\ 0 \\ 0 \\ 0 \end{pmatrix}$$

$$C_0 = \begin{pmatrix} 0.1116 & 0.0043 & 0.1088 & 0.0014 \end{pmatrix} \times 10^{-5}, \tag{9.59}$$

This system is unstable, with the following poles:

$$0.9075 \pm j0.4330, \quad 0.9503 \pm j0.2249,$$

whose norms are

$$1.0055, \; 1.0055, \; 0.9766, \; 0.9766.$$

It is desired to control this system by a pole placement strategy using state estimate feedback in such a way that the closed loop poles are:

$$\{\lambda(A_0 - B_0 K_0)\} = 0.9844 \pm j0.0357, \quad 0.9643 \pm j0.0145.$$

The norms of these closed loop poles are:

$$0.9851, \; 0.9851, \; 0.9644, \; 0.9644.$$

We have purposefully chosen desired closed loop poles that are near the unit circle in order to better illustrate the FWL effects obtained with different compensator realizations. The magnitude frequency response of this desired closed loop system is given by the full line in Figure 9.2.

As described above, when the states of the system are not measured, this control strategy can be realized by inserting a state observer into the loop. We consider that the following are the desired locations for the observer poles:

$$\{\lambda(A_0 - L_0 C_0)\} = \{0.7152 \pm j0.6348, \; 0.3522 \pm j0.2857\}.$$

The desired poles for the closed loop system and the desired observer poles dictate the following choices for the control gain vector K_0 and the observer gain vector L_0:

$$K_0 = \begin{pmatrix} -0.1818 & 0.2831 & -0.0500 & -0.0617 \end{pmatrix},$$

and
$$L_0 = \begin{pmatrix} 1.0963 & 0.6385 & 0.3027 & 0.0744 \end{pmatrix}^T \times 10^6.$$

These are both the infinite precision gains corresponding to the realization (9.59) in control canonical form. The corresponding state estimate feedback controller realization is then given by (9.8)-(9.9), with (Φ_0, B_0, K_0, L_0) given by:

$$\Phi_0 = \begin{pmatrix} 2.6743 & -5.7443 & 2.5096 & -0.9176 \\ 0.2877 & -0.0273 & -0.6947 & -0.0088 \\ -0.3377 & 0.9871 & -0.3294 & -0.0042 \\ -0.0830 & -0.0032 & 0.9190 & -0.0010 \end{pmatrix}$$

$$B_0 = \begin{pmatrix} 1 \\ 0 \\ 0 \\ 0 \end{pmatrix}, L_0 = \begin{pmatrix} 1.0963 \\ 0.6385 \\ 0.3027 \\ 0.0744 \end{pmatrix} \times 10^6$$

$$K_0 = \begin{pmatrix} -0.1818 & 0.2831 & -0.0500 & -0.0617 \end{pmatrix}. \quad (9.60)$$

Taking (9.59) and (9.60) as the initial realizations for the system and the compensator, respectively, one can compute the four weighted Gramians defined in (9.29) by using the algorithm described in Appendix 4.B.

Minimizing the sensitivity upper bound (9.32) with the algorithm given in Chapter 4 yields an optimal similarity transformation matrix $T^{opt}(\bar{M}_{cl,L_{12}})$. Applying this transformation to the controllable realization then yields the optimal compensator realization $R^{opt}(\bar{M}_{cl,L_{12}})$ is:

$\boxed{R^{opt}(\bar{M}_{cl,L_{12}}):}$

$$\Phi^{opt}(\bar{M}_{cl,L_{12}}) = \begin{pmatrix} -0.2106 & -0.1393 & 0.0110 & -0.0006 \\ 3.3882 & 0.9948 & -0.0943 & 0.0038 \\ 0.4976 & 0.9588 & 0.6010 & -0.0673 \\ 5.0339 & -2.9354 & 0.6403 & 0.9314 \end{pmatrix}$$

$$B^{opt}(\bar{M}_{cl,L_{12}}) = \begin{pmatrix} 0.0000 \\ -0.0001 \\ 0.0013 \\ -0.0398 \end{pmatrix} \quad L^{opt}(\bar{M}_{cl,L_{12}}) = \begin{pmatrix} 37.8156 \\ -100.9617 \\ -35.3556 \\ -90.3173 \end{pmatrix}$$

$$K^{opt}(\bar{M}_{cl,L_{12}}) = \begin{pmatrix} -341.1329 & 273.5402 & -67.5640 & 1.4450 \end{pmatrix}$$

The optimal controller realization above has been derived by minimizing the sensitivity upper bound of the closed loop transfer function with respect to the controller parameters, that is following strategy (ii) as described in the introduction. We have compared its performance with those of

- that of a FWL implementation of the controller (Φ_0, B_0, K_0, L_0), which we denote by R_c because it is derived from a representation of the system in *control canonical form*;

- that of a FWL implementation of a *balanced realization* of the controller transfer function $(C_1(z) \; C_2(z))$, which we denote by R_b: this balanced realization optimizes the L_1/L_2 transfer function sensitivity measure of the controller itself.

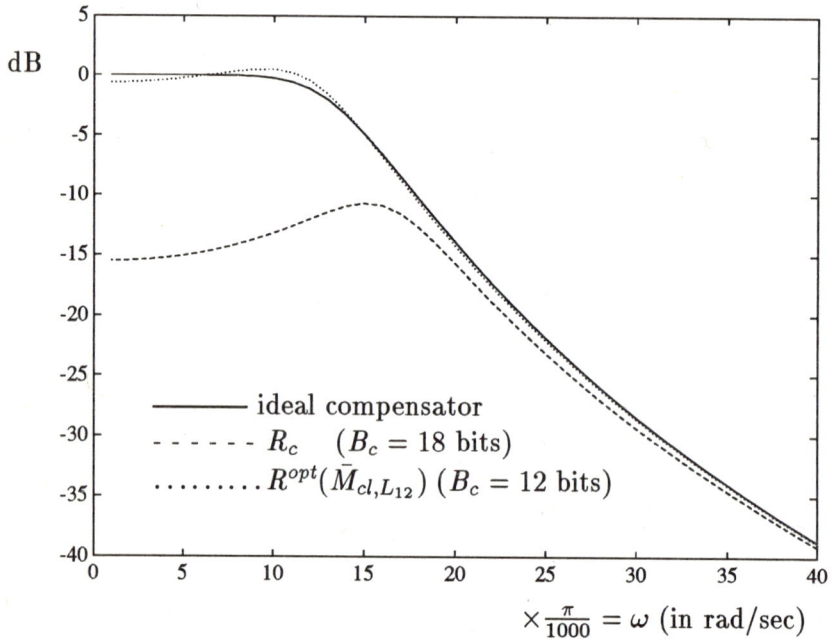

FIGURE 9.2. Closed loop magnitude frequency responses with ideal and with 2 FWL compensators

The values of the sensitivity upper bound for the realizations R_c, R_b and $R^{opt}(\bar{M}_{cl,L_{12}})$ are given by

$$R_c \quad : \quad \bar{M}_{cl,L_{12}} = 9.4094 \times 10^9$$
$$R_b \quad : \quad \bar{M}_{cl,L_{12}} = 8.5863 \times 10^4$$
$$R^{opt}(\bar{M}_{cl,L_{12}}) \quad : \quad \bar{M}_{cl,L_{12}} = 4.5441 \times 10^2$$

The values of these sensitivity measure upper bounds are quite eloquent, but they only give one synthetic number that does perhaps not accurately reflect the different behaviours of these three different FWL controllers as a function of frequency. Therefore we have also computed the magnitudes of the frequency responses of the closed loops for the truncated versions of these three controllers. In these computations, the fractional part of every coefficient of the compensator under investigation has been rounded off to B_c bits for various values of B_c. The results are presented in Figures 9.2 and 9.3. Figure 9.2 illustrates the magnitude of the frequency response of the ideal closed loop system in full line, of the system controlled by an 18-bit coefficient implementation of R_c and of the system controlled by a 12-bit coefficient implementation of $R^{opt}(\bar{M}_{cl,L_{12}})$, while Figure 9.3 compares the ideal response and that of the closed-loop-optimal $R^{opt}(\bar{M}_{cl,L_{12}})$ with the response of the system controlled by a 12-bit coefficient implementation of the realization R_b. These computations portray only the effects of coefficient errors, not those of roundoff errors.

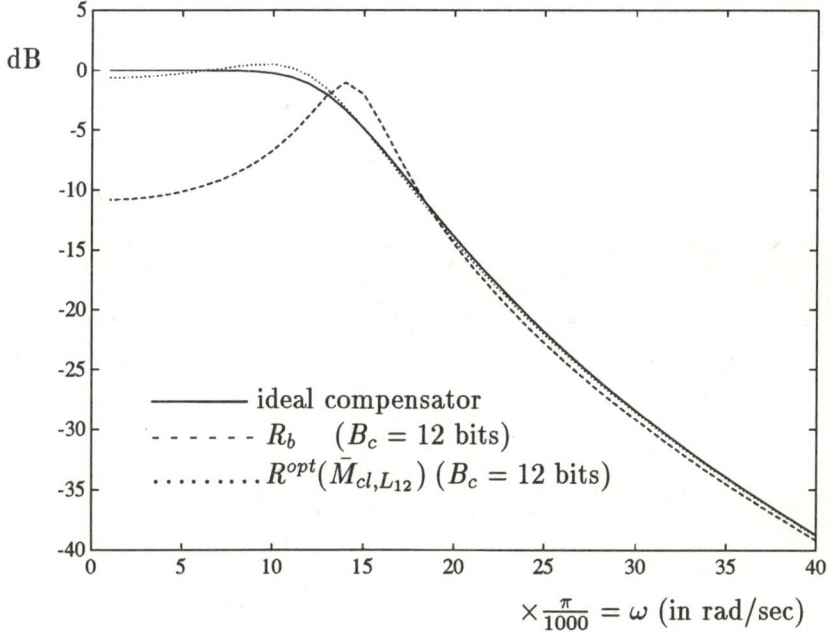

FIGURE 9.3. Closed loop magnitude frequency responses with ideal and with 2 FWL compensators

Comments:

1. With only 12 bits, the optimal realization $R^{opt}(\bar{M}_{cl,L_{12}})$ yields a closed loop performance that is almost as good as the infinite wordlength compensator.

2. Figure 9.2 shows the very poor performance of the realization R_c, particularly at low frequencies, and the clear superiority of the optimal realization: with only 12 bits, $R^{opt}(\bar{M}_{cl,L_{12}})$ performs much better than R_c with 18 bits. This confirms our theoretical computations.

3. Figure 9.3 shows that, with the same number of bits (12), the optimal realization $R^{opt}(\bar{M}_{cl,L_{12}})$ also performs much better than R_b. Recall that the internally balanced realization R_b minimizes the sensitivity measure of the compensator viewed as an open loop system. This computation is probably the most striking: it strongly confirms and supports our argument that one should optimize the compensator realization in terms of optimizing a closed loop performance measure.

We have also computed the poles of the closed loop transfer function for each of the compensator realizations described above. They are given in Figures 9.4 and 9.5, where '+', 'x' and '*' indicate the closed loop pole locations for the ideal closed loop, and for the closed loops with compensators R_c (using 18 bits) and $R^{opt}(\bar{M}_{cl,L_{12}})$ (using 12 bits) for Figure 9.4, and with R_b (using 12 bits) and $R^{opt}(\bar{M}_{cl,L_{12}})$ (using 12 bits) for Figure 9.5, respectively.[3]

We recall that the observer dynamics, which disappear from the input/output transfer function by pole-zero cancellation when the compensator is implemented with infinite precision, do not cancel when the compensator is implemented with finite precision. Thus, the closed loop systems with FWL compensators become 8th-order systems, and this explains why 8 different poles appear in Figures 9.4 and 9.5. Observe that the closed loop system becomes unstable when the compensator is implemented with R_c using 18 bits, and even when the balanced realization R_b (using 12 bits) is used. On the other hand, the closed-loop-optimal compensator $R^{opt}(\bar{M}_{cl,L_{12}})$, implemented with only 12 bits, produces a stable closed loop.

---

[3] The bit lengths given here are for the fractional parts of each coefficient of the compensator. No roundoff error is assumed.

9. Parametrizations in control problems 241

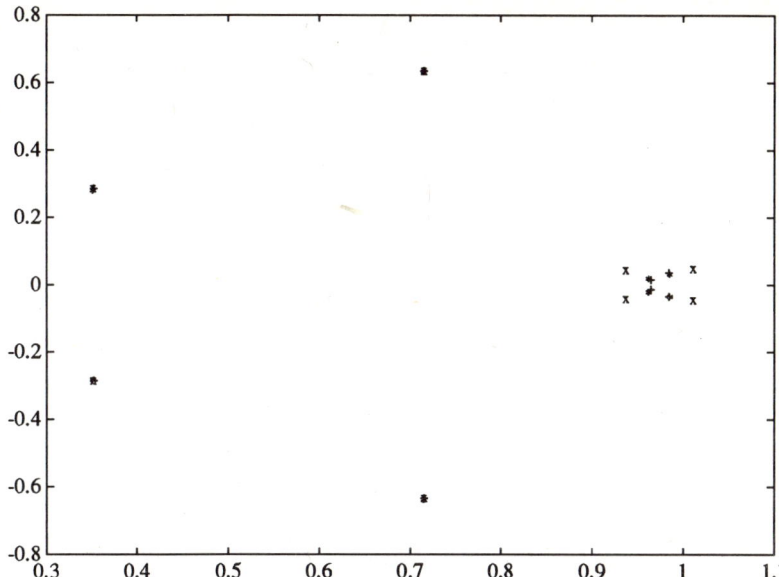

FIGURE 9.4. Closed loop pole locations for the ideal compensator (+), and for the FWL compensators R_c (x) and $R^{opt}(\bar{M}_{cl,L_{12}})$ (*).

To illustrate this point more fully, we present the poles of the closed loop system with the ideal compensator, with R_c using 18 bits, with R_b using 12 bits and with $R^{opt}(M_{cl,L_{12}})$ using 12 bits. The vectors of poles are, respectively, and with obvious notations:

$$V_{P_0} = \begin{pmatrix} 0.9844 \pm j0.0357 \\ 0.9643 \pm j0.0145 \\ 0.7152 \pm j0.6348 \\ 0.3522 \pm j0.2857 \end{pmatrix}, \quad V_{P_c} = \begin{pmatrix} 1.0115 \pm j0.0469 \\ 0.9372 \pm j0.0428 \\ 0.7152 \pm j0.6348 \\ 0.3522 \pm j0.2857 \end{pmatrix}$$

$$V_{P_b} = \begin{pmatrix} 1.0049 \pm j0.0411 \\ 0.9435 \pm j0.0363 \\ 0.7152 \pm j0.6347 \\ 0.3522 \pm j0.2857 \end{pmatrix}, \quad V_{P_{opt}} = \begin{pmatrix} 0.9861 \pm j0.0334 \\ 0.9622 \pm j0.0205 \\ 0.7155 \pm j0.6353 \\ 0.3520 \pm j0.2821 \end{pmatrix}$$

242   9. Parametrizations in control problems

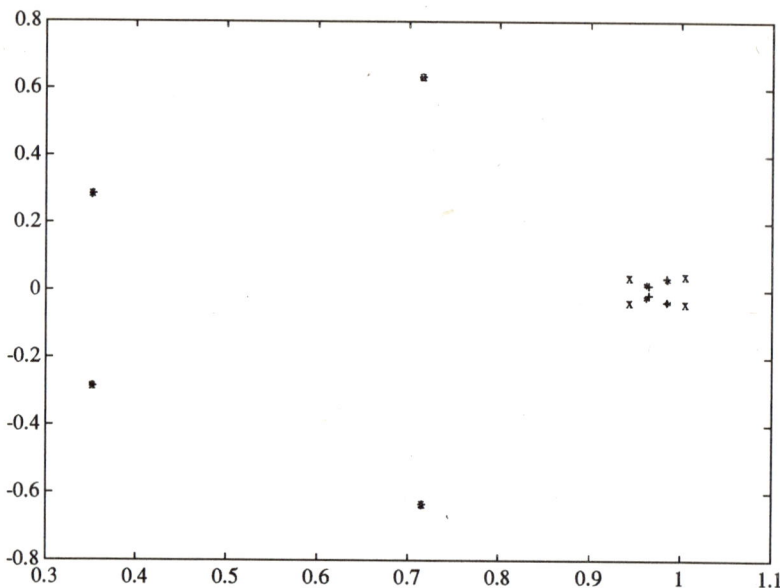

FIGURE 9.5. Closed loop pole locations for the ideal compensator (+), and for the FWL compensators R_b (x) and $R^{opt}(\bar{M}_{cl,L_{12}})$ (*).

The corresponding norms are:

$$\bar{V}_{P_0} = \begin{pmatrix} 0.9851 \\ 0.9644 \\ 0.9563 \\ 0.4535 \end{pmatrix}, \quad \bar{V}_{P_c} = \begin{pmatrix} 1.0126 \\ 0.9382 \\ 0.9563 \\ 0.4535 \end{pmatrix}$$

$$\bar{V}_{P_b} = \begin{pmatrix} 1.0058 \\ 0.9442 \\ 0.9563 \\ 0.4535 \end{pmatrix}, \quad \bar{V}_{P^{opt}} = \begin{pmatrix} 0.9867 \\ 0.9624 \\ 0.9569 \\ 0.4511 \end{pmatrix}$$

With the same initial realization (9.59) and (9.60), we compute the controllability and observability Gramians of the state space realization of the closed loop system that are needed to perform the roundoff noise minimization design (9.56)-(9.58). We have computed an optimal transformation $T^{opt}(G)$ from the controller realization R_c, and hence the corresponding optimal realization $R^{opt}(G)$:

$\boxed{R^{opt}(G):}$

$$\Phi^{opt}(R^{opt}(G)) = \begin{pmatrix} 0.0817 & -2.2239 & 1.9883 & -1.2443 \\ 0.7069 & 2.7454 & -1.5575 & 1.1983 \\ 1.6778 & 3.0280 & 0.5877 & 2.7439 \\ -1.2346 & -2.0275 & -0.0673 & -1.0982 \end{pmatrix}$$

$$B^{opt}(R^{opt}(G)) = \begin{pmatrix} -0.0085 \\ 0.0094 \\ 0.0038 \\ -0.0047 \end{pmatrix} \quad L^{opt}(R^{opt}(G)) = \begin{pmatrix} -8.1371 \\ 8.3577 \\ 39.0678 \\ -31.1646 \end{pmatrix}$$

$$K^{opt}(R^{opt}(G)) = \begin{pmatrix} 360.0162 & 842.5326 & -649.1845 & 545.7773 \end{pmatrix}.$$
(9.61)

We have also examined the l_2-scaled version of R_c, denoted R_c^{sc}, and the l_2-scaled MRH realization, which minimizes the roundoff noise gain of the compensator itself considered as a simple open loop system, denoted R_{MRH}^{sc}. The coordinate dependent contributions, $G(T)$, of the closed loop roundoff noise gains (see (9.53)) and the sensitivity measure upper bounds, $\bar{M}_{cl,L_{12}}$, for these realizations are as follows:

$$\begin{aligned} R_c^{sc} : G &= 3.3713 \times 10^8 & \bar{M}_{cl,L_{12}} &= 1.9805 \times 10^9 \\ R_{MRH}^{sc} : G &= 4.9859 \times 10^2 & \bar{M}_{cl,L_{12}} &= 2.8888 \times 10^3 \\ R^{opt}(G) : G &= 1.0216 \times 10^2 & \bar{M}_{cl,L_{12}} &= 5.9377 \times 10^2 \end{aligned}$$

Comments:

1. Note the huge difference between the gain $G(T)$ of the closed loop system using the l_2-scaled compensator realization R_c^{sc} derived from the control canonical realization (9.59) of the system to be controlled, and that obtained when the optimal realization $R^{opt}(G)$ is used. This shows that a poor choice of parametrization for the compensator can cause a very large roundoff error propagation in the closed loop system. This certainly motivates spending some effort on a proper design of parametrization for the compensator.

2. The roundoff noise gain analysis again shows that the standard optimal FWL state space realization design can not be applied

directly to the controller implementation, even though the difference in roundoff noise gain between open loop and closed loop design is not quite so dramatic as with the sensitivities. The l_2-scaled realization that minimizes the roundoff noise gain of the compensator, R_{MRH}^{sc}, yields an error propagation gain that is almost 4 times higher than that of the optimal realization which minimizes the error propagation gain of the closed loop system, $R^{opt}(G)$.

3. We mention that the MRH structure R_{MRH} without l_2-scaling is even worse:

$$R_{MRH} : G = 1.9738 \times 10^6 \quad \bar{M}_{cl,L_{12}} = 3.7824 \times 10^6.$$

We note that the sensitivity performance of $R^{opt}(G)$ is almost as nice as that of $R^{opt}(\bar{M}_{cl,L_{12}})$. This confirms similar findings in the filter design (or open loop design) problem. However, the proximity of the performances of $R^{opt}(G)$ and $R^{opt}(\bar{M}_{cl,L_{12}})$ is in part due to the adoption of an unnatural L_1/L_2 sensitivity measure, which makes the minimization problems of these two measures very close, and even identical in the presence of l_2 scaling constraints.

9.3 FWL LQG controller design

The results of the previous section have been presented in the context of model reference control design. An ideal (infinite wordlength) controller was computed first to achieve a predefined (i.e. model reference) closed loop transfer function, and the subsequent FWL optimal design was there to guarantee that, with a FWL implementation of this controller, the closed loop system does not deviate too much from its ideal performance. This FWL design clearly follows strategy (ii) according to our classification in the introduction of this chapter. A design along strategy (iii) could be envisaged in so far as coefficient errors are concerned: one could, for example, minimize an H_∞ norm of the difference between the ideal closed loop transfer function and the closed loop transfer function with FWL compensator, where the minimization would be carried over all observer gains L and controller gains K as well as over all possible coordinates in which the controller is represented. The optimization would probably not be easy but, more importantly, such procedure would only take into account the coefficient errors. It is not at all clear what performance criterion might best be used to minimize the effects of roundoff errors in a model reference control design context.

In a regulation context, on the other hand, where a Linear Quadratic Gaussian (LQG) criterion is optimized, the criterion itself lends itself naturally to an incorporation of any errors that can be modelled as stochastic disturbances. It is therefore not surprising that the recent results on FWL design of LQG controllers have been on the minimization of roundoff error effects [WK89], [LSG92] and that they have led to a solution following the more interesting strategy (iii) of our introduction. In fact, by adopting a stochastic description of coefficient errors, one could probably easily incorporate the effects of coefficient errors in the optimal FWL LQG design method of [LSG92]. These coefficient error effects have been examined in [MWH80] using precisely such stochastic approach, but only in terms of estimating their impact on the global LQG cost, and with a view of deciding which coefficient wordlength is required to keep the performance degradation within a predetermined level.

Because the LQG regulator, with its in-built criterion for FWL performance degradation, lends itself to a development of strategy (iii) for the computation of an optimal FWL controller, it is worthy of a special presentation. The present section is thus devoted to it, and we follow basically the developments of [LSG92].

9.3.1 The 'ideal' LQG controller

We consider that the system is again described by a linear time invariant strictly proper discrete time state space model, but that it is now subject to stochastic disturbances:

$$\begin{aligned} x(t+1) &= A_0 x(t) + B_0 u(t) + D_0 w(t) \\ y(t) &= C_0 x(t) + v(t) \end{aligned} \quad (9.62)$$

with $A_0 \in \mathbb{R}^{n \times n}, B_0 \in \mathbb{R}^n, D_0 \in \mathbb{R}^n, C_0^T \in \mathbb{R}^n$. The scalar stochastic disturbances $w(t)$ and $v(t)$ are assumed to have zero mean with:

$$\begin{aligned} E[w(t)w(s)] &= \sigma_w^2 \delta_{ts} \\ E[v(t)v(s)] &= \sigma_v^2 \delta_{ts} \\ E[w(t)v(s)] &= 0 \quad \forall t, s. \end{aligned} \quad (9.63)$$

9. Parametrizations in control problems

We assume as in the previous section that this system is minimal, that is the pair $[A_0, B_0]$ is completely reachable and the pair $[C_0, A_0]$ is completely observable.

In the simplest possible LQG regulation problem, it is desired to construct a control sequence $\{u(t), u(t+1), \ldots\}$ such that the following criterion is minimized:

$$J = \lim_{N \to \infty} \frac{1}{N} \sum_{t=0}^{N} E[y^2(t) + Ru^2(t)], \qquad (9.64)$$

where R is a positive scalar. It is a classical result of optimal control theory (see e.g. [AM89], [KS72]) that the solution of this infinite horizon optimization problem takes the form of a linear time invariant state estimate feedback regulator:

$$\begin{aligned} \hat{x}(t+1) &= A_0\hat{x}(t) + B_0 u(t) + L_0[y(t) - C_0\hat{x}(t)] \\ u(t) &= -K_0\hat{x}(t), \end{aligned} \qquad (9.65)$$

where the gains K_0 and L_0 are obtained from the solutions of two Riccati equations:

$$\begin{aligned} \text{a) } K_0 &= (R + B_0^T P B_0)^{-1} B_0^T P A_0^T \\ P &= A_0^T P A_0 + C_0^T C_0 - A_0^T P B_0 (R + B_0^T P B_0)^{-1} B_0^T P A_0 \quad (9.66) \\ \text{b) } L_0 &= A_0 \Sigma C_0^T (C_0 \Sigma C_0^T + \sigma_v^2)^{-1} \\ \Sigma &= A_0 \Sigma A_0^T + \sigma_w^2 D_0 D_0^T - A_0 \Sigma C_0^T (A_0 \Sigma A_0^T + \sigma_v^2)^{-1} C_0 \Sigma A_0^T. \end{aligned}$$
$$(9.67)$$

We observe that, except for the absence of a reference signal $r(t)$, the ideal LQG regulator has exactly the same structure as the pole placement controller (9.6). The only (but significant) difference is that the gains K_0 and L_0 are determined not by direct pole location considerations, but by the solution of a criterion minimization. The regulator can be rewritten as:

$$\begin{aligned} \hat{x}(t+1) &= \Phi_0 \hat{x}(t) + L_0 y(t) \\ u(t) &= -K_0 \hat{x}(t), \end{aligned} \qquad (9.68)$$

with Φ_0 defined as in (9.9).

The controller transfer function therefore takes the same form as for the pole placement controller (see (9.10)) except for the absence of a reference signal:

$$u(t) = -K_0(zI - \Phi_0)^{-1} L_0 y(t) = -C_2(z) y(t). \qquad (9.69)$$

Combining the equations of the system and of the controller yields the following augmented state equations:

$$\begin{pmatrix} x(t+1) \\ \hat{x}(t+1) \end{pmatrix} = \bar{A}_0 \begin{pmatrix} x(t) \\ \hat{x}(t) \end{pmatrix} + G \begin{pmatrix} w(t) \\ v(t) \end{pmatrix}$$

$$y(t) = \bar{C}_0 \begin{pmatrix} x(t) \\ \hat{x}(t) \end{pmatrix} + v(t), \qquad (9.70)$$

where $\bar{A}_0$ and $\bar{C}_0$ have been defined in (9.39), and where G is defined as

$$G \triangleq \begin{pmatrix} D_0 & 0 \\ 0 & L_0 \end{pmatrix}. \qquad (9.71)$$

Using these equations and the expression (9.68) of $u(t)$, we can compute the optimal cost:

$$\begin{aligned} J &\triangleq \lim_{N \to \infty} \frac{1}{N} \sum_{t=0}^{N} E[y^2(t) + Ru^2(t)] \\ &= \lim_{N \to \infty} \frac{1}{N} \sum_{t=0}^{N} E[\begin{pmatrix} x^T(t) & \hat{x}^T(t) \end{pmatrix} \begin{pmatrix} C_0^T C_0 & 0 \\ 0 & K_0^T R K_0 \end{pmatrix} \begin{pmatrix} x(t) \\ \hat{x}(t) \end{pmatrix} \\ &\quad + \sigma_v^2] \\ &= tr\{QP + \sigma_v^2\}, \qquad (9.72) \end{aligned}$$

where $Q \in \mathbb{R}^{2n \times 2n}$ is defined as

$$Q \triangleq \begin{pmatrix} C_0^T C_0 & 0 \\ 0 & K_0^T R K_0 \end{pmatrix} \qquad (9.73)$$

and where $P \in \mathbb{R}^{2n \times 2n}$ is the solution of the Lyapunov equation

$$P = \bar{A}_0 P \bar{A}_0^T + G \Omega G^T, \qquad (9.74)$$

with

$$\Omega \triangleq \begin{pmatrix} \sigma_w^2 & 0 \\ 0 & \sigma_v^2 \end{pmatrix}. \qquad (9.75)$$

Having briefly recalled the fundamentals of theoretical LQG regulator design, we now examine the FWL effects. As explained above, the presently available optimal design studies are limited to roundoff noise errors and we thus limit our analysis to this case.

9.3.2 ROUNDOFF NOISE STUDY OF AN LQG CONTROLLER

In this subsection we consider the case where the gains K_0 and L_0 (and hence also the closed loop matrix Φ_0) have been computed on the basis of the expressions (9.66)-(9.67) of the 'ideal' LQG controller.[4] We assume that the coefficients in (Φ_0, B_0, K_0, L_0) have an exact FWL expression, and we examine the deterioration in the performance index when roundoff errors occur in the regulator due to the encoding of the signals with a finite number, B_s, of bits. We shall be able to compute explicitly the increase in the achieved cost due to these roundoff errors. Our analysis in the next subsection will show that, when roundoff errors occur in the regulator, the gains K_0 and L_0 computed from the ideal LQG theory are in fact not optimal, and that *optimal FWL LQG gains* can be computed.

If we assume again that the regulator is implemented with *Roundoff Before Multiplication*, then the equations of the *actual* controller are:

$$\begin{aligned} \hat{x}^*(t+1) &= \Phi_0 \hat{x}^*(t) + L_0 y^*(t) - M^s(t) \\ u^*(t) &= -K_0 \hat{x}^*(t) - n^s(t) \end{aligned} \quad (9.76)$$

where

$$\begin{aligned} M^s(t) &\triangleq \Phi_0 e_x(t) + L_0 e_y(t) \\ n^s(t) &\triangleq -K_0 e_x(t) \end{aligned} \quad (9.77)$$

with $e_x(t)$ and $e_y(t)$ defined as in (9.36), and where $y^*(t)$ is the output of the true system driven by $Q[u^*(t)]$:

$$\begin{aligned} x^*(t+1) &= A_0 x^*(t) + B_0 Q[u^*(t)] + D_0 w(t) \\ y^*(t) &= C_0 x^*(t) + v(t). \end{aligned} \quad (9.78)$$

Note that the roundoff noise analysis for this LQG regulator completely parallels that of subsection 9.2.4 for the pole placement controller. In particular, compare expressions (9.76) with (9.34), and (9.78) with (9.37). The state space equations of the actual closed loop system with roundoff before multiplication and with exact coefficients are:

$$\begin{pmatrix} x^*(t+1) \\ \hat{x}^*(t+1) \end{pmatrix} = \bar{A}_0 \begin{pmatrix} x^*(t) \\ \hat{x}^*(t) \end{pmatrix} + G \begin{pmatrix} w(t) \\ v(t) \end{pmatrix} - N(t)$$

---

[4] The word 'ideal' refers to the fact that the gains are those of the 'ideal' infinite wordlength LQG controller. We shall see later that these gains are not optimal for a FWL LQG regulator.

$$y^*(t) = \bar{C}_0 \begin{pmatrix} x^*(t) \\ \hat{x}^*(t) \end{pmatrix} + v(t) \tag{9.79}$$

where $N(t)$ is given by

$$\begin{aligned} N(t) &\triangleq \begin{pmatrix} B_0(n^s(t) + e_u(t)) \\ M^s(t) \end{pmatrix} \\ &= \begin{pmatrix} -B_0 K_0 & 0 & B_0 \\ \Phi_0 & L_0 & 0 \end{pmatrix} \begin{pmatrix} e_x(t) \\ e_y(t) \\ e_u(t) \end{pmatrix} \\ &\triangleq S_N \begin{pmatrix} e_x(t) \\ e_y(t) \\ e_u(t) \end{pmatrix} \end{aligned} \tag{9.80}$$

with $e_u(t) \triangleq u^*(t) - Q[u^*(t)]$. Comparing these equations of the closed loop system with the corresponding equations (9.70) of the system with an ideal (roundoff-error-free) regulator, we notice that the difference lies in the additional roundoff noise driving term $N(t)$. Again, these expressions are to be compared with (9.38)-(9.40) in the roundoff noise analysis of the pole placement controller.

We now compute the achieved LQG cost when the system is controlled by this regulator in which roundoff errors occur. We follow the same steps as in the computations of (9.72)-(9.74), but with $u^*(t) = -K_0 \hat{x}^*(t) - n^s(t)$. This yields:

$$J = tr\{QP + K_0^T R K_0 \sigma^2\} + \sigma_v^2, \tag{9.81}$$

where Q has been defined in (9.73), σ^2 is again the unit roundoff noise variance of e_x, e_y, e_u, and P is the covariance matrix of the state of the closed loop system (9.79). It is the solution of a modified Lyapunov equation:

$$P = \bar{A}_0 P \bar{A}_0^T + G\Omega G^T + S_N S_N^T \sigma^2. \tag{9.82}$$

The matrices $\bar{A}_0$, G, Ω and S_N have been defined in (9.39), (9.71), (9.75) and (9.80), respectively. Since the Lyapunov equation is linear in P, we can decompose P to separate the respective effects of the disturbances $w(.)$ and $v(.)$, and of the roundoff errors $e_x(.)$, $e_y(.)$ and $e_u(.)$ in the total cost. We write

$$P = P_{wv} + \sigma^2 P_e,$$

where
$$P_{wv} = \bar{A}_0 P_{wv} \bar{A}_0^T + G\Omega G^T \tag{9.83}$$
$$P_e = \bar{A}_0 P_e \bar{A}_0^T + S_N S_N^T. \tag{9.84}$$

Thus, the total cost J can be decomposed into
$$J = J_{wv} + \sigma^2 J_e, \tag{9.85}$$
where
$$J_{wv} = tr\{QP_{wv}\} + \sigma_v^2 \tag{9.86}$$
$$J_e = tr\{QP_e + K_0^T R K_0\}. \tag{9.87}$$

Comparing (9.86) with (9.72), we observe that J_{wv} is identical to the ideal cost in the absence of roundoff errors in the regulator: it is the contribution to the total cost of the stochastic disturbances $w(.)$ and $v(.)$. As for $\sigma^2 J_e$, it is the additional cost due solely to these roundoff errors. Hence, J_e can be regarded as the roundoff noise gain of this LQG controller, since it is the factor by which the unit roundoff noise variance σ^2 in the regulator is multiplied to penalize the total LQG cost. In the next subsection, we examine how these various contributions to the total cost are affected by the implementation of the controller, that is we study their dependence on the coordinates in which the controller is represented.

9.3.3 OPTIMAL IMPLEMENTATIONS OF FWL LQG CONTROLLERS

We recall that we examine the FWL implementation of an 'ideal' LQG regulator, that is one whose gains have been computed on the basis of the infinite precision expressions (9.66)-(9.67). Here we examine how the total cost, $J = J_{wv} + \sigma^2 J_e$, changes when a coordinate transformation T is applied to the representation of the controller, the aim being to then formulate an optimal controller realization problem.

If a similarity transformation matrix T is applied to the state $x(t)$ and the state estimate $\hat{x}(t)$, then the matrices A_0, B_0, C_0, D_0, Φ_0, K_0, L_0 are changed, respectively, to $T^{-1}A_0T$, $T^{-1}B_0$, C_0T, $T^{-1}D_0$, $T^{-1}\Phi_0 T$, $K_0 T$, $T^{-1} L_0$, while the matrices $\bar{A}_0$, Q and S_N are changed, respectively, to $\bar{T}^{-1}\bar{A}_0\bar{T}$, $\bar{T}^T Q \bar{T}$, and
$$\bar{T}^{-1} S_N \begin{pmatrix} T & 0 & 0 \\ 0 & I & 0 \\ 0 & 0 & I \end{pmatrix},$$

where
$$\bar{T} \triangleq \begin{pmatrix} T & 0 \\ 0 & T \end{pmatrix}.$$

It then follows that P_{wv} is transformed to $\bar{T}^{-1} P_{wv} \bar{T}^{-T}$ and hence $J_{wv} \triangleq tr\{Q P_{wv}\} + \sigma_v^2$ is coordinate independent.[5] As for J_e, the term $K_0^T R K_0$ is transformed to $T^T K_0^T R K_0 T$, while the term $tr\{Q P_e\}$ contains the sum of coordinate independent terms (the contributions to the state covariance of e_u and e_y) and of a coordinate dependent term (the contribution of e_x). Indeed, it follows from the linearity of P_e in (9.84) and the form of S_N (see (9.80)) that P_e can again be decomposed into the sum of two terms, $P_e = P_{e1} + P_{e2}$, where

$$P_{e1} = \bar{A}_0 P_{e1} \bar{A}_0^T + \begin{pmatrix} B_0 B_0^T & 0 \\ 0 & L_0 L_0^T \end{pmatrix}$$

$$P_{e2} = \bar{A}_0 P_{e2} \bar{A}_0^T + \begin{pmatrix} -B_0 K_0 \\ \Phi_0 \end{pmatrix} \begin{pmatrix} -K_0^T B_0^T & \Phi_0^T \end{pmatrix}. \quad (9.88)$$

With this decomposition we get

$$J_e = tr\{Q P_{e1} + Q P_{e2} + K_0^T R K_0\}.$$

Clearly, a coordinate transformation transforms P_{e1} into $\bar{T}^{-1} P_{e1} \bar{T}^{-T}$, and hence $tr\{Q P_{e1}\}$ is coordinate independent. We denote $\eta^2 \triangleq tr\{Q P_{e1}\}$ and we now examine the term $tr\{Q P_{e2} + K_0^T R K_0\}$. Let Π denote the solution of the Lyapunov equation

$$\Pi = \bar{A}_0^T \Pi \bar{A}_0 + Q. \quad (9.89)$$

It follows from (9.88) and (9.89) that

$$tr\{\Pi \begin{pmatrix} -B_0 K_0 \\ \Phi_0 \end{pmatrix} \begin{pmatrix} -K_0^T B_0^T & \Phi_0^T \end{pmatrix}\} = tr\{Q P_{e2}\}.$$

In addition it follows from (9.39), (9.73) and (9.89) that

$$\Pi(2,2) = \begin{pmatrix} -K_0^T B_0^T & \Phi_0^T \end{pmatrix} \Pi \begin{pmatrix} -B_0 K_0 \\ \Phi_0 \end{pmatrix} + K_0^T R K_0. \quad (9.90)$$

---

[5] Since we have shown above that J_{wv} is identical to the optimal cost of the ideal LQG regulator, the coordinate independence of J_{wv} is well known and is a direct consequence of the fact that the performance index (9.64) is expressed in terms of the coordinate independent variables $u(.)$ and $y(.)$.

Therefore,

$$\begin{aligned}
tr\{QP_{e2} + K_0^T R K_0\} &= tr\{\begin{pmatrix} -K_0^T B_0^T & \Phi_0^T \end{pmatrix} \Pi \begin{pmatrix} -B_0 K_0 \\ \Phi_0 \end{pmatrix} \\
&\quad + K_0^T R K_0\} \\
&= tr\{\Pi(2,2)\}.
\end{aligned} \qquad (9.91)$$

Now, it follows immediately from the definition (9.89) that a coordinate transformation transforms Π into $\bar{T}^T \Pi \bar{T}$, and thus $\Pi(2,2)$ into $T^T \Pi(2,2) T$.

Collecting all these results, it follows that the dependence of the total cost on a coordinate transformation T of the regulator is as follows:

$$\begin{aligned}
J(T) &= J_{wv} + \sigma^2 J_e(T) \\
&= J_{wv} + \sigma^2 [\eta^2 + tr\{T^T \Pi(2,2) T\}].
\end{aligned} \qquad (9.92)$$

In this decomposition, J_{wv} represents the contribution to the total cost of the stochastic disturbances w and v, $\sigma^2 \eta^2$ represents the contribution of the roundoff noises e_u and e_y in the D/A and A/D converters, while the third, T-dependent term, represents the contribution of the roundoff noise at the states of the digital regulator. It is worth comparing the expression (9.92) with (9.49)-(9.50) in the case of a pole placement controller with roundoff.

The expression (9.92) shows that, unless some constraint is imposed on the transformation matrices T, the roundoff noise effect in the regulator can make the LQG cost become infinitely large. Note that the same conclusion applied to the error variance $\sigma^2_{\Delta y}(T)$ in the pole placement controller. When an l_2 constraint is imposed on the states $\hat{x}(t)$ of the controller, T cannot vary arbitrarily and this cannot happen. This leads us to the formulation of an optimal FWL regulator implementation problem, where the FWL effects account only for the roundoff noise. Traditionally, this optimal realization problem is posed with the l_2 constraint imposed not on the states $\hat{x}^*(t)$ of the FWL regulator, but on the states $\hat{x}(t)$ of the infinite wordlength regulator. Thus, the optimization problem can be formulated as follows:

$$\min_{T:\,\det T\neq 0} tr\{T^T\Pi(2,2)T\}, \qquad (9.93)$$

where Π satisfies (9.89), subject to

$$(T^{-1}P_{wv}(2,2)T^{-T})_{i,i} = 1 \text{ for } i = 1, 2, \ldots, n. \qquad (9.94)$$

Here $P_{wv}(2,2)$ is the (2,2) block of the solution P_{wv} of the Lyapunov equation (9.83): hence, it is the covariance of the state $\hat{x}(t)$ of the regulator without taking into account roundoff noise errors.

The optimization problem (9.93)-(9.94) is identical to the optimization problem (9.56)-(9.58), and its solution is again obtained by the algorithmic procedure of Hwang explained in Chapter 3.

9.3.4 OPTIMAL FWL LQG CONTROLLERS

The solution of the optimization problem (9.93)-(9.94) yields the best possible FWL *implementation* Φ_0, K_0, L_0 of an LQG regulator (see (9.68)) whose gains K_0 and L_0 have been pre-computed from the 'ideal' infinite wordlength expressions (9.66)-(9.67) and whose transition matrix Φ_0 is then obtained from K_0 and L_0 via $\Phi_0 = A_0 - B_0K_0 - L_0C_0$. In other words, in the approach of the last subsection, the gains are computed *first* from the two LQG Riccati equations, and this is *followed* by a coordinate transformation whose purpose is to realize this fixed structure controller in a coordinate space that minimizes the effects of roundoff errors. This procedure follows strategy (ii) and is thus clearly *suboptimal*.

The *optimal* solution to the FWL LQG regulator design is to minimize the cost criterion *directly* with respect to the regulator quantities Φ_0, K_0 and L_0, without imposing any structure on these quantities, and in such a way that the cost criterion incorporates the effects of roundoff errors. This optimization problem therefore involves two steps:

- the first step is to compute the expression of the LQG cost - including the roundoff error effects - as a function of an arbitrary realization (Φ_0, K_0, L_0) of a regulator;

- the second step is to minimize this cost with respect to the regulator realization, that is with respect to Φ_0, K_0 and L_0.

Now we observe that the first step has already been accomplished. Indeed, the total LQG cost, including the effects of the roundoff errors in the regulator, is expressed by (9.85), together with (9.86)-(9.87) and (9.83)-(9.84). Recall that

$$\bar{A}_0 = \begin{pmatrix} A_0 & -B_0 K_0 \\ L_0 C_0 & \Phi_0 \end{pmatrix}.$$

We observe that the dependence of this total cost on the regulator parameters Φ_0, K_0, L_0 is via $\bar{A}_0$ in the two Lyapunov equations, and that this expression does not depend on K_0 and L_0 having been obtained via the solutions of two Riccati equations, or on Φ_0 being related to the other quantities by the expression (9.9). In other words, (9.83) to (9.87) express the total LQG cost obtained for any controller of the form (9.68) applied to the system (9.62) when roundoff errors with variance σ^2 occur in the regulator. The optimal FWL LQG regulator is thus obtained by optimizing the expression (9.85) of J with respect to Φ_0, K_0 and L_0, while satisfying an l_2 constraint on the states of the regulator. As explained above, this l_2 constraint is usually imposed on the states of the ideal regulator (i.e. without taking account of the roundoff noise). Thus, a first formulation of the optimization problem could be:

$$\min_{\Phi_0, K_0, L_0} \{J_{wv} + J_e\} \quad (9.95)$$

subject to

$$(P_{wv}(2,2))_{i,i} = 1 \quad \text{for} \quad i = 1, 2, \ldots, n. \quad (9.96)$$

Now, we have seen in the previous subsection that J_{wv} is independent of the coordinate space of the regulator realization, whereas P_{wv} and J_e both depend on this coordinate space. On the basis of our analysis of that subsection, we can therefore reformulate the optimization problem as:

$$\min_{\Phi_0, K_0, L_0, T} \{J_{wv} + J_e(T)\} = \min_{\Phi_0, K_0, L_0} \{J_{wv} + \min_T [J_e(T)]\} \quad (9.97)$$

subject to

$$(T^{-1} P_{wv}(2,2) T^{-T})_{i,i} = 1 \quad \text{for} \quad i = 1, 2, \ldots, n. \quad (9.98)$$

The dependence of $J_e(T)$ on T was expressed in (9.92). The optimization problem (9.97)-(9.98) has been solved in [LSG92], where some interesting simulations are also presented. These results are remarkable in that they provide a solution to a FWL controller design problem that follows the precepts of strategy (iii), as defined in the introduction of this chapter, at least as far as the roundoff noise errors are concerned. The simulations of Liu and Skelton show that the optimal FWL LQG regulator can yield a dramatically smaller cost than the 'ideal' LQG regulator that is only optimized with respect to its realization. Thus, these simulations clearly illustrate the superiority of strategy (iii) over strategy (ii).

9.4 Conclusions

In this chapter, we have shown that the techniques of optimal design of FWL filters developed in the first part of this book can be extended to the optimal design of FWL digital controllers. At first sight, it might appear that the optimal controller design problem is a particular case of optimal filter design, since the controller transfer function can be considered as a filter to be implemented in a way that minimizes FWL effects. We hope to have convinced the reader by now, both by our theoretical developments and our simulations, that this naive strategy, which we labelled strategy (i), is very far from being optimal.

One of the contributions of this chapter has been to show that the optimal design of FWL controllers is in fact much richer than the FWL filter design problem in that it lends itself to a range of ever more sophisticated design strategies, depending on how locally the optimization is being performed. We have proposed a classification of strategies, labelled (i) to (iv), where strategy (i) refers to an optimization at the most local level, that is optimization of the controller realization taken in isolation, and where strategy (iv) refers to a global optimization of the continuous-discrete closed loop system with respect to the set of all FWL controllers. Whereas in strategies (i) and (ii) the optimization is only with respect to the *realization* of a controller that has already been designed on the assumption of an infinite precision implementation, in strategy (iii) - and possibly (iv) - the controller *design* takes account of the finite wordlength of the controller, that is the optimization is performed with respect to all FWL controllers. In addition to this local-to-global hierarchy of strategies, the FWL controller design problem also allows for a wide choice of performance criteria by which different FWL controllers are judged and compared. These choices are to a certain extent guided by the application at hand, as we have explained in the introduction. Thus, in a regulation application it makes more sense to design a FWL regulator that minimizes the effects of *roundoff errors*, since it is the variance of the output signal (or of the output signal and the control signal in LQG control) that needs to be minimized, whereas the exact closed loop transfer function may not matter that much. On the other hand, in a tracking application one may want to have an *actual* closed loop system that behaves as closely as possible to some *ideal* one, in which case the effects of controller coefficient errors may be more important than roundoff error effects, and one may want to minimize some *sensitivity function* of the closed loop transfer function with respect to the controller parameters.

At the time of writing of this book, only some combinations of strategies and of performance criteria have been examined and resolved. In this chapter, we have presented some solution schemes:

- a model reference scheme where a separate analysis and controller design scheme has been proposed for closed loop sensitivity minimization and for closed loop roundoff noise gain minimization;

- an LQG regulation scheme where an optimal FWL controller design has been discussed for roundoff noise minimization.

The controller design schemes presented in this chapter minimize either some roundoff noise measure, or some sensitivity measure, but not both. Even though some intuitive arguments may sometimes suggest that one of the causes of FWL errors dominates the other in specific applications, as explained above, and even though simulations suggest that a controller realization that is designed to minimize the closed loop roundoff noise gain often has low closed loop transfer function sensitivity, this is not always the case. It is therefore interesting to develop a controller realization synthesis procedure that minimizes the combined effects of roundoff and coefficient errors in a synthetic way. Such combined analysis should also provide analysis tools to select the best distribution between coefficient and signal bit lengths. This is the topic of the next chapter.

Appendix 9: Computation of sensitivity functions of the closed loop system.

We derive the expressions (9.20)-(9.23) of the sensitivity functions of the closed loop transfer function $H_c(z)$ with respect to the controller coefficient matrices Φ, B, L and K, evaluated at the infinite precision values Φ_0, B_0, L_0 and K_0.

We recall that the realization of the closed loop system is (see (9.16)-(9.17)):

$$\bar{A} = \begin{pmatrix} A_0 & -B_0 K \\ LC_0 & \Phi \end{pmatrix}, \bar{B} = \begin{pmatrix} B_0 \\ B \end{pmatrix}, \bar{C} = \begin{pmatrix} C_0 & 0 \end{pmatrix} \quad (9.A.1)$$

and the closed loop transfer function is

$$H_c(z) = \bar{C}(zI - \bar{A})^{-1}\bar{B}. \quad (9.A.2)$$

Denote

$$F(z) \triangleq (zI - \bar{A})^{-1}\bar{B} = \begin{pmatrix} F_1(z) \\ F_2(z) \end{pmatrix}$$

$$G(z) \triangleq (zI - \bar{A}^T)^{-1}\bar{C}^T = \begin{pmatrix} G_1(z) \\ G_2(z) \end{pmatrix}. \quad (9.A.3)$$

Noting that

$$\frac{\partial (zI - \bar{A})^{-1}}{\partial x} = (zI - \bar{A})^{-1}\frac{\partial \bar{A}}{\partial x}(zI - \bar{A})^{-1},$$

one has

$$\frac{\partial H_c(z)}{\partial \Phi(i,j)} = \bar{C}\frac{\partial (zI - \bar{A})^{-1}}{\partial \Phi(i,j)}\bar{B}$$

$$= G^T(z)\begin{pmatrix} 0 & 0 \\ 0 & E_{ij} \end{pmatrix} F(z) = G_{2,i}(z)F_{2,j}(z),$$

where $E_{ij} \triangleq e_i e_j^T$ with e_k a column vector made up of zeros except for a 1 in the k-th position. It then follows that

$$\frac{\partial H_c(z)}{\partial \Phi} = G_2(z)F_2^T(z). \quad (9.A.4)$$

Similarly, one can show that

$$\frac{\partial H_c(z)}{\partial B} = G_2(z) \tag{9.A.5}$$

$$\frac{\partial H_c(z)}{\partial K^T} = -G_1^T(z) B_0 F_2(z) \tag{9.A.6}$$

$$\frac{\partial H_c(z)}{\partial L} = C_0 F_1(z) G_2(z). \tag{9.A.7}$$

Now, it is easy to verify the following equality

$$(zI - \bar{A}_0) = TMT^{-1}$$

with

$$M = \begin{pmatrix} M_1 & M_3 \\ 0 & M_2 \end{pmatrix}$$
$$\triangleq \begin{pmatrix} zI - (A_0 - B_0 K_0) & B_0 K_0 \\ 0 & zI - (A_0 - L_0 C_0) \end{pmatrix} \tag{9.A.8}$$
$$T = \begin{pmatrix} I & 0 \\ I & I \end{pmatrix}$$

by noting that

$$\Phi_0 = A_0 - B_0 K_0 - L_0 C_0.$$

Therefore,

$$(zI - \bar{A}_0)^{-1} = TM^{-1}T^{-1}$$
$$= \begin{pmatrix} M_1^{-1}(I + M_3 M_2^{-1}) & -M_1^{-1} M_3 M_2^{-1} \\ M_1^{-1}(I + M_3 M_2^{-1}) - M_2^{-1} & (I - M_1^{-1} M_3) M_2^{-1} \end{pmatrix} \tag{9.A.9}$$

and hence $F(z)$ and $G(z)$, evaluated at the infinite precision coefficient matrices Φ_0, B_0, L_0, K_0, are

$$F(z) = \begin{pmatrix} F_1(z) \\ F_2(z) \end{pmatrix} = \begin{pmatrix} M_1^{-1} B_0 \\ M_1^{-1} B_0 \end{pmatrix}$$
$$G(z) = \begin{pmatrix} G_1(z) \\ G_2(z) \end{pmatrix} = \begin{pmatrix} (M_1^{-1}(I + M_3 M_2^{-1}))^T C_0^T \\ -(M_1^{-1} M_3 M_2^{-1})^T C_0^T \end{pmatrix}.$$
$$\tag{9.A.10}$$

With M_i for $i = 1, 2, 3$ given by (9.A.8), one has

$$\begin{aligned}
F_1(z) &= F_2(z) = M_1^{-1} B_0 \\
&= (zI - (A_0 - B_0 K_0))^{-1} B_0 \triangleq F_{c2}(z) \\
G_1(z) &= (I + M_3 M_2^{-1})^T M_1^{-T} C_0^T \\
&= M_1^{-T} C_0^T + (zI - (A_0 - L_0 C_0))^{-T} K_0^T B_0^T M_1^{-T} C_0^T \\
G_2(z) &= -(zI - (A_0 - L_0 C_0))^{-T} K_0^T B_0^T (zI - (A_0 - B_0 K_0))^{-T} C_0^T \\
&= -C_0 (zI - (A_0 - B_0 K_0))^{-1} B_0 (zI - (A_0 - L_0 C_0))^{-T} K_0^T \\
&\triangleq -H_c^0(z) G_o(z).
\end{aligned}$$

(9.A.11)

It then follows from (9.A.4)-(9.A.7) that

$$\begin{aligned}
\frac{\partial H_c(z)}{\partial \Phi} &= -H_c^0(z) G_o(z) F_{c2}^T(z) \triangleq G_{o1}(z) F_{c2}^T(z) \\
\frac{\partial H_c(z)}{\partial B} &= -H_c^0(z) G_o(z) \triangleq G_{o1}(z) \\
\frac{\partial H_c(z)}{\partial K^T} &= -G_1^T(z) B_0 F_{c2}(z) = -H_c^0(z)(1 + H_{c0}(z)) F_{c2}(z) \\
&\triangleq F_{c4}(z) \\
\frac{\partial H_c(z)}{\partial L} &= C_0 F_{c2}(z) G_2(z) = -H_c^0(z) H_c^0(z) G_o(z) \triangleq G_{o3}(z)
\end{aligned}$$

(9.A.12)

where $H_c^0(z)$ is the closed loop transfer function evaluated at the infinite precision realization, while $H_{c0}(z)$ is defined in (9.26). The quantities $G_{o1}(z)$, $F_{c2}(z)$, $G_{o3}(z)$, $F_{c4}(z)$, $G_o(z)$ have all been defined in (9.20)-(9.25). We have thus demonstrated the formulæ (9.20)-(9.23).

10

Synthetic FWL compensator design

10.1 Introduction

In the previous chapter, we have introduced the FWL controller realization problem and we have shown that several features distinguish this problem from the more classical FWL filter realization problem. To make this distinction more transparent, we have introduced a hierarchy of design strategies, denoted strategies (i) to (iv), and in the conclusion of Chapter 9 we have distinguished between optimal FWL *controller realization* and optimal FWL *controller design*. Depending on the combination of performance criterion and optimization strategy, a wide variety of optimal design schemes can be envisaged, and so far only a few have been studied. Four of these have been presented in Chapter 9:

- an optimal controller *realization* scheme for the minimization of the effects of controller coefficient errors in pole placement controllers (in subsection 9.2.3);

- an optimal controller *realization* scheme for the minimization of the effects of signal roundoff errors in pole placement controllers (in subsection 9.2.4);

- an optimal controller *realization* scheme for the minimization of the effects of signal roundoff errors in LQG regulators (in subsection 9.3.3);

- an optimal controller *design* scheme for the minimization of the effects of signal roundoff errors in LQG regulators (in subsection 9.3.4).

In these four optimal schemes, the performance criterion incorporates either the effects of coefficient errors, or those of roundoff errors, but never both. In the present chapter, we build on the assets of Chapter 7 to construct a synthetic optimal controller realization scheme that incorporates the combined effects of coefficient errors and of signal roundoff errors. Thus, our performance criterion is more general than any of those of Chapter 9. On the other hand, our strategy is clearly an *optimal realization* strategy (i.e. strategy (ii)) rather than an *optimal design* strategy. Thus, we assume throughout this chapter that an ideal (that is infinite precision controller) has been computed on the basis of some design criterion, and that the task is to optimize the *implementation* of this controller in order to minimize some measure of performance degradation of the ideal closed loop system due to FWL effects. This approach is suboptimal in comparison with one that would directly optimize over the set of FWL implemented controllers, as in strategy (iii). However, the direct optimization over the set of FWL controllers is presently not possible for all design criteria, even though we have discussed such scheme for the case of an LQG regulator in the previous chapter. On the other hand, the synthetic controller design that we present in this chapter has the advantage that, just like the synthetic filter design of Chapter 7, it allows for clear design choices for the respective bitlengths of coefficients and signals in the controller implementation.

In addition to a more general performance criterion, our synthetic design is also applicable to any general control law, encompassing pole placement, LQG, H_∞, or any other control design scheme as a special case. Indeed, our starting point is a general two-degree-of-freedom controller[1], obtained as the result of any infinite precision design scheme.

To obtain a synthetic performance measure, we shall again adopt a statistical model for the coefficient errors, a concept pioneered by Knowles and Olcayto [KO68] and Crochiere [Cro75] in the FWL filter analysis context (see our application of this statistical model for design in Chapter 7), and later extended to the FWL controller analysis context by Moroney, Willsky and Houpt [MWH80]. Our performance criterion will then be in the form of an output error variance which expresses the performance degradation due to all FWL effects in the compensator. After normalization, this criterion can be seen as a closed loop gain. One contribution of this chapter is to show that this gain is the sum of a closed loop L_2 sensitivity measure and of

---

[1]To the control 'ignorandus', we apologize for this abrupt introduction of a perhaps weird sounding expression. The concept of a two-degree-of-freedom controller will be made clear in the next section.

a closed loop roundoff noise gain. Another contribution is to find the optimal compensator realizations in terms of minimizing this closed loop performance criterion under the usual dynamic range constraint on the compensator states. The necessary and sufficient optimality conditions are derived and an algorithm is given, which leads to an optimal compensator realization. The superiority of this optimal realization is illustrated with a numerical example.

An outline of this chapter is as follows. In Section 10.2 we present the state space description of a closed loop system for a general two-degree-of-freedom controller and we illustrate the difference between controller sensitivity minimization and closed loop sensitivity minimization by examining the relationships between controller and closed loop sensitivities. This serves as further motivation for performing the FWL controller realization problem in closed loop. Our first new contribution is in Section 10.3, where we give a synthetic analysis of the FWL effects on the desired plant output and derive a computable expression of what we call the Total Noise Gain, which is a measure of the degradation of the closed loop system. In Section 10.4 we formulate and solve the optimal compensator realization problem. Finally a computational example in Section 10.5 confirms the superiority of this optimal FWL compensator realization. Some concluding remarks are given in Section 10.6.

10.2 State space description of a compensator

In this section we describe a two-degree-of-freedom controller in state space form and discuss the FWL design options in general terms.

Thus, we assume from now on that some design procedure (e.g. model reference, or LQG, or H_∞) has led to the specification of a two-degree-of-freedom controller $(C_1(z), C_2(z))$ as represented in Figure 10.1. Here $C_1(z)$ and $C_2(z)$ denote the ideal (infinite precision) controller transfer functions obtained as a result of the control design calculations, while $H_0(z)$ denotes the exact (infinite precision) model transfer function.

The closed loop transfer function from $r(t)$ to $y(t)$ is

$$H_c(z) = \frac{C_1(z)H_0(z)}{1 + H_0(z)C_2(z)}. \qquad (10.1)$$

The two-degree-of-freedom controller of Figure 10.1 originates as the result of several controller design schemes. For example, an LQG

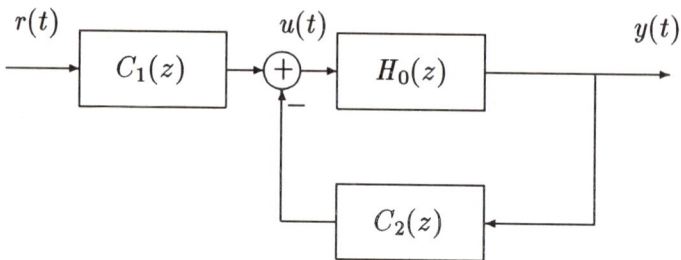

FIGURE 10.1. Closed loop system with a two-degree-of-freedom controller

tracking problem results in a two-degree of freedom controller (see e.g. [BGW90]) with

$$C_1(z) = [1 - K(zI - A_0)^{-1}B_0]^{-1}$$
$$C_2(z) = K(zI - A_0 + LC_0 + B_0K)^{-1}L,$$

where (A_0, B_0, C_0) is an exact realization of the plant $H_0(z)$, and K and L are the corresponding optimal control gain and optimal Kalman filter gain, respectively, both of which are obtained by solving a Riccati equation (see Chapter 9).

Alternatively, a polynomial design of a model reference controller with $H_0(z) = B_0(z)/A_0(z)$ also leads to a two-degree-of-freedom controller with $C_1(z) = T(z)/R(z)$ and $C_2(z) = S(z)/R(z)$ (see [ÅW84]). If $H_m(z) = B_m(z)/A_m(z)$ is the desired closed loop transfer function, then the controller polynomials $T(z), R(z)$ and $S(z)$ are obtained as the solution of a Diophantine equation in such a way that the closed loop transfer function is given by

$$H_c(z) = \frac{B_0(z)T(z)}{A_0(z)R(z) + B_0(z)S(z)} = \frac{B_m(z)}{A_m(z)}.$$

For our analysis of FWL effects, it is useful to write a state space description of the closed loop system of Figure 10.1. Even though the plant is not implemented in a computer, it will be useful to think of $H_0(z)$ as being implemented by an infinite precision state variable realization (A_0, B_0, C_0, d_0) in some coordinate system:

$$\begin{aligned} x(t+1) &= A_0 x(t) + B_0 u(t) \\ y(t) &= C_0 x(t) + d_0 u(t) \end{aligned} \quad (10.2)$$

where

$$H_0(z) = d_0 + C_0(zI - A_0)^{-1}B_0. \quad (10.3)$$

10. Synthetic FWL compensator design

As for the compensator, it is a multiple input single output (MISO) system which can be rewritten as

$$u(t) = C_1(z)r(t) - C_2(z)y(t) \triangleq C(z) \begin{pmatrix} r(t) \\ y(t) \end{pmatrix}. \quad (10.4)$$

The implementation of the compensator corresponds to a particular state space realization (A_r, B_r, C_r, D_r):

$$x_r(t+1) = A_r x_r(t) + B_r \begin{pmatrix} r(t) & y(t) \end{pmatrix}^T \quad (10.5)$$

$$u(t) = C_r x_r(t) + D_r \begin{pmatrix} r(t) & y(t) \end{pmatrix}^T \quad (10.6)$$

where $A_r \in \mathbb{R}^{m \times m}, B_r \in \mathbb{R}^{m \times 2}, C_r \in \mathbb{R}^{1 \times m}, D_r \in \mathbb{R}^{1 \times 2}$, m is the order of $C(z)$,

$$B_r = \begin{pmatrix} B_1 & B_2 \end{pmatrix}, D_r = \begin{pmatrix} d_1 & 0 \end{pmatrix} \quad (10.7)$$

and

$$\begin{aligned} C(z) &= D_r + C_r(zI - A_r)^{-1} B_r \\ &= \begin{pmatrix} C_1(z) & -C_2(z) \end{pmatrix}. \end{aligned} \quad (10.8)$$

The actual implementation of the compensator is a FWL implementation of the realization (10.5)-(10.6).

Comments:

1. In order for a closed loop system to be 'well-posed', there needs to be a delay somewhere in the loop. This prevents the presence of 'algebraic loops' in which information circulates around the closed loop infinitely fast. This means that either the plant or the controller needs to have a delay. Without loss of generality, we have assumed a delay in the controller here. We could also have taken $d_0 = 0$ and $D_r = (d_1 \quad d_2)$. This would of course slightly alter the closed loop formulae below.

2. The two control schemes studied in Chapter 9 were state estimate feedback controllers, so that the state, x_r, of the controller was in fact an estimate of the state of the system. Here we study general dynamic output feedback controllers, of which observer based controllers are a special case.

It is easy to understand that there is a set of equivalent realizations for this compensator, which is given by

$$S_c = \{(A_r, B_r, C_r, D_r) : C(z) = D_r + C_r(zI - A_r)^{-1}B_r\}.$$

Clearly, if (A_r, B_r, C_r, D_r) belongs to S_c, so does $(T^{-1}A_rT, T^{-1}B_r, C_rT, D_r)$ for any similarity transformation matrix T. The optimal FWL state space compensator realization problem is to identify those realizations in the set S_c which optimize some specified performance criterion on the closed loop system in the face of FWL errors. For reasons that have already been amply justified, this performance criterion should be on the closed loop system rather than on the compensator seen as a filter.

The state model of the closed loop system is given by

$$\begin{aligned} \bar{x}(t+1) &= \bar{A}\bar{x}(t) + \bar{B}r(t) \\ y(t) &= \bar{C}\bar{x}(t) + \bar{d}r(t) \end{aligned} \quad (10.9)$$

where $\bar{A}$, $\bar{B}$, $\bar{C}$ and $\bar{d}$ are the following functions of (A_0, B_0, C_0, d_0) and (A_r, B_r, C_r, D_r):

$$\bar{A} \triangleq \begin{pmatrix} A_0 & B_0 C_r \\ B_2 C_0 & A_r + d_0 B_2 C_r \end{pmatrix}, \quad \bar{B} \triangleq \begin{pmatrix} d_1 B_0 \\ B_1 + d_0 d_1 B_2 \end{pmatrix},$$

$$\bar{C} \triangleq \begin{pmatrix} C_0 & d_0 C_r \end{pmatrix}, \quad \bar{d} \triangleq d_0 d_1.$$

The ideal closed loop transfer function is then given by

$$H_c(z) = \bar{d} + \bar{C}(zI - \bar{A})^{-1}\bar{B}. \quad (10.10)$$

In practice, the compensator (10.5)-(10.6) is implemented with Finite Word Length. Thus the ideal realization (A_r, B_r, C_r, D_r) of the compensator is replaced in (10.10) by its FWL version $(A_r^*, B_r^*, C_r^*, D_r^*)$. In addition, the input signal of the controller, $y(t)$, is quantized in the A/D converter, and the state and output signals of the controller, $x_r(t)$ and $u(t)$, are rounded off. As a result, the actual control signal and hence the actual output of the plant differ from the desired ones: this causes a degradation of the performance of the closed loop system with respect to the designed performance.

Since the closed loop transfer function is a function of the realization (A_r, B_r, C_r, D_r) of the compensator, this performance degradation is a function of the particular realization of the compensator, that is it depends on the choice of coordinates used for this state space realization.

10. Synthetic FWL compensator design

Once again, the simple strategy would be to optimize the realization of the compensator in such a way as to minimize its roundoff noise gain, some sensitivity measure of its transfer function, or better still some synthetic measure of its FWL performance degradation. For example, if an L_1/L_2 sensitivity measure is used, this would lead one to implement the controller in balanced form. But this is clearly suboptimal, since the quantity of interest is the closed loop transfer function. To illustrate the connection between the minimization of the sensitivity of the controller transfer function and that of the closed loop transfer function, we establish the relationships between their respective sensitivity functions. Recall that the closed loop transfer function $H_c(z)$ has been defined in (10.1), where $C_1(z)$ and $C_2(z)$ are the following functions of the controller realization (see (10.7)-(10.8)):

$$C_1(z) = d_1 + C_r(zI - A_r)^{-1} B_1$$

and

$$C_2(z) = -C_r(zI - A_r)^{-1} B_2.$$

This immediately leads to the relationships between closed loop and controller sensitivities:

$$\begin{aligned}
\frac{\partial H_c}{\partial A_r} &= H_0 \left[\frac{\partial C_1}{\partial A_r} \frac{1}{1 + H_0 C_2} - \frac{C_1 H_0}{(1 + H_0 C_2)^2} \frac{\partial C_2}{\partial A_r} \right] \\
&\triangleq W_1(z) \frac{\partial C_1(z)}{\partial A_r} + W_2(z) \frac{\partial C_2(z)}{\partial A_r} \\
\frac{\partial H_c}{\partial B_1} &= \frac{\partial C_1}{\partial B_1} \frac{H_0}{1 + H_0 C_2} \triangleq W_1(z) \frac{\partial C_1(z)}{\partial B_1} \\
\frac{\partial H_c}{\partial B_2} &= -\frac{H_0 C_1}{(1 + H_0 C_2)^2} H_0 \frac{\partial C_2}{\partial B_2} \triangleq W_2(z) \frac{\partial C_2(z)}{\partial B_2} \\
\frac{\partial H_c}{\partial C_r} &= W_1(z) \frac{\partial C_1(z)}{\partial C_r} + W_2(z) \frac{\partial C_2(z)}{\partial C_r} \\
\frac{\partial H_c}{\partial d_1} &= W_1(z) \frac{\partial C_1(z)}{\partial d_1},
\end{aligned} \tag{10.11}$$

where

$$\begin{aligned}
W_1(z) &\triangleq \frac{H_0}{1 + H_0 C_2} = \frac{H_c(z)}{C_1(z)}, \\
W_2(z) &\triangleq -\frac{C_1 H_0^2}{(1 + H_0 C_2)^2} = -\frac{H_c^2(z)}{C_1(z)}.
\end{aligned} \tag{10.12}$$

One now observes that the sensitivity functions of the closed loop transfer function with respect to the controller parameters are linear combinations of weighted sensitivity functions of the two controller transfer functions with respect to these same parameters. These frequency weightings, $W_1(z)$ and $W_2(z)$, both contain the quantity $\frac{1}{1+H_0 C_2}$ as a factor. In control, this quantity plays a fundamental role and is known as the sensitivity of the closed loop. It expresses, among other things, how much a relative change (or error) in the open loop frequency response $H_0(z)$ at some frequency is reflected in a corresponding relative change of the closed loop frequency response $H_c(z)$. The presence of this sensitivity function in W_1 and W_2 means that, as is to be expected, the effects of controller transfer function errors on the closed loop transfer function are emphasized in the frequency ranges where the closed loop sensitivity is high, that is around the closed loop crossover frequency. On the other hand, these same controller transfer function errors do not matter as much in the frequency ranges where the closed loop sensitivity is low, as is typically the case at low frequencies.

The expressions (10.11) also clearly indicate that optimizing a measure of the controller sensitivities will not, in general, optimize the closed loop transfer function sensitivity measure. Therefore, we argue again here that it is natural and reasonable to carry out the optimal FWL compensator implementation design in terms of optimizing a closed loop performance rather than an open loop performance.

Classically, the FWL effects due to signal roundoff in the controller and to controller coefficient errors are separately analysed. This can not describe what happens in an actual implementation where these two FWL errors appear simultaneously. In the next section, we follow the approach developed in Chapter 7 for filter design, and adapt it to our present controller design situation. We define a unified measure of all FWL effects of the controller implementation on the closed loop system. In Section 10.4 we then show that this measure can actually be minimized subject to a normalizing l_2 constraint on the controller.

10.3 Analysis of FWL effects of a compensator

In this section we analyse the performance degradation of the closed loop system due to a FWL implementation of the controller. With (A_r, B_r, C_r, D_r) replaced by its FWL version $(A_r^*, B_r^*, C_r^*, D_r^*)$, (10.5)-(10.6) becomes

$$x_r^*(t+1) = A_r^* x_r^*(t) + B_r^* \begin{pmatrix} r(t) & y^*(t) \end{pmatrix}^T \quad (10.13)$$

$$u^*(t) = C_r^* x_r^*(t) + D_r^* \begin{pmatrix} r(t) & y^*(t) \end{pmatrix}^T \quad (10.14)$$

where $y^*(t)$ is the output $y(t)$ of the plant with $u(t)$ replaced by $u^*(t)$. The closed loop system is then given by

$$\begin{pmatrix} x^*(t+1) \\ x_r^*(t+1) \end{pmatrix} = \bar{A}^* \begin{pmatrix} x^*(t) \\ x_r^*(t) \end{pmatrix} + \bar{B}^* r(t)$$

$$y^*(t) = \bar{C}^* \begin{pmatrix} x^*(t) \\ x_r^*(t) \end{pmatrix} + \bar{d}^* r(t) \quad (10.15)$$

where $\bar{A}^*, \bar{B}^*, \bar{C}^*$ and $\bar{d}^*$ are the closed loop system matrices $\bar{A}, \bar{B}, \bar{C}$ and $\bar{d}$ given by (10.10) but with (A_r, B_r, C_r, D_r) replaced by its FWL version $(A_r^*, B_r^*, C_r^*, D_r^*)$.

When, additionally, the roundoff in arithmetic operations is also considered, the above model has to be modified. A fixed point implementation with *Roundoff Before Multiplication* (RBM) leads to the following actual computation model of the compensator:

$$x_r'(t+1) = A_r^* Q[x_r'(t)] + B_r^* \begin{pmatrix} Q[r(t)] & Q[y'(t)] \end{pmatrix}^T \quad (10.16)$$

$$u'(t) = C_r^* Q[x_r'(t)] + D_r^* \begin{pmatrix} Q[r(t)] & Q[y'(t)] \end{pmatrix}^T \quad (10.17)$$

where $y'(t)$ is the actual output of the plant with $u(t)$ replaced by $Q[u'(t)]$. As usual, $Q[\,.\,]$ denotes the quantization operation, the signals are assumed to be implemented with B_s bits, while the parameters in $(A_r^*, B_r^*, C_r^*, D_r^*)$ are implemented with B_c bits.

We denote the roundoff errors by

$$\begin{array}{ll} e_x(t) = x_r'(t) - Q[x_r'(t)], & e_r(t) = r(t) - Q[r(t)] \\ e_y(t) = y'(t) - Q[y'(t)], & e_u(t) = u'(t) - Q[u'(t)]. \end{array} \quad (10.18)$$

As explained in previous chapters, these error processes can all be modelled as independent white noise having zero mean and a variance $\sigma_n^2 = (1/12)2^{-2B_s}$. Using (10.18), we can rewrite (10.16)-(10.17) as

$$x_r'(t+1) = A_r^* x_r'(t) + B_r^* \begin{pmatrix} r(t) & y'(t) \end{pmatrix}^T$$

$$u'(t) = \begin{array}{l} -A_r^* e_x(t) - B_r^* \begin{pmatrix} e_r(t) & e_y(t) \end{pmatrix}^T \quad (10.19) \\ C_r^* x_r'(t) + D_r^* \begin{pmatrix} r(t) & y'(t) \end{pmatrix}^T \\ -C_r^* e_x(t) - D_r^* \begin{pmatrix} e_r(t) & e_y(t) \end{pmatrix}^T. \quad (10.20) \end{array}$$

Now, the degradation of the output of the closed loop system due to a FWL implementation of the compensator and to a quantization of the signals in the compensator can be measured by the difference between the desired and the actual output of the plant:

$$\begin{aligned} \Delta y(t) &\triangleq y(t) - y'(t) = [y(t) - y^*(t)] + [y^*(t) - y'(t)] \\ &\triangleq \Delta y^*(t) + \Delta y'(t). \end{aligned} \quad (10.21)$$

Just as in Chapter 7, the overall output error process $\Delta y(t)$ can be separated into two processes $\Delta y^*(t) = y(t) - y^*(t)$ and $\Delta y'(t) = y^*(t) - y'(t)$, which stand for the errors on $y(t)$ due to the FWL effects on coefficients and arithmetic operations, respectively.

Note that $y^*(t)$ is the output of the plant when the control signal $u^*(t)$ is computed with (10.13)-(10.14). It is thus produced by the closed loop system (10.15). As for $y'(t)$, it is the plant output when the actual input to the plant is $Q[u'(t)]$ with $u'(t)$ computed from (10.19)-(10.20):

$$\begin{aligned} x'(t+1) &= A_0 x'(t) + B_0 Q[u'(t)] \\ y'(t) &= C_0 x'(t) + d_0 Q[u'(t)] \end{aligned} \quad (10.22)$$

The closed loop system is then given by

$$\begin{aligned} \begin{pmatrix} x'(t+1) \\ x_r'(t+1) \end{pmatrix} &= \bar{A}^* \begin{pmatrix} x'(t) \\ x_r'(t) \end{pmatrix} + \bar{B}^* r(t) - N(t) \\ y'(t) &= \bar{C}^* \begin{pmatrix} x'(t) \\ x_r'(t) \end{pmatrix} + \bar{d}^* r(t) - M(t) \end{aligned} \quad (10.23)$$

where

$$N(t) \triangleq \begin{pmatrix} B_0[C_r^* e_x(t) + e_u(t) + d_1^* e_r(t)] \\ \\ (A_r^* + d_0 B_2^* C_r^*) e_x(t) + B_2^* e_y(t) \\ + d_0 B_2^* e_u + (B_1^* + d_0 d_1^* B_2^*) e_r(t) \end{pmatrix}$$

$$M(t) \triangleq d_0[C_r^* e_x(t) + d_1^* e_r(t) + e_u(t)]. \quad (10.24)$$

10. Synthetic FWL compensator design

Denote
$$E_x(t) \triangleq \begin{pmatrix} x^*(t) \\ x_r^*(t) \end{pmatrix} - \begin{pmatrix} x'(t) \\ x_r'(t) \end{pmatrix}.$$

It follows from (10.15) and (10.23) that

$$E_x(t+1) = \bar{A}^* E_x(t) + N(t) \qquad (10.25)$$
$$\Delta y'(t) = \bar{C}^* E_x(t) + M(t). \qquad (10.26)$$

These state equations will allow us to compute the variance of $\Delta y'(t)$.

Now, consider the term $\Delta y^*(t)$. Let $\{h_i\}$ and $\{h_i^*\}$ be the impulse responses of the closed loop system with infinite and finite coefficient implementation of the compensator, respectively. In addition, let $\{p_i\}$ denote the theoretical parameters of the realization of the compensator and $\{p_i^*\}$ their FWL version with $p_i = p_i^* + \Delta p_i$. Clearly,

$$h_i - h_i^* = \frac{1}{2\pi j} \oint_{|z|=1} [H_c(z) - H_c^*(z)] z^{i-1} dz. \qquad (10.27)$$

With a first order approximation, one has

$$\Delta H_c(z) = H_c(z) - H_c^*(z) = \sum_i \frac{\partial H_c}{\partial p_i} \Delta p_i. \qquad (10.28)$$

Therefore, the error $\Delta y^*(t)$ can be written as a function of $\{\Delta p_i\}$ and r as follows:

$$\Delta y^*(t) \triangleq y(t) - y^*(t) = \sum_{k=0}^{t-1} [h_k - h_k^*] r(t-k)$$
$$= \sum_{k=0}^{t-1} [\sum_i \frac{1}{2\pi j} \oint_{|z|=1} \frac{\partial H_c}{\partial p_i} z^{k-1} dz \Delta p_i] r(t-k). \qquad (10.29)$$

We adopt the same stochastic assumption on the coefficient errors as in Chapter 7, that is the coefficient errors $\{\Delta p_i\}$ are assumed to be uniformly distributed zero mean uncorrelated random variables. The roundoff errors e_x, e_y, e_u and e_r are also assumed to be uniformly distributed zero mean uncorrelated random variables, uncorrelated with the coefficient errors. Finally, for simplicity, $r(t)$ is taken to be unit variance white noise. The steady state variance of the total plant output error, $\Delta y(t)$, is then the sum of the variances of $\Delta y^*(t)$ and $\Delta y'(t)$:

$$\sigma^2 = \lim_{t \to \infty} E[\Delta y(t)^2] = \sigma_1^2 + \sigma_2^2 \qquad (10.30)$$

where
$$\sigma_1^2 = \lim_{t\to\infty} E[\Delta y^*(t)^2], \quad \sigma_2^2 = \lim_{t\to\infty} E[\Delta y'(t)^2].$$

Computation of σ_1^2

We first compute the variance of the error due to coefficient wordlength errors, $\sigma_1^2 = \lim_{t\to\infty} E[\Delta y^*(t)^2]$. Assuming that the $\{\Delta p_i\}$ all have variance $\sigma_c^2 = \frac{1}{12}2^{-2B_c}$, and that they are independent of the input signal $r(t)$, itself a unit variance white noise, then

$$\sigma_1^2 = \lim_{t\to\infty} E[\Delta y^*(t)^2] = \sum_{i=1}^{N_0}\left\{\frac{1}{2\pi j}\oint_{|z|=1}\left(\frac{\partial H_c}{\partial p_i}\right)\left(\frac{\partial H_c}{\partial p_i}\right)^H z^{-1}dz\right\}\sigma_c^2 \tag{10.31}$$

where N_0 is the number of parameters of finite wordlength in the compensator realization (A_r, B_r, C_r, D_r). To compute (10.31), one needs the sensitivity functions $\{\frac{\partial H_c}{\partial p_i}\}$ of the closed loop transfer function with respect to the controller parameters. These computations have already been performed in (10.11). However, these expressions do not lend themselves easily to the computation of the corresponding frequency independent measures. Starting from (10.10), and after some manipulations, one obtains the following sensitivity matrix functions:

$$\begin{aligned}\frac{\partial H_c}{\partial A_r} &= G_2(z)F_2^T(z) \\ \frac{\partial H_c}{\partial B_1} &= G_2(z), \quad \frac{\partial H_c}{\partial B_2} = H_c(z)G_2(z) \\ \frac{\partial H_c}{\partial C_r^T} &= \{d_0 + G^T(z)[B_0^T \ d_0 B_2^T]^T\}F_2(z) \\ \frac{\partial H_c}{\partial d_1} &= \frac{H_0(z)}{1 + H_0(z)C_2(z)},\end{aligned} \tag{10.32}$$

where

$$\begin{aligned}F(z) &= \left(F_1^T(z) \ F_2^T(z)\right)^T = (zI - \bar{A})^{-1}\bar{B} \\ G^T(z) &= \left(G_1^T(z) \ G_2^T(z)\right) = \bar{C}(zI - \bar{A})^{-1}.\end{aligned} \tag{10.33}$$

So, for a given compensator realization with N_0 nontrivial parameters, σ_1^2 can be computed using (10.31)-(10.32). Assuming that the compensator realization is fully parametrized, then (10.31) can be written as follows:

$$\begin{aligned}\sigma_1^2 &= [\|\frac{\partial H_c}{\partial A_r}\|_2^2 + \|\frac{\partial H_c}{\partial B_1}\|_2^2 + \|\frac{\partial H_c}{\partial B_2}\|_2^2 \\ &\quad + \|\frac{\partial H_c}{\partial C_r^T}\|_2^2 + \|\frac{\partial H_c}{\partial d_1}\|_2^2]\sigma_c^2 \\ &= \{tr[W_{A_r} + W_{B_1} + W_{B_2} + W_{C_r}] + W_{d_1}\}\sigma_c^2 \end{aligned}$$
(10.34)

where

$$W_X = \frac{1}{2\pi j}\oint_{|z|=1}\left(\frac{\partial H_c}{\partial X}\right)\left(\frac{\partial H_c}{\partial X}\right)^H z^{-1}dz \qquad (10.35)$$

with $X = A_r, B_1, B_2, C_r$ and d_1. We note that W_{d_1} is a coordinate independent scalar quantity. It is easy to see that the matrices W_{B_1}, W_{B_2} and W_{C_r} can each be computed as a submatrix of a weighted Gramian of the closed loop system with a particular weighting function. For example, if M is the observability Gramian of the closed loop system weighted by $H_c(z)$, then its 2×2 block submatrix $M(2,2)$ corresponding to the partition in (10.33) is W_{B_2}.

As to W_{A_r}, it can be computed in the following way. Denote

$$F(z) = \begin{pmatrix} F_1^T(z) & F_2^T(z) \end{pmatrix}^T = (zI - \bar{A})^{-1}\bar{B} = \sum_{i=0}^{\infty} \bar{f}(i)z^{-(i+1)}$$

and

$$G^T(z) = \begin{pmatrix} G_1^T(z) & G_2^T(z) \end{pmatrix} = \bar{C}(zI - \bar{A})^{-1} = \sum_{i=0}^{\infty} \bar{g}^T(i)z^{-(i+1)}$$

where

$$\bar{f}(i) \triangleq \begin{pmatrix} \bar{f}_1^T(i) & \bar{f}_2^T(i) \end{pmatrix}^T = \bar{A}^i\bar{B},$$
$$\bar{g}(i) \triangleq \begin{pmatrix} \bar{g}_1^T(i) & \bar{g}_2^T(i) \end{pmatrix}^T = (\bar{A}^i)^T\bar{C}^T.$$

Therefore,

$$\frac{\partial H_c}{\partial A_r} = G_2(z)F_2^T(z) = \sum_{i,j\geq 0}\bar{g}_2(i)\bar{f}_2^T(j)z^{-(i+j+2)} = \sum_{k=0}^{\infty}H(k)z^{-(k+2)}$$

where

$$H(k) \triangleq \sum_{i+j=k}\bar{g}_2(i)\bar{f}_2^T(j). \qquad (10.36)$$

274   10. Synthetic FWL compensator design

Thus, one has

$$W_{Ar} = \sum_{k=0}^{\infty} H(k)H^T(k). \qquad (10.37)$$

Finally, W_{d_1} can easily be computed with residue theory, but it is coordinate independent.

Computation of σ_2^2

We now compute the second term $\sigma_2^2 = \lim_{t\to\infty} E[\Delta y'(t)^2]$. From (10.25), one gets the following expression for the state error, assuming $E(0) = 0$:

$$E_x(t) = \sum_{i=0}^{t-1} (\bar{A}^*)^i N(t-1-i).$$

Since

$$N(t) = \begin{pmatrix} B_0 C_r^* & 0 & B_0 & d_1^* B_0 \\ A_r^* + d_0 B_2^* C_r^* & B_2^* & d_0 B_2^* & B_1^* + d_0 d_1^* B_2^* \end{pmatrix} \begin{pmatrix} e_x(t) \\ e_y(t) \\ e_u(t) \\ e_r(t) \end{pmatrix}$$

$$\triangleq PE(t),$$

$\Delta y'(t)$ can be expressed as

$$\Delta y'(t) = \bar{C}^* \sum_{i=0}^{t-1} (\bar{A}^*)^i PE(t-1-i) + d_0 \begin{pmatrix} C_r^* & 0 & 1 & d_1^* \end{pmatrix} E(t).$$

Since all roundoff errors have been assumed uncorrelated, we then have $E\{E(t)E^T(\tau)\} = \sigma_n^2 I \delta_{t\tau}$. Thus, the steady state output error variance becomes

$$\sigma_2^2 = \{tr[P^T W_o P] + d_0^2 [C_r^* (C_r^*)^T + 1 + (d_1^*)^2]\} \sigma_n^2 \qquad (10.38)$$

where W_o is the observability Gramian of the FWL closed loop system:

$$W_o \triangleq \sum_{i=0}^{\infty} (\bar{A}^{*T})^i (\bar{C}^*)^T \bar{C}^* (\bar{A}^*)^i = \begin{pmatrix} W_o(1,1) & W_o(1,2) \\ W_o(2,1) & W_o(2,2) \end{pmatrix}. \qquad (10.39)$$

With some manipulations, the variance in (10.38) can be separated into two parts:

$$\sigma_2^2 = \{tr(W) + \eta^2\} \sigma_n^2 \qquad (10.40)$$

where $tr(W)$ is the coordinate dependent part,

$$\begin{aligned}tr(W) &= [d_0^2 + B_0^T W_o(1,1) B_0] C_r^* (C_r^*)^T \\ &+ tr\{2(C_r^*)^T B_0^T W_o(1,2)[A_r^* + d_0 B_2^* C_r^*] \\ &+ (A_r^* + d_0 B_2^* C_r^*)^T W_o(2,2)(A_r^* + d_0 B_2^* C_r^*)\}, \end{aligned} \quad (10.41)$$

while η^2 is coordinate independent,

$$\begin{aligned}\eta^2 &\triangleq (1 + (d_1^*)^2)[d_0^2 + B_0^T W_o(1,1) B_0] \\ &+ 2[1 + (d_1^*)^2] d_0 B_0^T W_o(1,2) B_2^* + 2 d_1^* B_0^T W_0(1,2) B_1^* \\ &+ [1 + (1 + (d_1^*)^2) d_0^2] (B_2^*)^T W_o(2,2) B_2^* \\ &+ 2 d_0 d_1^* (B_1^*)^T W_0(2,2) B_2^* + (B_1^*)^T W_o(2,2) B_1^*. \end{aligned} \quad (10.42)$$

If an initial compensator realization (A_r, B_r, C_r, D_r) yields W^0 for the matrix W of (10.41), then a realization obtained from (A_r, B_r, C_r, D_r) by a similarity transformation matrix T applied to the compensator only will correspond to $W = T^T W^0 T$. We have now done all the footwork towards the computation of the total output error variance σ^2 due to both coefficient errors and roundoff of the signals in the compensator:

$$\sigma^2 = \{tr(W) + \eta^2\}\sigma_n^2 + \{tr[W_{A_r} + W_{B_1} + W_{B_2} + W_{C_r}] + W_{d_1}\}\sigma_c^2. \quad (10.43)$$

With a given compensator realization, one can evaluate the degradation of the closed loop performance due to the FWL implementation of the compensator, as measured by the variance σ^2 of $\Delta y(t)$. Recall that σ_n^2 and σ_c^2 are the variances of the roundoff noise and of the statistical coefficient random noise; they depend on the signal and coefficient wordlengths, B_s and B_c, respectively. In line with our machinations of Chapter 7, we can define a *Total Noise Gain*, $G_{cl,T}$, of the closed loop system due to FWL effects in the compensator:

$$G_{cl,T} = \{tr[W_{A_r} + W_{B_1} + W_{B_2} + W_{C_r}] + W_{d_1}\} + \rho^2\{tr[W] + \eta^2\}, \quad (10.44)$$

where $\rho^2 \triangleq \sigma_n^2/\sigma_c^2 = 2^{2(B_c - B_s)}$ depends on the wordlengths of signals and coefficients. The Total Noise Gain is a function of the particular coordinates in which the compensator is represented via the terms W_{A_r}, W_{B_1}, W_{B_2}, W_{C_r} and W.

Comment:

The Total Noise Gain $G_{cl,T}$ can be rewritten in short as

$$G_{cl,T} = M_{cl,L_2} + \rho^2 G_{cl}, \qquad (10.45)$$

where M_{cl,L_2} is the L_2 sensitivity measure of the *closed loop transfer function* with respect to the parameters in a compensator realization while G_{cl} is the classical roundoff noise gain of the *closed loop system* when the computations in the compensator are performed with finite precision. In classical FWL effect analysis, these two effects are studied separately. Since in a FWL implementation of a compensator these two effects act simultaneously, our global measure obviously gives a better account of the total FWL error effects in the closed loop system. The discussion we have had in Chapter 7 concerning the respective weights of M_{cl,L_2} and $\rho^2 G_{cl}$ in the total gain, and the help this provides in selecting the respective bit lengths, apply of course to our controller realization design case in exactly the same way.

The most interesting use we can make of the expression (10.44) of the gain $G_{cl,T}$ is to minimize it with respect to all equivalent realizations (A_r, B_r, C_r, D_r) of the compensator. In order to minimize the probability of overflow, the components of the state $x'_r(t)$ in (10.16) must be maintained within the same dynamic range. This is achieved by the now familiar method of l_2 scaling. If

$$W_c = \begin{pmatrix} W_c(1,1) & W_c(1,2) \\ W_c(2,1) & W_c(2,2) \end{pmatrix} = \sum_{i=0}^{\infty} \bar{A}^i \bar{B} \bar{B}^T (\bar{A})^i \qquad (10.46)$$

is the controllability Gramian of the closed loop system (10.10), then this l_2-scaling on the state $x'_r(t)$ of the compensator can approximately be achieved with the following constraint:

$$(W_c(2,2))_{ii} = 1 \text{ for all } i. \qquad (10.47)$$

In the next section, we formulate and solve the optimal l_2-constrained compensator realization problem.

10.4 Optimal FWL compensator realizations

With our synthetic gain measure $G_{cl,T}$, the optimal l_2-constrained FWL compensator realization problem can be formulated as the following constrained minimization problem:

$$\min_{(A_r,B_r,C_r,D_r)\in S_c} G_{cl,T} = M_{cl,L_2} + \rho^2 G_{cl} \qquad (10.48)$$

subject to
$$(W_c(2,2))_{ii} = 1 \quad i = 1,\ldots,n. \qquad (10.49)$$

To solve this problem, we first examine the coordinate dependence of $G_{cl,T}$ and W_c. If an initial compensator realization $(A_r^0, B_r^0, C_r^0, D_r)$ in S_c corresponds to
$$\{W^0, W_{A_r}^0, W_{B_1}^0, W_{B_2}^0, W_{C_r}^0, W_c^0(2,2)\},$$
then a realization (A_r, B_r, C_r, D_r) obtained from this initial one by a similarity transformation T will have $\{T^T W^0 T, W_{A_r}(T), T^T W_{B_1}^0 T, T^T W_{B_2}^0 T, T^{-1} W_{C_r}^0 T^{-T}, T^{-1} W_c^0(2,2) T^{-T}\}$. The only term that causes some difficulty is $W_{A_r}(T)$, but as it happens the dependence of this term on the transformation matrix T has exactly the same form as the dependence of the corresponding term $W_A(T)$ of (7.27) in the synthetic filter design problem:

$$W_{A_r}(T) = \sum_{k=0}^{\infty} T^T H^0(k) T^{-T} T^{-1} H^0(k)^T T. \qquad (10.50)$$

Comment: Since the terms W_{d_1} and $\rho^2 \eta^2$ in (10.44) are coordinate independent, they do not affect the solutions of (10.48)-(10.49). Therefore, the minimization problem of the Total Noise Gain $G_{cl,T}$ can be performed as if these two terms were nil. This will be assumed in the sequel.

With the definitions $M_1^0 \triangleq W_{C_r}^0$, $M_2^0 \triangleq W_{B_1}^0 + W_{B_2}^0 + \rho^2 W^0$ and $M_3^0 \triangleq W_c^0(2,2)$, the Total Noise Gain $G_{cl,T}$ of the closed loop system can be rewritten as

$$G_{cl,T} = \text{tr}\left[\sum_{k=0}^{\infty} P H^0(k) P^{-1} H^0(k)^T\right]$$
$$+ \text{tr}[P^{-1} M_1^0] + \text{tr}[P M_2^0] \triangleq R(P) \qquad (10.51)$$

with $P = TT^T$. The optimal design problem can thus be restated as

$$\min_{T: \det T \neq 0} G_{cl,T} \iff \min_{P: P > 0} R(P) \qquad (10.52)$$

both with the constraint
$$(T^{-1} M_3^0 T^{-T})_{ii} = 1 \quad \forall i. \qquad (10.53)$$

278   10. Synthetic FWL compensator design

The structure of this optimization problem is now identical to that of problem (7.30)-(7.32) in Section 7.3. The solution is thus identical, with the proper substitutions. We define

$$L(P,\lambda) = R(P) + \lambda[tr(M_3^0 P^{-1}) - n]. \qquad (10.54)$$

Using Lagrange's method, we obtain the following equations:

$$\frac{\partial L}{\partial P} = \sum_{k=0}^{\infty} \{H^0(k) P^{-1} H^0(k)^T - P^{-1} H^0(k)^T P H^0(k) P^{-1}\}$$
$$+ M_2^0 - P^{-1} M_1^0 P^{-1} - \lambda P^{-1} M_3^0 P^{-1} = 0, \qquad (10.55)$$

$$\frac{\partial L}{\partial \lambda} = tr[M_3^0 P^{-1}] - n = 0. \qquad (10.56)$$

Since P is nonsingular, (10.55) is equivalent with

$$P\left[M_2^0 + \sum_{k=0}^{\infty} H^0(k) P^{-1} H^0(k)^T\right] P = (M_1^0 + \lambda M_3^0) + \sum_{k=0}^{\infty} H^0(k)^T P H^0(k) \qquad (10.57)$$

Theorem 7.1 applies exactly to the present problem, and we therefore have the following result for the optimal state space FWL compensator design.

Theorem 10.1 : *The optimization problem (10.52)-(10.53) has a unique solution: it is the unique positive definite solution P of (10.56)-(10.57).*

An explicit solution of (10.56)-(10.57) appears beyond reach (at least for the authors of this book). To compute the solution P^{opt}, the following iterative algorithm can be used:

$$P(k+1) = P(k) - \mu_1 \frac{\partial L(P,\lambda)}{\partial P}\Big|_{P=P(k),\lambda=\lambda(k)},$$
$$\lambda(k+1) = \lambda(k) - \mu_2 \frac{\partial L(P,\lambda)}{\partial \lambda}\Big|_{P=P(k),\lambda=\lambda(k)} \qquad (10.58)$$

where μ_1 and μ_2 are positive step sizes. When these step sizes are chosen small enough, the algorithm (10.58) converges to the unique solution of (10.55)-(10.56) or, equivalently, (10.56)-(10.57).

Once an optimal solution P^{opt} has been obtained, the set of optimal transformations T^{opt} can be computed from P^{opt} using an SVD under the constraint of l_2 normalization. The procedure has been explained at the end of Section 7.3 to which we refer the reader.

10.5 A design example

We illustrate our optimal synthetic compensator realization procedure with the following example. We consider a discrete (or sampled data) system to be controlled, described in controllable canonical (i.e. direct) form as follows:

$$A_0 = \begin{pmatrix} 3.7156 & -5.4143 & 3.6525 & -0.9642 \\ 1.0000 & 0 & 0 & 0 \\ 0 & 1.0000 & 0 & 0 \\ 0 & 0 & 1.0000 & 0 \end{pmatrix} \quad B_0 = \begin{pmatrix} 1 \\ 0 \\ 0 \\ 0 \end{pmatrix}$$

$$C_0 = \begin{pmatrix} 0.2991 & 0.0891 & 0.2546 & 0.0234 \end{pmatrix} \times 10^{-2}. \tag{10.59}$$

This system has the following open loop poles:

$$0.9075 \pm j0.4330, \quad 0.9503 \pm j0.2249$$

The control strategy is to perform pole placement by state estimate feedback, leaving the zeros unchanged and moving the closed loop poles to the following desired locations:

$$\{\lambda(A_0 - B_0 K_0)\} = \{0.8490 \pm j0.2570, \ 0.7413 \pm j0.0859\}.$$

Suppose in addition that the desired observer poles are:

$$\{\lambda(A_0 - L_0 C_0)\} = \{0.7692 \pm j0.3425, \ 0.6448 \pm j0.2344\}.$$

From these design requirements one can compute the ideal (infinite precision) state estimate feedback gain and the infinite precision observer (or Kalman) gain corresponding to the realization (10.59):

$$K_0 = \begin{pmatrix} 0.5350 & -1.5531 & 1.5403 & -0.5260 \end{pmatrix},$$

and

$$L_0 = \begin{pmatrix} 1.6549 & 1.3882 & 1.0038 & 0.5714 \end{pmatrix}^T \times 10^2.$$

We note by comparing the state variable realization (9.8) of the state estimate feedback controller with the state variable realization (10.5)-(10.6) that pole placement with state estimate feedback is a special case of a two-degree-of-freedom controller with $A_r = \Phi_0$, $B_r = (B_0 \ L_0)$, $C_r = -K_0$ and $D_r = (1 \ 0)$. The plant to be controlled has been given by its controllable realization as (10.59). The corresponding realization R_c of the two-degree-of-freedom controller is given as[2]

$\boxed{R_c:}$

$$A_c = \begin{pmatrix} 2.6856 & -4.0087 & 1.6909 & -0.4769 \\ 0.5848 & -0.1237 & -0.3534 & -0.0325 \\ -0.3002 & 0.9106 & -0.2556 & -0.0235 \\ -0.1709 & -0.0509 & 0.8545 & -0.0134 \end{pmatrix}$$

$$B_c = \begin{pmatrix} 1 & 165.49 \\ 0 & 138.82 \\ 0 & 100.38 \\ 0 & 57.14 \end{pmatrix}$$

$$C_c = \begin{pmatrix} -0.5350 & 1.5531 & -1.5403 & 0.5260 \end{pmatrix}.$$

We first compute a controller that minimizes $G_{cl,T}$ under l_2 scaling and with $\rho^2 = 0$, that is a controller that optimizes the L_2 sensitivity measure M_{cl,L_2} of the closed loop system subject to an l_2 scaling on the controller states. This can be done with the algorithm (10.58). An optimal similarity transformation from R_c is given by

$$T^{opt}(M_{cl,L_2}) = \begin{pmatrix} -43.3707 & -46.4892 & -25.8687 & 55.5262 \\ -26.2039 & -33.6042 & -41.2598 & 55.1077 \\ -6.5950 & -21.7176 & -51.0795 & 48.7035 \\ 13.2411 & -11.5886 & -54.8527 & 38.6034 \end{pmatrix}.$$

The corresponding realization is given by

$\boxed{R^{opt}(M_{cl,L_2}):}$

$$A^{opt}(M_{cl,L_2}) = \begin{pmatrix} 1.0941 & -0.3182 & -0.2101 & -0.3437 \\ 0.0993 & 0.8350 & -0.0183 & -0.0073 \\ 0.7191 & -0.0702 & 1.4030 & -1.3379 \\ 0.7523 & 0.0335 & 1.1174 & -1.0392 \end{pmatrix}$$

---

[2]We caution that all state variable realizations described hereafter are realizations of the *controller* only.

10. Synthetic FWL compensator design

$$B^{opt}(M_{cl,L_2}) = \begin{pmatrix} -0.3164 & -0.8065 \\ -0.3116 & -0.5712 \\ -0.5288 & 0.3004 \\ -0.7364 & 2.0121 \end{pmatrix}$$

$$C^{opt}(M_{cl,L_2}) = \begin{pmatrix} -0.3708 & 0.0371 & -0.4156 & 1.1687 \end{pmatrix}.$$

For this sensitivity optimal realization $R^{opt}(M_{cl,L_2})$, the closed loop sensitivity measure is

$$M_{cl,L_2}(R^{opt}(M_{cl,L_2})) = 19.2212 = G_{cl,T}(R^{opt}(M_{cl,L_2})).$$

The corresponding roundoff noise gain of the closed loop system is also computed: $G_{cl}(R^{opt}(M_{cl,L_2})) = 0.6895$.

We now turn to an optimal controller realization that minimizes the roundoff noise gain G_{cl} of the closed loop system, that is we solve (10.48)-(10.49) with $\rho^2 = \infty$. An optimal transformation matrix from R_c is

$$T^{opt}(G_{cl}) = \begin{pmatrix} -57.8829 & -45.0098 & -8.4623 & 7.3129 \\ -49.6854 & -45.7821 & 4.4561 & -11.1804 \\ -37.6134 & -41.8993 & 12.6644 & -27.7561 \\ -23.1264 & -35.3665 & 16.2494 & -40.6423 \end{pmatrix}$$

and the corresponding optimal realization is

$$\boxed{R^{opt}(G_{cl}):}$$

$$A^{opt}(G_{cl}) = \begin{pmatrix} 0.8495 & 0.3604 & -0.3329 & -0.4830 \\ -0.7901 & -0.4957 & 0.7812 & -0.0673 \\ 0.0406 & -0.1818 & 0.9603 & 0.1464 \\ 0.6979 & 0.7763 & -0.3974 & 0.9790 \end{pmatrix}$$

$$B^{opt}(G_{cl}) = \begin{pmatrix} -0.4890 & -1.4851 \\ 0.5876 & -1.6299 \\ -0.1537 & 0.0184 \\ -0.2946 & 0.8648 \end{pmatrix}$$

$$C^{opt}(G_{cl}) = \begin{pmatrix} -0.4276 & -1.0893 & 0.4883 & 0.0982 \end{pmatrix}.$$

The L_2 sensitivity measure and the roundoff noise gain of the closed loop system for this roundoff-noise-optimal controller realization are:

$$M_{cl,L_2}(R^{opt}(G_{cl})) = 20.9444, \quad G_{cl}(R^{opt}(G_{cl})) = 0.5903$$

10. Synthetic FWL compensator design

Comment: For both of these optimal realizations the output error variance due to errors on the coefficients of the controller are significantly higher (by a factor of about 30) than the variance due to roundoff in the controller. This indicates that more bits should be used for coefficient implementation than for the storage of the signals.

We therefore solve the optimization problem (10.48)-(10.49) again with $B_c - B_s = 2$, that is $\rho^2 = 16$. One optimal solution obtained from the iterative algorithm is

$$T^{opt}(G_{cl,T}) = \begin{pmatrix} 17.5492 & -22.1273 & -70.2202 & -22.8375 \\ 30.5583 & -3.1222 & -61.5354 & -25.8695 \\ 38.0369 & 15.4649 & -49.5421 & -24.1579 \\ 39.7126 & 31.9777 & -35.9587 & -19.4315 \end{pmatrix}.$$

The corresponding optimal realization is

$\boxed{R^{opt}(G_{cl,T}) :}$

$$A^{opt}(G_{cl,T}) = \begin{pmatrix} 1.3300 & -0.2444 & -0.5876 & -1.0871 \\ 0.2182 & 0.9747 & -0.2950 & 0.2157 \\ 0.3739 & 0.1642 & 0.4241 & -0.3516 \\ 0.9743 & -0.0602 & -1.0961 & -0.4358 \end{pmatrix}$$

$$B^{opt}(G_{cl,T}) = \begin{pmatrix} 0.4505 & -0.1725 \\ -0.1908 & -0.8053 \\ -0.0976 & -1.6175 \\ 0.7873 & -1.6251 \end{pmatrix}$$

$$C^{opt}(G_{cl,T}) = \begin{pmatrix} 0.3718 & -0.0114 & -0.6074 & -0.9704 \end{pmatrix}$$

The L_2 sensitivity measure and the roundoff noise gain for the closed loop system with this controller realization are

$$M_{cl,L_2}(R^{opt}(G_{cl,T})) = 19.4393 \quad G_{cl}(R^{opt}(G_{cl,T})) = 0.6209.$$

For comparison, we have also computed the L_2 sensitivity measure and the roundoff noise gain of the closed loop system with a controller obtained from R_c above by l_2 scaling, denoted R_c^{sc}, and with a controller in l_2 scaled MRH structure, R_{MRH}^{sc}. This last structure is obtained by scaling an MRH structure that is a roundoff noise optimal realization of the compensator seen as a filter rather than as a component of a closed loop system. The l_2 scaling is achieved by applying a diagonal similarity transformation to R_c and R_{MRH}, respectively, such that R_c^{sc} and R_{MRH}^{sc} satisfy (10.49). The results are as follows:

$$M_{cl,L_2}(R_c^{sc}) = 2.5908 \times 10^5 \qquad G_{cl}(R_c^{sc}) = 1.7239$$
$$M_{cl,L_2}(R_{MRH}^{sc}) = 29.3389 \qquad G_{cl}(R_{MRH}^{sc}) = 0.9097$$

For ease of comparison, we present all the computation results in the following table:

$Realization/Measure$	M_{cl,L_2}	G_{cl}
R_c^{sc}	2.5908×10^5	1.7239
R_{MRH}^{sc}	29.3389	0.9097
$R^{opt}(M_{cl,L2})$	19.2212	0.6895
$R^{opt}(G_{cl})$	20.9444	0.5903
$R^{opt}(G_{cl,T})$	19.4393	0.6209

Comments:

1. Our computations show that a controller obtained from a system described in controller form yields FWL performances that are significantly worse than those of any of the optimal realizations. This is particularly true for the sensitivity of the closed loop transfer function with respect to the controller parameters: this sensitivity is several orders of magnitude larger for the realization R_c^{sc} than for any of the other realizations.

2. It is also clear that any of the realizations that optimize some closed loop performance criterion achieve a significantly better result than the realization R_{MRH}^{sc} that optimizes the controller roundoff noise gain irrespective of its effect on the closed loop system.

Numbers, gains, sensitivities are all compact pieces of information, but nothing is more appealing than a good looking curve. We have thus performed two simulations. The first one illustrates rather clearly the mediocre performance of the non-optimal realization R_c^{sc} when implemented with finite wordlength. In Figure 10.2 we have illustrated the response of the controlled system to a step change on the reference signal, for an ideal (infinite wordlength) implementation of the compensator, for a FWL version of the realization R_c^{sc} with $B_c = B_s = 8$ bits, and for a FWL version of the optimal realization $R^{opt}(G_{cl,T})$ given above with $B_c = B_s = 6$ bits. The numbers B_c and B_s indicate, as

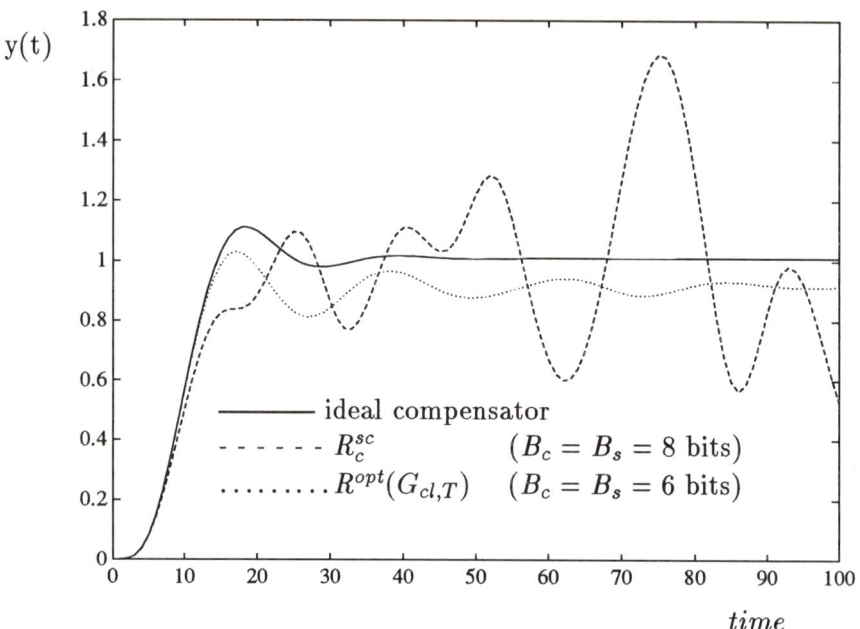

FIGURE 10.2. Output signal $y(t)$ due to a reference step change

usual, the number of bits for the fractional parts of coefficients and signals, respectively.

The signals have been rounded off *before multiplication*, i.e. an RBM model. The corresponding control signals are given in Figure 10.3. The superiority of the optimal realization $R^{opt}(G_{cl,T})$ over R_c^{sc} is so clearly shown in these figures that no additional comment is needed.

In the second simulation we compare the performance of different optimal FWL controller realizations, where the differences lie in the respective choices of coefficient and signal bitlengths. In one case, the optimal compensator realization $R^{opt}(G_{cl,T})$ obtained above is implemented with the choice $\{B_c = 6, B_s = 8\}$, that is more bits are used for the encoding of the signals than for the controller coefficients, while in the second case the implementation specifications $\{B_c = 8, B_s = 6\}$ are made for the same optimal realization $R^{opt}(G_{cl,T})$. Again a step change in the reference signal is imposed and the corresponding output and control signals are computed for the ideal realization of the controller and for the optimal realization $R^{opt}(G_{cl,T})$ with these two kinds of implementation specifications, respectively.

The objective of this simulation is to show the fact that the FWL

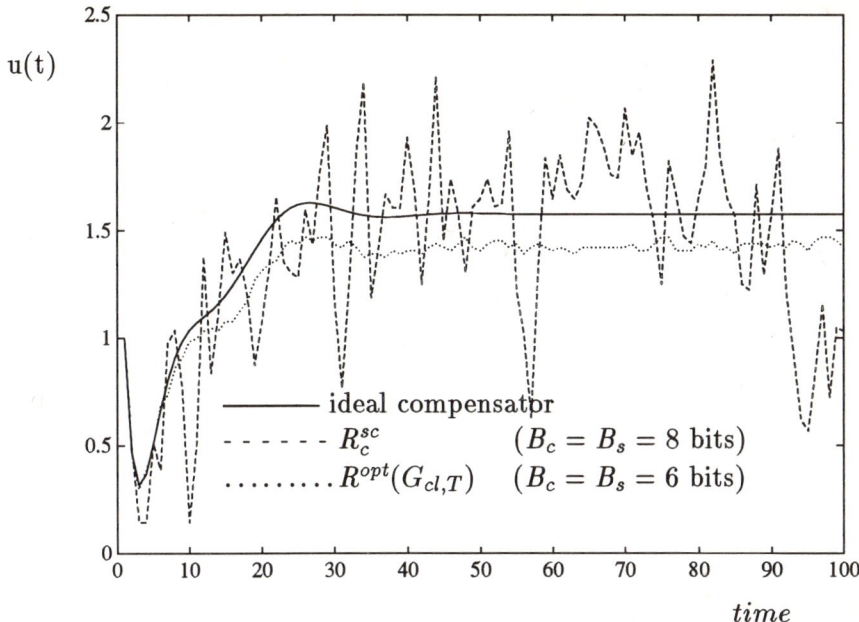

FIGURE 10.3. Control signal $u(t)$ due to a reference step change

error effect of the coefficients on the output of the closed loop system as well as on the control signal is more serious than that of the signals for this example. The results are presented in Figure 10.4 for the output signal response, and in Figure 10.5 for the control signal. The simulation results clearly confirm the arguments given just above on the importance of the coefficient quantization error for this example.

10.6 Conclusion

In this chapter, we have considered the optimal parametrization problem for the FWL implementation of the most general form of compensator, namely a two-degree-of-freedom compensator. We have extended to this control realization problem the synthetic design ideas of Chapter 7, in which the coefficient errors due to finite wordlength are treated as random. This allows one to adopt a global performance measure, which is the variance of the output of the closed loop system due to both the random controller coefficient errors and the controller roundoff errors. Even though this synthetic performance measure can be criticized on the grounds of its stochastic assumption on the coefficients,

286   10. Synthetic FWL compensator design

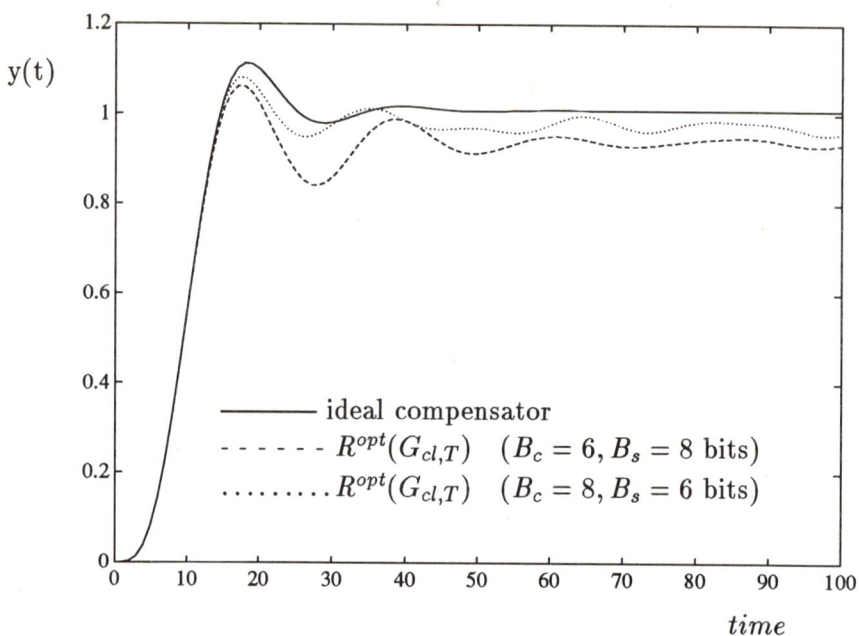

FIGURE 10.4. Output signal $y(t)$ due to a reference step change

we believe that it has the following merits:

- the variance contribution of the random coefficient errors naturally introduces a closed loop L_2 sensitivity measure;

- the synthetic measure allows for a natural comparison of the respective weights of coefficient error and roundoff error contribution in the total measure, thereby allowing for an easy design choice for the tradeoff between coefficient and signal bitlengths.

The contribution of this chapter has been twofold. The first was to derive a computable expression for this synthetic performance measure, called Total Noise Gain, and to show its dependence on the coordinate basis of the controller realization. The second contribution was to compute the set of optimal realizations, i.e. the set of realizations of the compensator that minimize the closed loop system performance degradation due to the FWL implementation of the compensator under dynamic range constraint. The optimal controller realizations have been characterized by a set of necessary and sufficient conditions, and an iterative algorithm for finding the optimal realizations has been given. We have illustrated with a numerical example the nice performance

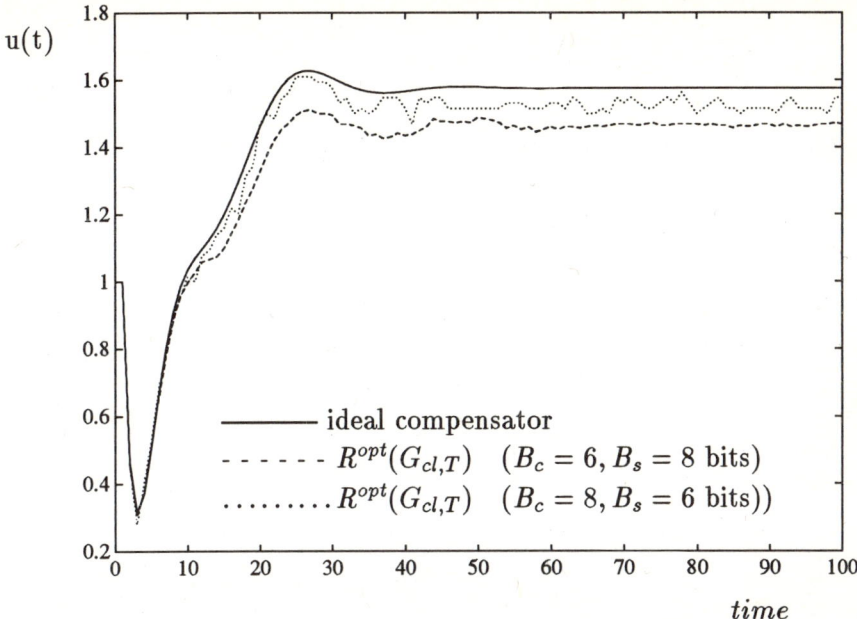

FIGURE 10.5. Control signal $u(t)$ due to a reference step change

that can be achieved by the optimal realizations as compared to realizations that optimize the FWL performance of the controller treated as an isolated filter.

This ends our treatment of FWL controller design. We are fully aware that many open problems remain. We believe that the distinctions we have outlined in the introduction of Chapter 9 between various possible performance criteria on the one hand, and a hierarchy of optimization strategies on the other hand, in some ways delineate a work program of possible FWL design problems to be addressed. In addition, we believe that some more thought needs to be given, in the control context, to the choice of the best FWL performance criterion in connection with each particular control application.

11
Parametrizations in the Delta operator

11.1 Introduction and motivation

So far in this book we have followed common practice by describing discrete time transfer functions with the usual shift operator z. This operator is easy to implement since it consists simply of a shift. However, a disadvantage of the shift operator is that it does not at all approximate the continuous time operator $d(.)/dt$. Heuristically, it might be suspected that a better correspondence is obtained between continuous and discrete time representations if the shift operator is replaced by a difference operator that is more like a derivative. Such operator, nowadays often called delta operator, is defined as $\delta = \frac{z-1}{T_s}$, where T_s is the sampling period. The delta operator approximates the time derivative and converges to it when the sampling period tends towards zero.

The concept and use of the delta operator is not entirely new, even though the name has been coined only recently by Middleton and Goodwin [MG90]. It has been known in numerical analysis as the first divided-difference operator. In the control literature, it has been used as a way of motivating z-transformations [Kar79] and as a way of relating continuous and discrete transfer functions [Gaw80], while the analysis of the numerical properties of this operator in digital filtering appears to have originated with the work of Agarwal and Burrus [AB75]. The technique introduced in [AB75] was later called 'delay replacement' in [OM84] and [Wil88].

In the last few years, both Peterka [Pet86] and Middleton and Goodwin [MG86] and [MG90] have promoted the use of the delta operator as opposed to the shift operator in estimation and control applications. Two major advantages are claimed for the use of delta operator parametrizations: a theoretically interesting unified formulation of continuous time and discrete time filtering and control theory which, in particular, entails a better understanding of discrete time control under fast sampling, and a range of practically interesting numerical advantages connected with finite wordlength (FWL) effects. These numerical

11. Parametrizations in the Delta operator

advantages can be summarized as follows (see e.g.[Goo88]):

1. superior finite word length coefficient representation;

2. superior finite word length rounding error performance;

3. superior numerical properties in calculations with discrete time models, e.g. control system design by pole assignment.

The usual delta operator is defined as $(z-1)/T_s$ where T_s is the sampling period. In this chapter, we define the delta operator as above but with a positive factor Δ replacing the sampling period T_s. This replacement allows us to study the numerical properties of this operator from a purely mathematical point of view. Before going into details, we first give some motivations for the use of the delta operator.

In a transfer function model, a system is characterized by its numerator and denominator polynomials. One can analyse the numerical properties of this model by studying the corresponding polynomials. Now, consider a real polynomial $p(z)$ in the shift operator z,

$$p(z) = \prod_{i=1}^{n}(z - \lambda_i) = \sum_{k=0}^{n} a_k z^{n-k}, a_0 = 1 \qquad (11.1)$$

where $\{\lambda_i\}$ are the roots of this polynomial. Suppose that λ_i is a simple root. By the same techniques as in Chapter 6, one can compute the sensitivity of this root with respect to the j-th coefficient a_j:

$$\frac{d\lambda_i}{da_j} = -\frac{\frac{\partial p(z)}{\partial a_j}}{\frac{\partial p(z)}{\partial \lambda_i}}\bigg|_{z=\lambda_i}$$

$$= \frac{\lambda_i^{n-j}}{\prod_{k \neq i}(\lambda_i - \lambda_k)}. \qquad (11.2)$$

It is noted that

1. the greater the absolute value of $|\lambda_i|$, the higher the sensitivity measure $|\frac{\partial \lambda_i}{\partial a_j}|$ is;

2. the smaller the difference between λ_i and a neighbouring root, the higher $|\frac{\partial \lambda_i}{\partial a_j}|$ is.

In many control applications where fast sampling is used, these two factors, which make the root sensitivity high, occur simultaneously since with fast sampling the poles and zeros of the controller are clustered around $z = +1$ [ÅSH84], [MG90]. This is one of the reasons why the transfer function model yields undesirable performance and why the redundancy in state space models has to be explored in order to, hopefully, find state space realizations that have better root sensitivity. It is also interesting to note that the higher the order of the system is, the higher the pole sensitivity is. This is another reason why higher order systems are often decomposed as a cascade of first and second order subsystems, as we have seen in Chapter 8. The pole sensitivity of each subsystem is then substantially reduced.

Now, consider the following algebraic transformation,

$$\delta = \frac{z-1}{\Delta}, \quad (11.3)$$

which will from now on serve as our definition for the delta operator. With this transformation, we get a polynomial $P(\delta)$ corresponding to $p(z)$,

$$\begin{aligned} P(\delta) &= p(z)|_{z=1+\Delta\delta} = \rho_0 \prod_{i=1}^{n}(\delta - \mu_i) \\ &= \rho_0 \sum_{k=0}^{n} \alpha_k \delta^{n-k}, \alpha_0 = 1 \end{aligned} \quad (11.4)$$

where $\{\mu_i = (\lambda_i - 1)/\Delta\}$ are the roots of the new polynomial $P(\delta)$.
We note that

1. $|\mu_i - \mu_j| > |\lambda_i - \lambda_j|$ as long as $0 < \Delta < 1$;

2.
$$\left|\frac{\partial \mu_i}{\partial \alpha_j}\right| = \Delta^{j-1} \frac{|1-\lambda_i|^{n-j}}{\prod_{k \neq i}|\lambda_i - \lambda_k|} < \left|\frac{\partial \lambda_i}{\partial a_j}\right|$$
if $0 < \Delta < 1$ and $|1 - \lambda_i| < |\lambda_i|$,

where the last inequality means that the root λ_i of $p(z)$ is closer to $z = +1$ than to $z = 0$.

This signifies that by a simple algebraic transformation such as (11.3) one can make the new polynomial $P(\delta)$ have better numerical properties in the case where the roots are closer to $z = +1$ than to $z = 0$. This observation motivates us to give a further study of this transformation.

Comment: In [MG90], Middleton and Goodwin defined the delta operator as (11.3) but with the sampling period T_s replacing Δ. Here our Δ is any positive number and has no physical meaning. The choice of a value for Δ and its role in improving some numerical properties will be discussed later.

Any polynomial of degree n in z is exactly equivalent to another polynomial of degree n in the δ-operator using the transformation (11.3). Thus, any discrete rational transfer function in z is equivalent to another rational function in δ (with numerator and denominator of the same degree as those of the rational transfer function in z). In [MG90], the numerical properties of the transfer functions in δ-operator have been investigated in some depth and compared to those of z-operator transfer functions.

Just as a shift operator transfer function, a delta operator transfer function can be represented in a state variable realization, and the state variable realization of a filter directly in terms of its δ-operator transfer function coefficients corresponds to a δ-operator direct form realization. Similarly, there exists an equivalence class, say S_δ, of δ-operator state variable realizations, that are all related by similarity transformations. The problems not studied in [MG90] are those of examining the optimal realizations, within this equivalence class S_δ, with respect to sensitivity of the transfer functions vis-à-vis coefficient errors or with respect to roundoff noise gain. Another problem not studied in [MG90] is that of comparing the numerical performance (sensitivity or roundoff noise gain) of optimal shift operator realizations with those of optimal delta operator realizations. The examination of the numerical properties of the realization set S_δ and a comparison of these realizations with those in S_z is the main object of this chapter.

An outline of this chapter is as follows. We first establish in Section 11.2 the relationships between shift and delta operator parametrizations, both in transfer function and in state space form. Section 11.3 is concerned with the study of the L_1/L_2 absolute transfer function sensitivity measure defined and discussed in Chapter 3. Our main new results of this section consist in computing an expression for the best achievable upper bound for the absolute sensitivity of all equivalent delta state space realizations. This expression is compared with the similar expression for shift operator realizations. More interestingly, we show that the set of optimal delta realizations can be connected in a simple way and therefore derived from the set of optimal shift realizations. It is then shown that, by a proper choice of the degree of freedom available in the definition of the delta operator, the absolute sensitivity measure with the delta operator state space models can normally be made smaller than that of shift operator state space models. In Section 11.4, we examine the possible advantages of using delta operator state space realizations rather than shift operator realizations of transfer functions in terms of minimizing the roundoff noise gain of the realization. In [Wil86], Williamson proposed the use of residue feedback to reduce the roundoff noise gain for shift operator realizations and compared it with the optimal gain for realizations without residue feedback. He introduced the concept of residue modes and showed that the superiority of the optimal realizations with residue feedback over the optimal realizations without residue feedback hinged on whether the sum of the residue modes was smaller than the sum of the Hankel singular values. We first show in this section that the δ-operator implementation is in fact a special case of residue feedback. We then give several conditions under which the optimal roundoff noise gain for delta operator realizations is smaller than the optimal gain for shift operator realizations. Two numerical examples are given in this section. Some concluding remarks are given in Section 11.5. Finally we note that some of the material of this chapter was presented in [LG90c].

11.2 Delta operator parametrizations

Throughout this chapter we consider scalar strictly proper time invariant discrete time transfer functions. In the old days (i.e. before Middleton and Goodwin [MG90]) it was customary to represent such transfer functions as follows:

$$H(z) = \frac{\sum_{i=1}^{n} b_i z^{n-i}}{z^n + \sum_{i=1}^{n} a_i z^{n-i}}. \qquad (11.5)$$

Such discrete time transfer function is often obtained from a continuous time transfer function $H_s(s)$ as the result of a discretization procedure with a sampling period T_s. Here, we consider that the starting point is a discrete time (rather than continuous time) transfer function, and we use (11.3) as our definition for the δ-operator, where Δ is any positive number, not necessarily a sampling period. Thus (11.3) should be seen purely as a linear operator. With this definition, the transfer function $H(z)$ of (11.5) can be re-expressed in δ-form as follows:

$$H(z) = \frac{\sum_{i=1}^{n} \beta_i \delta^{n-i}}{\delta^n + \sum_{i=1}^{n} \alpha_i \delta^{n-i}} = H_\delta(\delta). \tag{11.6}$$

We note that the coefficients $\{\alpha_i, \beta_i\}$ are obtained from the $\{a_i, b_i\}$ by substituting $z = 1 + \Delta\delta$ in $H(z)$. This yields the following relationships:

$$\begin{pmatrix} 1 \\ a_1 \\ \cdot \\ \cdot \\ \cdot \\ a_n \end{pmatrix} = \bar{T}_\delta \begin{pmatrix} 1 \\ \alpha_1 \\ \cdot \\ \cdot \\ \cdot \\ \alpha_n \end{pmatrix}, \quad \begin{pmatrix} 0 \\ b_1 \\ \cdot \\ \cdot \\ \cdot \\ b_n \end{pmatrix} = \bar{T}_\delta \begin{pmatrix} 0 \\ \beta_1 \\ \cdot \\ \cdot \\ \cdot \\ \beta_n \end{pmatrix}, \tag{11.7}$$

where $\bar{T}_\delta \in \mathbb{R}^{(n+1)\times(n+1)}$ is upper triangular,

$$\bar{T}_\delta \triangleq \begin{pmatrix} 1 & t_{12} & \cdot & \cdot & t_{1(n+1)} \\ 0 & t_{22} & \cdot & \cdot & t_{2(n+1)} \\ \cdot & \cdot & \cdot & \cdot & \cdot \\ 0 & 0 & \cdot & 0 & t_{(n+1)(n+1)} \end{pmatrix}$$

$$\triangleq \begin{pmatrix} 1 & \tau^T \\ 0 & T_\delta \end{pmatrix} \tag{11.8}$$

with $T_\delta \in \mathbb{R}^{n\times n}$ and

$$t_{ij} = (-1)^{(j-i)} C_{n+1-i}^{j-i} \Delta^{(i-1)}, j \geq i; \quad C_m^k = \frac{m!}{(m-k)!k!}.$$

Going back to (11.6), we observe that $H(z)$ and $H_\delta(\delta)$ are two different but equivalent parametrizations representing the same object. These two input-output relationships can be represented by a shift operator (resp. δ-operator) state space model as follows:

$$\begin{aligned} zx_t^{(1)} &= A_z x_t^{(1)} + B_z u_t \\ y_t &= C_z x_t^{(1)} \end{aligned} \tag{11.9}$$

and

$$\begin{aligned} \delta x_t^{(2)} &= A_\delta x_t^{(2)} + B_\delta u_t \\ y_t &= C_\delta x_t^{(2)}. \end{aligned} \tag{11.10}$$

The following relationships relate the internal and external representations:

$$H(z) = C_z(zI - A_z)^{-1}B_z, H_\delta(\delta) = C_\delta(\delta I - A_\delta)^{-1}B_\delta. \quad (11.11)$$

We shall for future use introduce the notion of a realization set S_ρ. We define:

$$S_\rho = \{(A_\rho, B_\rho, C_\rho) : H_\rho(\rho) = C_\rho(\rho I - A_\rho)^{-1}B_\rho\} \quad (11.12)$$

where $\rho = z$ or δ is called generalized operator, and $H_z(z) = H(z)$. Hence if $(A_\rho, B_\rho, C_\rho) \in S_\rho$, $(T^{-1}A_\rho T, T^{-1}B_\rho, C_\rho T) \in S_\rho$ if and only if T is nonsingular.

Substituting (11.3) in (11.10), it is straightforward to establish that the following relationship exists between the state space realizations $(A_z, B_z, C_z) \in S_z$ and $(A_\delta, B_\delta, C_\delta) \in S_\delta$:

$$\begin{aligned} A_z &= \Delta A_\delta + I \\ B_z &= \Delta B_\delta \\ C_z &= C_\delta. \end{aligned} \quad (11.13)$$

This means that if $(A_\delta, B_\delta, C_\delta) \in S_\delta$, one can find a corresponding realization $(A_z, B_z, C_z) \in S_z$ and vice-versa by the one-to-one mapping (11.13).

Before going into details for analysis, we give a brief discussion about the implementation of a δ-operator model. A state space realization (11.10) may be directly evaluated via

$$\begin{aligned} x_t^{(2)} &= \delta^{-1}(A_\delta x_t^{(2)} + B_\delta u_t) \\ y_t &= C_\delta x_t^{(2)} \end{aligned} \quad (11.14)$$

which uses the basic building block (δ^{-1}). This building block replaces the usual unit delay (z^{-1}) used in standard shift operator discrete time models.

The key operation in (11.14) is the operator δ^{-1}. It is defined as follows:

$$v(t) = \delta^{-1}w(t) \iff v(t+1) = v(t) + \Delta w(t). \quad (11.15)$$

The implementation of the operator δ^{-1} is illustrated in Figure 11.1.

11. Parametrizations in the Delta operator

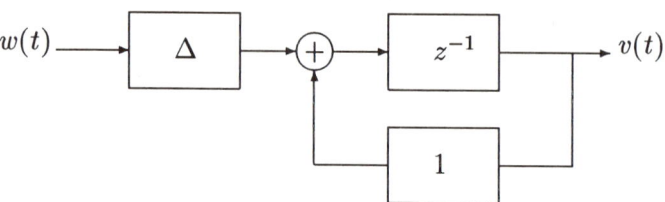

FIGURE 11.1. Implementation of the δ^{-1} operator

The hardware, or software, necessary to implement this building block is only marginally more complex than that necessary to implement the usual shift operator z^{-1}. The operation (11.15) contains one addition and one multiplication. Since Δ can be chosen in such a way that it has an exact FWL representation, there will be no coefficient error in implementing the δ-operator. However, a rounding error occurs in both fixed point arithmetic and floating point arithmetic when $\Delta \neq 1$. Some discussions on this issue can be found in [MG90]. We will show later that in fixed point implementation, this error is so small that it can be neglected (compared with other noises). Thus, for the moment the δ-operator is considered to be implemented perfectly.

11.3 Sensitivity of delta parametrizations

Our aim here is to compare the sensitivity of the transfer functions $H(z)$ and $H_\delta(\delta)$, respectively, with respect to numerical errors in the coefficients of their respective state space implementations. The choice of a value for Δ and its role in improving these sensitivities will be a central feature of this section.

Consider the generalized state space realization

$$\rho x_t = A_\rho x_t + B_\rho u_t$$
$$y_t = C_\rho x_t \qquad (11.16)$$

where ρ is z or δ (see (11.9)-(11.10)), and where (A_ρ, B_ρ, C_ρ) is an infinite precision implementation of a transfer function $H_\rho(\rho)$, $\rho = z$ or δ. Assume that B_c bits are available and denote by $(A_\rho^*, B_\rho^*, C_\rho^*)$ the implemented version of (A_ρ, B_ρ, C_ρ) where the coefficients have been truncated to B_c bits. The actually implemented state space model is then given by (11.16) with (A_ρ, B_ρ, C_ρ) replaced by $(A_\rho^*, B_\rho^*, C_\rho^*)$. It follows that the actual transfer function $H_\rho^*(\rho) = C_\rho^*(\rho I - A_\rho^*)^{-1} B_\rho^*$ and the ideal $H_\rho(\rho) = C_\rho(\rho I - A_\rho)^{-1} B_\rho$ will differ, and hence the output of the actually implemented filter to any input sequence will deviate from the output of the ideal filter. One way to evaluate this error is to compute a measure of the sensitivity of the transfer function $H_\rho(\rho)$ to errors on the matrices A_ρ, B_ρ, C_ρ. Here we use the definitions given in Chapter 3 for the sensitivity measure of the state space implementation of a transfer function in the generalized operator ρ. We then specialize these expressions to the case of shift and δ-operator representations.

11.3.1 Sensitivity measure

Using Definitions 3.1 and 3.2, one can define the L_1/L_2 sensitivity measure of the transfer function $H_\rho(\rho)$ with respect to the parameters in the realization (A_ρ, B_ρ, C_ρ) as

$$M_{\rho, L_{12}} = \|\frac{\partial H_\rho}{\partial A_\rho}\|_1^2 + \|\frac{\partial H_\rho}{\partial B_\rho}\|_2^2 + \|\frac{\partial H_\rho}{\partial C_\rho}\|_2^2 \qquad (11.17)$$

where

$$\frac{\partial H_\rho}{\partial A_\rho} = (\rho I - A_\rho^T)^{-1} C_\rho^T B_\rho^T (\rho I - A_\rho^T)^{-1}$$

$$\frac{\partial H_\rho}{\partial B_\rho} = (\rho I - A_\rho^T)^{-1} C^T$$

$$\frac{\partial H_\rho}{\partial C_\rho^T} = (\rho I - A_\rho)^{-1} B_\rho. \qquad (11.18)$$

It then easily follows from (11.3), (11.13) and (11.18) that:

$$\frac{\partial H_\delta}{\partial A_\delta} = \Delta \frac{\partial A_z}{\partial A_z}$$

$$\frac{\partial H_\delta}{\partial B_\delta} = \Delta \frac{\partial H_z}{\partial B_z}$$

$$\frac{\partial H_\delta}{\partial C_\delta} = \frac{\partial H_z}{\partial C_z}. \qquad (11.19)$$

Therefore, (11.17) specializes, for $\rho = z$ and $\rho = \delta$, to:

$$M_{z, L_{12}} = \|\frac{\partial H_z}{\partial A_z}\|_1^2 + \|\frac{\partial H_z}{\partial B_z}\|_2^2 + \|\frac{\partial H_z}{\partial C_z}\|_2^2, \qquad (11.20)$$

$$M_{\delta, L_{12}} = \|\frac{\partial H_\delta}{\partial A_\delta}\|_1^2 + \|\frac{\partial H_\delta}{\partial B_\delta}\|_2^2 + \|\frac{\partial H_\delta}{\partial C_\delta}\|_2^2$$

$$= \Delta^2 \|\frac{\partial H_z}{\partial A_z}\|_1^2 + \Delta^2 \|\frac{\partial H_z}{\partial B_z}\|_2^2 + \|\frac{\partial H_z}{\partial C_z}\|_2^2. \quad (11.21)$$

Similarly, we have the following expressions for the L_2 sensitivity measure:

$$M_{z,L_2} = \|\frac{\partial H_z}{\partial A_z}\|_2^2 + \|\frac{\partial H_z}{\partial B_z}\|_2^2 + \|\frac{\partial H_z}{\partial C_z}\|_2^2, \quad (11.22)$$

$$M_{\delta,L_2} = \|\frac{\partial H_\delta}{\partial A_\delta}\|_2^2 + \|\frac{\partial H_\delta}{\partial B_\delta}\|_2^2 + \|\frac{\partial H_\delta}{\partial C_\delta}\|_2^2$$

$$= \Delta^2 \|\frac{\partial H_z}{\partial A_z}\|_2^2 + \Delta^2 \|\frac{\partial H_z}{\partial B_z}\|_2^2 + \|\frac{\partial H_z}{\partial C_z}\|_2^2. \quad (11.23)$$

We have therefore the following result.

Theorem 11.1: *Consider two realizations* $(A_z, B_z, C_z) \in S_z$ *and* $(A_\delta, B_\delta, C_\delta) \in S_\delta$ *of the same transfer function, related by (11.13). Then*

$$M_{\delta,L_{12}} < M_{z,L_{12}} \quad \text{and} \quad M_{\delta,L_2} < M_{z,L_2}$$

if and only if $\Delta < 1$.

Proof: The proof follows immediately from (11.20) - (11.23). ∎

Comment 11.3.1: Theorem 11.1 shows that, for both the L_1/L_2 and the L_2 sensitivity measures, M_δ can be made smaller than M_z provided Δ can be chosen smaller than 1. We should note, however, that the value of Δ influences the range of the coefficients appearing in $(A_\delta, B_\delta, C_\delta)$ as is clear from (11.13). Therefore, the games we can play with Δ are limited by dynamic range considerations. To show this, let us consider the following two examples. In a fixed point arithmetic implementation, the absolute values of all implemented coefficients are constrained to be within some interval, say [0.001, 1]. If a realization in shift operator is given by

$$A_z = \begin{pmatrix} 0.99 & -0.01 \\ 0.01 & 0.99 \end{pmatrix}, B_z = \begin{pmatrix} 0.02 \\ 0.10 \end{pmatrix}, C_z^T = \begin{pmatrix} 0.50 \\ -0.67 \end{pmatrix},$$

then the maximal absolute value of the elements in $(A_z - I)$ is 0.01. So, the maximal absolute value of the elements in $(A_z - I)$ and B_z is 0.1. Therefore, according to (11.13), the choices of Δ that will keep the coefficients of A_δ, B_δ within the required range are any value between 0.1 and 1. Hence, by choosing a δ-operator implementation with $\Delta =$

0.1 we can significantly reduce the sensitivity with respect to a shift operator implementation while satisfying the dynamic range constraint.

Assume now that B_z and C_z are as before, but

$$A_z = \begin{pmatrix} 0.2314 & -0.0127 \\ 0.0231 & 0 \end{pmatrix}.$$

The maximal absolute value of the elements in $(A_z - I)$ is 1, the dynamic range constraint will then force $\Delta = 1$ and hence the δ-operator realization will have the same sensitivity performance as the shift operator realization. From these two examples, one can see that for a well scaled realization (for example, its largest absolute value is smaller than one) the choice of Δ is related to the poles of the system. For this issue, see also Comment 11.3.4 below and the numerical example in Section 11.4.

Just as in Chapter 5, we can discuss the optimal realization problem in terms of minimizing the L_2 sensitivity measure over all δ-operator realizations in the set S_δ. Evidently, one can get the necessary and sufficient conditions that optimal realizations must satisfy, but it is again impossible to construct explicit optimal solutions; they can only be obtained using the iterative algorithm proposed in Chapter 5. One advantage of the L_1/L_2 sensitivity measure is that the optimal realizations can be computed explicitly. Therefore, in the sequel we consider only the L_1/L_2 measure. However, the general conclusion that the optimal realizations in δ-operator have better L_1/L_2 sensitivity performance than the optimal shift operator realizations when the poles of the filter are clustered around $z = 1$ holds for the L_2 sensitivity measure as well.

11.3.2 Optimal realization set

The optimal filter realization problem in shift operator has been solved in Chapter 3. We now establish two new results. First we give an expression for the minimizing value of $M_{\delta, L_{12}}$ over all $(A_\delta, B_\delta, C_\delta) \in S_\delta$. Then we characterize the optimal set S_δ^{opt} by relating the optimal realizations in delta form to the optimal realizations in shift form.

Similarly to the procedure used by Thiele, we first replace $M_{\delta, L_{12}}$ by an upper bound $\bar{M}_{\delta, L_{12}}$ using the same Cauchy-Schwartz inequality:

$$M_{\delta, L_{12}} \leq \bar{M}_{\delta, L_{12}} = \|\frac{\partial H_\delta}{\partial B_\delta}\|_2^2 \|\frac{\partial H_\delta}{\partial C_\delta}\|_2^2 + \|\frac{\partial H_\delta}{\partial B_\delta}\|_2^2 + \|\frac{\partial H_\delta}{\partial C_\delta}\|_2^2. \quad (11.24)$$

Theorem 11.2 : *i) The minimal value of $M_{\delta,L_{12}}$ over all equivalent realizations $(A_\delta, B_\delta, C_\delta) \in S_\delta$ is*

$$\min_{S_\delta} M_{\delta,L_{12}} = \min_{S_\delta} \bar{M}_{\delta,L_{12}} = \Delta^2(\sum_{i=1}^n \sigma_i)^2 + 2\Delta \sum_{i=1}^n \sigma_i, \quad (11.25)$$

where $\sigma_i, i = 1, 2, ..., n$, are the Hankel singular values of the transfer function $H(z)$ defined in Chapter 3 (see (3.29)), which are invariant with respect to all equivalent realizations.

ii) The set of optimal realizations is characterized by

$$S_\delta^{opt} = \{(A_\delta, B_\delta, C_\delta) : W_c = \Delta^2 W_o\} \quad (11.26)$$

where W_c and W_o are the Gramians of the corresponding z-operator realization obtained from (11.13).

Proof: For every $(A_\delta, B_\delta, C_\delta)$ there exists a corresponding triple (A_z, B_z, C_z) defined by (11.13); it has a controllability Gramian W_c and an observability Gramian W_o. We now denote $\bar{W}_o = \Delta^2 W_o$. Therefore by (11.19), (11.24), (3.23) and (3.24), $\bar{M}_{\delta,L_{12}}$ can be expressed as

$$\begin{aligned}\bar{M}_{\delta,L_{12}} &= \Delta^2 tr W_c tr W_o + \Delta^2 tr W_o + tr W_c \\ &= tr \bar{W}_o tr W_c + tr \bar{W}_o + tr W_c \end{aligned} \quad (11.27)$$

Denote $\mu_i = [\lambda_i(W_c \bar{W}_o)]^{1/2} = [\lambda_i(\Delta^2 W_c W_o)]^{1/2} = \Delta \sigma_i$. It then follows from Theorem 3.2 in Chapter 3 that the minimizing value of $\bar{M}_{\delta,L_{12}}$ is

$$\min_{S_\delta} \bar{M}_{\delta,L_{12}} = (\sum_{i=1}^n \mu_i)^2 + 2\sum_{i=1}^n \mu_i, \quad (11.28)$$

that is (11.25), and that this value is achieved if and only if $W_c = \bar{W}_o$ or $W_c = \Delta^2 W_o$. Using the same procedure as used in [Thi86], one can show that

$$\min_{S_\delta} M_{\delta,L_{12}} = \min_{S_\delta} \bar{M}_{\delta,L_{12}}.$$

∎

As in the shift operator realization case, the optimal realizations in the δ-operator realization set are not unique and they form a set S_δ^{opt}.

Theorem 11.3 : Let $S_\delta^{opt} = \{A_\delta^{opt}, B_\delta^{opt}, C_\delta^{opt}\}$ denote the subset of S_δ that minimizes $M_{\delta,L_{12}}$ and let $S_z^{opt} = \{A_z^{opt}, B_z^{opt}, C_z^{opt}\}$ denote the subset of S_z that minimizes $M_{z,L_{12}}$. Then to each $(A_z^{opt}, B_z^{opt}, C_z^{opt})$ there corresponds a $(A_\delta^{opt}, B_\delta^{opt}, C_\delta^{opt})$ such that

$$\begin{aligned} A_\delta^{opt} &= \Delta^{-1}(A_z^{opt} - I) \\ B_\delta^{opt} &= \Delta^{-1/2} B_z^{opt} \\ C_\delta^{opt} &= \Delta^{-1/2} C_z^{opt}. \end{aligned} \quad (11.29)$$

Proof: Consider a member $(A_z^{opt}, B_z^{opt}, C_z^{opt})$ of S_z^{opt}. Then the corresponding Gramians satisfy $W_c = W_o$. Let $(A_\delta^{opt}, B_\delta^{opt}, C_\delta^{opt})$ be obtained from $(A_z^{opt}, B_z^{opt}, C_z^{opt})$ by (11.13):

$$A_\delta = \Delta^{-1}(A_z^{opt} - I), \quad B_\delta = \Delta^{-1} B_z^{opt}, \quad C_\delta = C_z^{opt}. \quad (11.30)$$

We know by Theorem 11.2 that optimality in S_δ requires $W_c = \Delta^2 W_o$. Now apply a similarity transformation $T = \Delta^{-1/2} I$ to $(A_z^{opt}, B_z^{opt}, C_z^{opt})$ and define

$$A_z^{(1)} = T^{-1} A_z^{opt} T, \quad B_z^{(1)} = T^{-1} B_z^{opt}, \quad C_z^{(1)} = C_z^{opt} T. \quad (11.31)$$

The Gramians of the realization $(A_z^{(1)}, B_z^{(1)}, C_z^{(1)})$ are $W_c^{(1)} = \Delta W_c$ and $W_o^{(1)} = \Delta^{-1} W_o$. Since $W_c = W_o$, it follows that $W_c^{(1)} = \Delta^2 W_o^{(1)}$, and hence, by Theorem 11.2 ii), the δ-realization corresponding to $(A_z^{(1)}, B_z^{(1)}, C_z^{(1)})$ is optimal in S_δ (see (11.26)). It now follows from (11.30), (11.31) and $T = \Delta^{-1/2} I$ that this δ-realization is expressed in terms of $(A_z^{opt}, B_z^{opt}, C_z^{opt})$ by (11.29). ∎

Comment 11.3.2: Theorem 11.3 is important in that it shows that the search for optimal realizations $(A_\delta, B_\delta, C_\delta)$ in δ-operator form does not require a new construction. The results of Thiele [Thi84] that characterize the optimal shift operator state variable forms also completely characterize the optimal delta operator forms via (11.29).

Comment 11.3.3: For non-optimal realizations, $\bar{M}_{\rho,L_{12}}$ is much easier to compute than $M_{\rho,L_{12}}$ since it only involves Gramians. It is therefore a reasonable measure of comparison between different realizations.

Comment 11.3.4: A balanced realization (A_b, B_b, C_b) is one of the optimal realizations in shift operator: see Section 3.4. From (11.13) one observes that the possible choices for Δ, compatible with dynamic range considerations, depend on the diagonal elements of $(A_b - I)$. When the largest absolute value of these elements is less than the largest absolute value of the elements of A_b, one can choose Δ smaller than one. This is the case when the poles of a system are clustered around $z = +1$, i.e. when fast sampling is used (see [MG90]). In this case, the diagonal elements of A_b will be near 1, and hence the diagonal elements of $(A_b - I)$ will be much smaller than 1, which yields the possibility of choosing Δ smaller than 1. See also Comment 11.4.1 for a further discussion on the choice of Δ. The following example also illustrates this point.

11.3.3 NUMERICAL EXAMPLE

We now illustrate our previous results and calculations on the optimal sensitivity measure with the following examples, already used in Chapter 3.

Example 11.3.1: Consider a system described in shift operator implementation by the following control canonical companion form:

$\boxed{R_{c,z}:}$

$$A_{c,z} = \begin{pmatrix} 1.9749 & -1.5562 & 0.4538 \\ 1.0000 & 0 & 0 \\ 0 & 1.0000 & 0 \end{pmatrix}$$

$$B_{c,z} = \begin{pmatrix} 1 \\ 0 \\ 0 \end{pmatrix}, \; C_{c,z} = \begin{pmatrix} 0.0793 \\ 0.0230 \\ 0.0232 \end{pmatrix}^T.$$

The poles are at 0.6579 and $0.6585 \pm j0.5061$.

A balanced form in S_z, one of the optimal realizations minimizing $M_{z,L_{12}}$, is given by

$\boxed{R_z^{opt}(M_{L_{12}}):}$

$$A_z^{opt}(M_{L_{12}}) = \begin{pmatrix} 0.8236 & 0.3999 & -0.0164 \\ -0.3999 & 0.5935 & 0.3425 \\ -0.0164 & -0.3425 & 0.5577 \end{pmatrix}$$

$$B_z^{opt}(M_{L_{12}}) = \begin{pmatrix} 0.4424 \\ 0.3799 \\ 0.1671 \end{pmatrix}, \quad C_z^{opt}(M_{L_{12}}) = \begin{pmatrix} 0.4424 \\ -0.3799 \\ 0.1671 \end{pmatrix}^T.$$

The largest number (in magnitude) in the realization $R_z^{opt}(M_{L_{12}})$ is 0.4424. Therefore we choose $\Delta = 2^{-1}$. The corresponding companion and optimal realizations in S_δ are, respectively,

$\boxed{R_{c,\delta}:}$

$$A_{c,\delta} = \begin{pmatrix} -2.0503 & -2.4258 & -1.0203 \\ 1.0000 & 0 & 0 \\ 0 & 1.0000 & 0 \end{pmatrix}$$

$$B_{c,\delta} = \begin{pmatrix} 1 \\ 0 \\ 0 \end{pmatrix}, \quad C_{c,\delta} = \begin{pmatrix} 0.1586 \\ 0.7265 \\ 1.0040 \end{pmatrix}^T.$$

and

$\boxed{R_\delta^{opt}(M_{L_{12}}):}$

$$A_\delta^{opt}(M_{L_{12}}) = \begin{pmatrix} -0.3527 & 0.7999 & -0.0329 \\ -0.7999 & -0.8130 & 0.6849 \\ -0.0329 & -0.6849 & -0.8846 \end{pmatrix}$$

$$B_\delta^{opt}(M_{L_{12}}) = \begin{pmatrix} 0.6256 \\ 0.5373 \\ 0.2363 \end{pmatrix}, \quad C_\delta^{opt}(M_{L_{12}}) = \begin{pmatrix} 0.6256 \\ -0.5373 \\ 0.2363 \end{pmatrix}^T.$$

We note that the coefficient ranges of $R_z^{opt}(M_{L_{12}})$ and $R_\delta^{opt}(M_{L_{12}})$ are roughly the same: in both cases the coefficients are between 2^{-6} and 1. The ratios between maximal and minimal coefficient in the absolute value sense are, respectively, $0.8236/0.0165 \approx 50$ for the shift form, and $0.8846/0.0329 \approx 27$ for the δ form. The optimal values of the sensitivity measures are, respectively,

$$M_{z,L_{12}}^{min} = 4.7560 = \bar{M}_{z,L_{12}}^{min} = M_{z,L_{12}}(R_z^{opt}(M_{L_{12}}))$$

$$M_{\delta,L_{12}}^{min} = 1.8886 = \bar{M}_{\delta,L_{12}}^{min} = M_{\delta,L_{12}}(R_\delta^{opt}(M_{L_{12}})).$$

304   11. Parametrizations in the Delta operator

We have also computed $\bar{M}_{L_{12}}$ for the shift operator and delta operator control canonical forms. These are, respectively, $\bar{M}_{z,L_{12}}(R_{c,z}) = 81.9891$ and $\bar{M}_{\delta,L_{12}}(R_{c,\delta}) = 5.1605$.

These theoretical results will now be confirmed by a numerical simulation on the same example. For both the optimal z-form realization $R_z^{opt}(M_{L_{12}})$ and the optimal δ-form realization $R_\delta^{opt}(M_{L_{12}})$ presented above, we compute the corresponding frequency response $H_{fwl}^{B_c}(\omega)$ obtained when the coefficients are implemented in fixed point with B_c bits for the fractional part of the coefficients, with B_c ranging from 5 to 30. We compare this with the ideal frequency response $H_{id}(\omega)$ implemented with infinite precision by computing the worst deviation over the frequency range, i.e. the H_∞ error:

$$R(B_c) \triangleq log[\max_{\omega \in (0, 2\pi)} |H_{id}(\omega) - H_{fwl}^{B_c}(\omega)|].$$

The results for the example described above are shown in Figure 11.2, which clearly shows the superiority of the optimal δ-form realization over the optimal z-form realization whatever the number of bits.

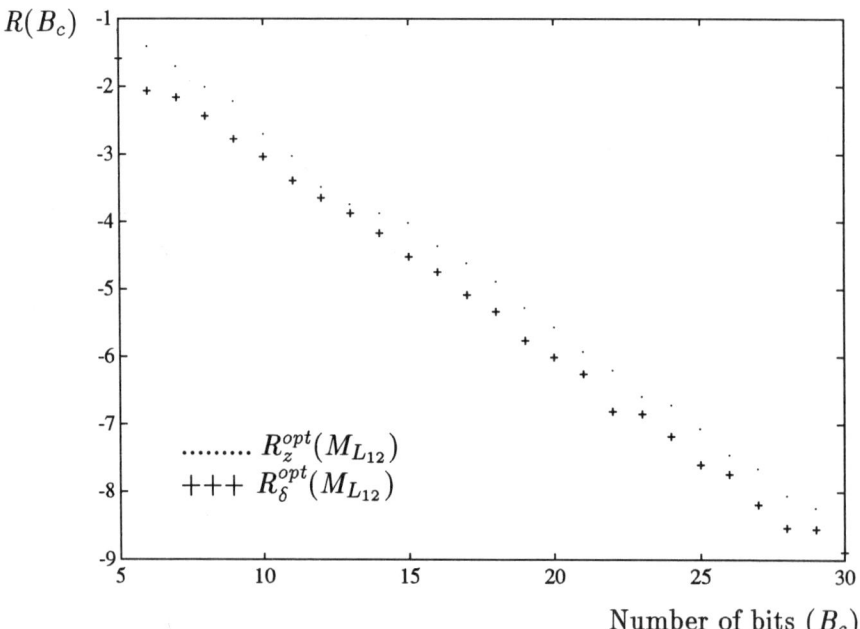

FIGURE 11.2. H_∞ error between ideal and FWL responses for $R_z^{opt}(M_{L_{12}})$ and $R_\delta^{opt}(M_{L_{12}})$

11.4 Roundoff noise analysis

11.4.1 INTRODUCTION

In Chapter 3, the problem of searching for implementations that minimize the roundoff noise gain of the realization has been studied. Within the class of usual shift operator state space realizations, the roundoff noise gain G_z is equal to the trace of the observability Gramian. Subject to a commonly used dynamic range constraint on the states of the realization, the set of realizations minimizing this roundoff noise gain has been completely characterized.

In [Wil86], Williamson proposed the use of residue feedback to reduce the roundoff noise gain for shift operator realizations and compared it with the optimal gain for realizations without residue feedback. He introduced the concept of residue modes and showed that the superiority of the optimal realizations with residue feedback over the optimal realizations without residue feedback hinged on whether the sum of the residue modes was smaller than the sum of the Hankel singular values.

In this section, we study the roundoff noise gain G_δ for state variable realizations implemented in the delta operator, with the aim of examining under what conditions the optimal δ-operator realization roundoff noise gain G_δ^{min} is smaller than the optimal shift operator realization roundoff noise gain G_z^{min}. We first show that the δ-operator implementation is in fact a special case of residue feedback. Therefore, following [Wil86], G_δ^{min} will be smaller than G_z^{min} if and only if the sum of the residue modes is smaller than the sum of the Hankel singular values. We give a few new conditions (i.e. sharper than those in [Wil86]) under which this holds.

11.4.2 THE ROUNDOFF NOISE GAIN OF SHIFT AND DELTA OPERATOR REALIZATIONS

We first recall a few basic facts concerning roundoff noise propagation in state variable realizations when the quantizations are carried out *before multiplication*: see Chapter 2. Assuming that the coefficients of (A_ρ, B_ρ, C_ρ) are represented exactly with B_c fractional bits and that the input signal has B_s fractional bits, the actual state $x^*(t)$ and the actual output $y^*(t)$ then have $B_c + B_s$ fractional bits. The Finite Word

Length (FWL) implementation of (11.16) is

$$\begin{cases} \rho x_t^* = A_\rho Q[x_t^*] + B_\rho u_t \\ y_t^* = C_\rho Q[x_t^*]. \end{cases} \quad (11.32)$$

Here Q represents the quantizer: it rounds the $(B_c + B_s)$-bit fraction x_t^* to B_s bits before multiplication. The roundoff noise

$$e_t \triangleq x_t^* - Q[x_t^*] \quad (11.33)$$

is usually modelled as white noise of zero mean with covariance $q^2 I$, with $q^2 = \frac{1}{12} 2^{-2B_s}$.

Comment 11.4.1: Expression (11.32) represents the FWL implementation of (11.10) under the assumption that the operator ρ is implemented exactly. This is the case when $\rho = z$, or $\rho = \delta$ with $\Delta = 1$. If $\rho = \frac{z-1}{\Delta}$ with $\Delta \neq 1$, the actual implementation of the δ-operator model (11.32) is as follows (see (11.15)):

$$\begin{cases} x_{t+1}^* = x_t^* + \Delta(A_\delta Q[x_t^*] + B_\delta u_t) \\ y_t^* = C_\delta Q[x_t^*] \end{cases} \quad (11.34)$$

We have shown in the previous section that, in order to minimize the sensitivity of the transfer function to coefficient errors, Δ should be chosen as small as possible, but compatible with the dynamic range constraints on the coefficients of A_δ, B_δ, and C_δ. This often allows one to choose values $\Delta < 1$ yielding a minimal sensitivity. Clearly, when $\Delta < 1$, an additional noise is introduced by the multiplication of Δ with $w_t \triangleq A_\delta Q[x_t^*] + B_\delta u_t$ in (11.34). Indeed, if the implementation of Δ requires B_Δ bits, then w_t must be rounded off to $B_c + B_s - B_\Delta$ bits to produce a $(B_c + B_s)$-bit number in (11.34). A complete analysis of this additional roundoff noise can easily be performed. In practice $B_\Delta \ll B_c$ (for example, $B_\Delta = 1$, $B_c = 8$ is typical), and the analysis then shows that this additional noise introduced by the multiplication with Δ in (11.34) can be neglected.

Comment 11.4.2: One procedure that has been advocated to reduce the effect of roundoff noise in digital filter realizations is the use of integer residue feedback [Wil86], [AP79] and [WS85]. In such case, the FWL shift operator state space realization (11.32) is modified according to

$$\begin{cases} x^*_{t+1} = A_z Q[x^*_t] + B_z u_t + Je_t \\ y^*_t = C_z Q[x^*_t] + he_t, \end{cases} \quad (11.35)$$

where all components of J and h are integers (see [Wil86]). We note that with the choices $J = I$ and $h = 0$ and using (11.13), (11.35) becomes identical to the δ-operator implementation (11.34) with $\Delta = 1$. We conclude that the FWL δ-realization is a special case of the residue feedback realization for the choices $J = I$ and $h = 0$.

Denoting $\epsilon y_t \triangleq y_t - y^*_t$, then the roundoff noise gain is usually defined as

$$G = \frac{1}{q^2} \lim_{t \to \infty} E[(\epsilon y_t)^2]. \quad (11.36)$$

To compute G in the special cases of z- and δ-operator realizations, we first write a state equation for the error $E_t \triangleq x_t - x^*_t$. It follows from (11.32)-(11.33) that

$$\begin{cases} \rho E_t = A_\rho E_t + A_\rho e_t \\ \epsilon y_t = C_\rho E_t + C_\rho e_t. \end{cases} \quad (11.37)$$

Replacing ρ, respectively, by z and δ in (11.37), it then follows that

$$G_z = tr(W_o) \quad (11.38)$$
$$G_\delta = tr(W), \quad (11.39)$$

where W_o is the observability Gramian of the realization (A_z, B_z, C_z),

$$W_o \triangleq \sum_{i=0}^{\infty} (A_z^i)^T C_z^T C_z A_z^i \quad (11.40)$$

and W is defined as

$$W \triangleq \Delta A_\delta^T W_o \Delta A_\delta + C_z^T C_z. \quad (11.41)$$

To obtain this last expression, note that (11.37) yields, using (11.13),

$$E_t = \sum_{k=-\infty}^{t-1} A_z^k \Delta A_\delta e_k,$$

and hence,

$$\epsilon y_t = C_\delta \sum_{k=-\infty}^{t-1} A_z^k \Delta A_\delta e_k + C_z e_t.$$

W_o is the solution of the Lyapunov equation

$$W_o = A_z^T W_o A_z + C_z^T C_z. \tag{11.42}$$

Using (11.42) and (11.13), alternative expressions for W can be obtained:

$$\begin{aligned} W &= (A_z - I)^T W_o (A_z - I) + C_z^T C_z \\ &= (I - A_z)^T W_o + W_o (I - A_z) = 2W_o - A_z^T W_o - W_o A_z. \end{aligned} \tag{11.43}$$

It should be clear that the previous expressions hold for matrices (A_δ, C_δ) and (A_z, C_z) that are related by (11.13).

11.4.3 MINIMIZATION OF THE ROUNDOFF NOISE GAIN

For any realization $(A_z, B_z, C_z) \in S_z$, whose corresponding realization in S_δ is $(A_\delta, B_\delta, C_\delta) \in S_\delta$ obtained through (11.13), one has

$$A'_z = T^{-1} A_z T, \; B'_z = T^{-1} B_z, \; C'_z = C_z T \tag{11.44}$$

and

$$A'_\delta = T^{-1} A_\delta T, \; B'_\delta = T^{-1} B_\delta, \; C'_\delta = C_\delta T, \tag{11.45}$$

where T is any nonsingular matrix of proper dimension. It is therefore clear that if (W_o, W) defined in (11.40) and (11.41) correspond to (A_z, B_z, C_z) (equivalently to $(A_\delta, B_\delta, C_\delta)$), the corresponding (W'_o, W') in the new coordinates satisfy the following transformation:

$$W'_o = T^T W_o T, \; W' = T^T W T. \tag{11.46}$$

It follows from (11.38) and (11.39) that different realizations in S_ρ ($\rho = z, \delta$) yield different roundoff noise gains. The interesting problem is to find the optimal realizations in S_ρ which minimize the roundoff noise gain:

$$\min_{(A'_z, B'_z, C'_z) \in S_z} G'_z = \min_{T: \det T \neq 0} tr(T^T W_o T) \tag{11.47}$$

$$\min_{(A_\delta, B_\delta, C_\delta) \in S_\delta} G'_\delta = \min_{T: \det T \neq 0} tr(T^T W T). \tag{11.48}$$

Note that the problems (11.47) and (11.48) do not make sense unless a scaling of the states is introduced since "smaller" T yielding smaller G'_ρ

would make the states larger. In order to maintain the amplitude of the states within an acceptable range, and hence to reduce the probability of overflow, an l_2 norm scaling on the states is introduced in practice, which is equivalent to the following constraint on the realizations in the new coordinates:

$$(W'_c)_{ii} = (T^{-1}W_c T^{-T})_{ii} = 1 \quad \forall i, \tag{11.49}$$

where

$$W_c \triangleq \sum_{i=0}^{\infty} A_z^i B_z B_z^T (A_z^i)^T \tag{11.50}$$

is the controllability Gramian of the system under the realization $(A_z, B_z, C_z) \in S_z$. So, now the problem of minimizing the roundoff noise gain G'_ρ under l_2 scaling can be formulated by combining (11.47)-(11.48) and (11.49)-(11.50). As we have shown in Chapter 3, the minimum achievable gain in S_z is

$$G_z^{min} = \frac{1}{n}(\sum_{i=1}^{n}\sigma_i)^2, \tag{11.51}$$

where $\{\sigma_i\}$ is the Hankel singular value set of the system defined by:

$$\{\sigma_i^2\} = \lambda(W_c W_o) = \lambda(W'_c W'_o). \tag{11.52}$$

This minimum is achieved by a set of realizations in S_z, all of which satisfy the dynamic range constraint (11.49). A constructive procedure for computing this optimal realization set has been given in Chapter 3.

The noise gain in S_δ is given by the exact same form as the noise gain in S_z, with W_o replaced by W and with the same dependence on T (see (11.47) and (11.48)), while the same l_2 norm scaling (11.49) applies. Therefore the minimum noise gain in S_δ is given by:

$$G_\delta^{min} = \frac{1}{n}(\sum_{i=1}^{n}\nu_i)^2, \tag{11.53}$$

where $\{\nu_i\}$ is called the residue mode set [Wil86] defined by

$$\{\nu_i^2\} = \lambda(W_c W) = \lambda(W'_c W'). \tag{11.54}$$

For the same reason, the optimal realizations in S_δ which achieve G_δ^{min} are obtained in the same way as in Chapter 3.

Comment 11.4.3: Since the residue feedback realization of [Wil86], in the case $J = I$ and $h = 0$, is identical to the δ-realization with $\Delta = 1$ (see Comment 11.4.2), it follows immediately that the optimal residue feedback realizations in this special case are identical to the optimal δ-realizations, and hence are also obtained by Hwang's procedure. Thus, there is no need to search for new optimal residue feedback realizations.

Now a reasonable question is under what conditions do we have

$$G_\delta^{min} < G_z^{min}. \tag{11.55}$$

Clearly (11.55) holds if and only if the sum of the residue modes is less than the sum of the Hankel singular values. It is thus interesting to produce simple conditions under which (11.55) holds without having to compute the Hankel singular values δ_i and the Residue Modes ν_i. This problem was studied by Williamson [Wil86] who gave some sufficient conditions for (11.55) to hold. In the next section, we will give new conditions and compare them with Williamson's.

11.4.4 NEW CONDITIONS FOR SUPERIORITY OF δ-REALIZATIONS

In this subsection, we give some conditions under which δ-operator implementations yield a smaller roundoff noise gain than shift operator realizations, that is conditions on the transfer function under which (11.55) can be achieved. In order to do so, we first present the following lemma.

Lemma 11.1 *Let $\{\rho_i^2, \rho_i \geq 0\}$ and $\{\theta_i^2, \theta_i \geq 0\}$ be the diagonal element and eigenvalue sets, respectively, of a nonnegative definite symmetric matrix M. Then*

$$\sum_{i=1}^{n} \rho_i \geq \sum_{i=1}^{n} \theta_i, \tag{11.56}$$

and equality is achieved if and only if the matrix M is diagonal, i.e. $\rho_i = \theta_i \; \forall i$.

Proof: The proof of this lemma is given by Hwang (see [Hwa77]) for the case where M is positive definite symmetric. It is easy to see that the conclusion can be extended to a nonnegative definite matrix. ∎

With this lemma we can prove the following theorem which gives a first new set of sufficient conditions under which (11.55) holds.

Theorem 11.4 *For any stable minimal SISO system (11.5), (11.55) holds if all the diagonal elements $\{a_{ii}\}$ of A_z^{in} satisfy*

$$\frac{1}{2} \leqslant a_{ii} \quad \forall i, \tag{11.57}$$

where $(A_z^{in}, B_z^{in}, C_z^{in})$ is the *input balanced realization* of $H(z)$, characterized by its Gramian matrices:

$$\begin{aligned} W_c^{in} = I, \; W_o^{in} &= \operatorname{diag}(\sigma_1^2, \sigma_2^2, \ldots, \sigma_n^2) \\ &\triangleq \Sigma^2 \end{aligned} \tag{11.58}$$

Proof: First, we note that the Hankel singular values $\{\sigma_i\}$ and Residue Modes $\{\nu_i\}$ are invariant under a coordinate transformation. So

$$\{\nu_i^2\} = \lambda(W_c W) = \lambda(W_c^{in} W^{in}) = \lambda(W^{in}) \tag{11.59}$$

where W^{in} is defined by (11.43) for the input balanced realization characterized by (11.58):

$$\begin{aligned} W^{in} &= (A_z^{in} - I)^T \Sigma^2 (A_z^{in} - I) + (C_z^{in})^T C_z^{in} \\ &= (I - A_z^{in})^T \Sigma^2 + \Sigma^2 (I - A_z^{in}) \end{aligned} \tag{11.60}$$

Denote $W^{in} \triangleq \{w_{ij}\}$ and $A_z^{in} \triangleq \{a_{ij}\}$. It is clear that

$$w_{ii} = 2(1 - a_{ii})\sigma_i^2 \; \forall i.$$

Since W^{in} is positive definite and symmetric, it follows that $w_{ii} > 0$. According to Lemma 11.1, one has

$$\sum_{i=1}^n \nu_i \leqslant \sum_{i=1}^n w_{ii}^{\frac{1}{2}} = \sum_{i=1}^n \sqrt{2(1 - a_{ii})} \sigma_i.$$

The theorem follows from the fact that (11.55) holds if (11.57) is satisfied. ■

In [Wil86], Williamson has given another sufficient condition under which (11.55) holds. This condition is on the *internally balanced realization* $(A^{ib}, B^{ib}, C^{ib}) \in S_z$, which is characterized by its Gramians:

$$W_c^{ib} = W_o^{ib} = \Sigma = \operatorname{diag}(\sigma_1, \sigma_2, \ldots, \sigma_n). \tag{11.61}$$

We now give a sharper result, also on the internally balanced realization.

Theorem 11.5 *For any stable minimal discrete linear time invariant SISO system, there exists an internally balanced form* $(A^{ib}, B^{ib}, C^{ib}) \in S_z$ *such that*

1.
$$A_z^{ib} = \begin{pmatrix} A_{11} & A_{12} \\ -A_{12}^T & A_{22} \end{pmatrix} \quad \text{with} \quad A_{ii} = A_{ii}^T \quad i = 1, 2 \quad (11.62)$$

2. *if* $\{\theta_i\} \triangleq$ *union of* $\lambda(A_{11})$ *and* $\lambda(A_{22})$ *and* λ_k *is any eigenvalue of* A^{ib}, *then*
$$\min_i \theta_i \leq Re(\lambda_k) \leq \max_i \theta_i \quad \forall k \quad (11.63)$$

3. *if*
$$\min \theta_i \geq \frac{1}{2}, \quad \text{then (11.55) holds.} \quad (11.64)$$

Proof: see Appendix 11.

We now compare our results with those of Williamson [Wil86].

1. The existence of (11.62) is always guaranteed while, in [Wil86], the Hankel singular values are assumed to be distinct for (11.62) to exist;

2. A sufficient condition for (11.55), given in [Wil86], is: $\min_i \theta_i \geq 1 - \frac{1}{2n}$ with n the order of system. Here we need only $\min_i \theta_i \geq \frac{1}{2}$, which is a sharper result.

Theorems 11.4 and 11.5 both yield sufficient conditions under which (11.55) holds. These conditions require the computation of (input or internally) balanced forms. Numerically well conditioned algorithms to compute balanced forms can be found in [Lau80] and [Moo81].

In our next Theorem, we give another sufficient condition for (11.55) to hold. This condition is on the poles of the system.

Theorem 11.6 *For any stable minimal discrete SISO linear time invariant system, if the poles* λ_i *of the system satisfy the condition*
$$\sum_{i=1}^n \lambda_i \geq n - \frac{1}{2}, \quad (11.65)$$
then (11.55) holds.

Proof: The internally balanced form defined by (11.61) satisfies the following Lyapunov matrix equation:

$$\Sigma = A_z^{ib} \Sigma A_z^{ibT} + B_z^{ib} B_z^{ibT}. \tag{11.66}$$

It follows that the diagonal elements $\{a_{ii}\}$ of A_z^{ib} satisfy

$$|a_{ii}| < 1. \tag{11.67}$$

From $\sum_{i=1}^{n} a_{ii} = \sum_{i=1}^{n} \lambda_i$ one obtains

$$\min_i a_{ii} + (n-1) > \sum_{i=1}^{n} \lambda_i$$

or

$$\min_i a_{ii} > 1 - (n - \sum_{i=1}^{n} \lambda_i).$$

Therefore, using (11.65), we have $\min_i a_{ii} > \frac{1}{2}$. The theorem follows directly by applying Theorem 11.4. ∎

The sufficient condition (11.65) can be rewritten as

$$\bar{\lambda} \triangleq \frac{1}{n} \sum_{i=1}^{n} \lambda_i \geq 1 - \frac{1}{2n} \triangleq \bar{\lambda}_{min} \tag{11.68}$$

Clearly, $\bar{\lambda}$ is the mean value of the poles.

Example 11.4.1: For $n = 4$, $\bar{\lambda}_{min} = 1 - \frac{1}{2n} = 0.875$. So, for any system of order 4, the optimal δ-operator implementation is superior to the shift operator implementation in terms of roundoff noise gain if the mean pole value $\bar{\lambda}$ is larger than 0.875.

Example 11.4.2: In [Wil88], a sixth-order narrow band low pass filter is considered, whose poles are $0.9723 \pm j0.1989$, $0.9389 \pm j0.1623$, $0.9152 \pm j0.0646$. For this filter, one has

$$\bar{\lambda} = 0.9441 \text{ and } \bar{\lambda}_{min} = 0.9167.$$

So, for this system, the optimal δ-operator implementation will have a better performance in terms of roundoff noise gain than the optimal shift operator implementation.

Theorem 11.6 yields a sufficient condition for (11.55) that is very easy to test, since the system is normally given by its transfer function from which the poles can be obtained easily.

The theorem guarantees the superiority of implementation in δ-operator over shift operator for a class of systems. In fact, it implies that for systems whose poles are clustered around $z = +1$, the δ-operator implementation yields a better performance in terms of minimizing the roundoff noise gain. The often used narrow band low pass filters belong to this class [Wil88]. In modern control, fast sampling is used in order to keep enough information [MG90]. The discrete time models used in practice come from the corresponding continuous time systems sampled with high frequency. With the sampling frequency chosen between 5 and 10 times the maximal frequency of interest as proposed by Middleton and Goodwin [MG90], the poles of the corresponding discrete time models and controllers are clustered around $z = +1$, and so here again the δ-operator models will typically perform better.

11.4.5 NUMERICAL EXAMPLES

In this subsection, we return to two numerical examples that have been used in earlier chapters.

Example 11.4.3: In Example 3.2, a low pass filter of order three was examined, with the following poles: $(0.6585 + j0.5061, 0.6585 - j0.5061, 0.6579)$. With Theorem 11.6 it is easy to verify that implementations in R_δ^{opt} yield a smaller roundoff noise gain for this filter than realizations in R_z^{opt}. In fact, $G_z^{min} = 0.6526$ and $G_\delta^{min} = 0.3106$.

Example 11.4.4: Now, let us consider the sixth-order narrow band filter of Example 11.4.2 again. According to Theorem 11.6, we have seen that implementations in R_δ^{opt} yield a better performance than in R_z^{opt}. This is confirmed by the computation of the noise gains: $G_z^{min} = 1.3329$ and $G_\delta^{min} = 0.0646$.

The optimal realizations in either S_z or S_δ, $(A_\rho^{opt}, B_\rho^{opt}, C_\rho^{opt})$, are full of non-zero-or-one elements. This is not very desirable since it maximizes the number of arithmetic operations. For those reasons, sparser realizations are preferred and some efforts have been made in this direction [SB88], [LH88] and [LAGP92]. We note that for the class of systems discussed just before, G_δ^{min} is smaller than G_z^{min}. This implies that some sparse realizations in S_δ could possibly have a noise gain G_δ that is near G_z^{min} (but of course larger than G_δ^{min}). For example, a companion form (direct form) realization in S_δ can possibly have a numerical performance that is better than that of a (fully parametrized) optimal shift operator realization, as illustrated in [Wil88]. In [LG93] we have produced alternative sparse realizations in S_δ, using some properties of Chebyshev polynomials. With the same numerical example as in Example 11.4.4, such Chebyshev based sparse realization yields a roundoff noise gain G_δ that is smaller than G_z^{min}: see [LG93].

11.5 Conclusions

Our aim in this chapter has been to investigate the numerical properties of the delta operator parametrizations by comparing them with those of the usual shift operator parametrizations in terms of transfer function sensitivity and roundoff noise gain.

We have examined both the L_1/L_2 and the L_2 transfer function sensitivity measures for δ-operator realizations and compared them with the corresponding measures for shift operator realizations. Our conclusion has been that δ-operator realizations can be made to have a smaller sensitivity measure if Δ is chosen smaller than 1. There is a limit, however, to how small Δ can be chosen, and this limit is determined by dynamic range considerations. The relationship between the optimal shift operator and δ-operator realization sets has been established: the optimal δ-operator realizations can thus be obtained directly from the optimal shift operator realizations (or, alternatively, they can be obtained directly by the same procedure) without the need for a new algorithm.

Similarly, we have analysed the roundoff noise effects in a FWL δ-operator state space implementation of a discrete time system. The expression of the roundoff noise gain has been derived. We have minimized this roundoff noise gain over the realization set of δ-operator models, S_δ, and have compared this expression, G_δ^{min} with the optimal gain for the shift operator realizations, G_z^{min}. Some new conditions for the superiority of optimal δ-operator implementations over optimal shift operator implementations have been given.

We have seen in this chapter that the use of δ-operator parametrizations can improve the numerical accuracy of the implementation of a digital filter, compared with the usual shift operator implementations. This improvement is particularly significant for narrow band low pass filters. From a purely algebraic point of view, $\delta = (z-1)/\Delta$ is nothing more than a linear transformation of the usual shift operator. We have also noted that this transformation corresponds to a *re-parametrization of the transfer function* itself, and thus δ-operator state space realizations are not obtainable within the equivalence class of shift operator realizations by similarity transformations. This motivates us to investigate more general operator transformations and to exploit new equivalence classes of transfer function parametrizations. This will be the topic of our next chapter.

Appendix 11: Proof of Theorem 11.5.

First, note that Kung [Kun78] has shown that for any stable minimal SISO discrete linear system, there exists an internally balanced form which satisfies the following symmetry property:

$$A_z^{ib} = Q A_z^{ib^T} Q, \quad B^{ib} = Q C_z^{ib^T} \qquad (11.A.1)$$

where Q is a sign matrix:

$$Q = diag(u_1, u_2, \ldots, u_n) \qquad u_i = \pm 1 \; \forall i.$$

Clearly, with a series of permutations, Q can be transformed to

$$\begin{pmatrix} I_1 & 0 \\ 0 & -I_2 \end{pmatrix} \text{ with } I_i = \begin{pmatrix} 1 & & 0 \\ & 1 & \\ & & \ddots \\ 0 & & & 1 \end{pmatrix}_{n_i \times n_i} \quad i = 1, 2 \qquad (11.A.2)$$

where $n_1 + n_2 = n$, while preserving the structure of (11.A.1). So, without loss of generality, Q in (11.A.1) can be assumed to be of the form (11.A.2). It follows from (A.1) that

$$A_z^{ib} Q = Q A_z^{ib^T} = (A_z^{ib} Q)^T = \begin{pmatrix} X_{11} & X_{12} \\ X_{12}^T & X_{22} \end{pmatrix}$$

with X_{ii}, $i = 1, 2$, real symmetric. It then follows that

$$A_z^{ib} = \begin{pmatrix} X_{11} & X_{12} \\ X_{12}^T & X_{22} \end{pmatrix} Q = \begin{pmatrix} X_{11} & X_{12} \\ X_{12}^T & X_{22} \end{pmatrix} \begin{pmatrix} I_1 & 0 \\ 0 & -I_2 \end{pmatrix}$$

$$= \begin{pmatrix} X_{11} & -X_{12} \\ X_{12}^T & -X_{22} \end{pmatrix}.$$

So, (11.62) is proved with the following identifications:

$$A_{11} = X_{11}, A_{22} = -X_{22}, A_{12} = -X_{12}$$

Now consider (11.63). It is well known that for any square real matrix A, the following decomposition holds

$$A = A_s + A_{sk} \text{ with } A_s = A_s^T, A_{sk}^T = -A_{sk}. \qquad (11.A.3)$$

If λ is an eigenvalue, $\lambda \in \lambda(A)$, with X the corresponding eigenvector, then

$$AX = \lambda X \Rightarrow X^H AX = \lambda \|X\|_2^2 \Rightarrow X^H A_s X = Re(\lambda)\|X\|_2^2.$$

By SVD decomposition, A_s can be written as

$$A_s = U \begin{pmatrix} \theta_1 & & 0 \\ & \ddots & \\ 0 & & \theta_n \end{pmatrix} U^T \text{ with } U \text{ orthogonal}.$$

Denote $Y = UX$, then

$$\sum_{i=0}^{n} \theta_i |y_i|^2 = Re(\lambda)\|X\|_2^2.$$

Since $\|Y\|_2^2 = \|X\|_2^2$, or $\sum_{i=1}^n |y_i|^2 = \sum_{i=1}^n |X_i|^2$, it follows that

$$\min_i \theta_i \leqslant Re(\lambda)) \leqslant \max_i \theta_i.$$

The inequality (11.63) follows with

$$A_s = \begin{pmatrix} A_{11} & 0 \\ 0 & A_{22} \end{pmatrix}, A_{sk} = \begin{pmatrix} 0 & A_{12} \\ -A_{12}^T & 0 \end{pmatrix}.$$

Finally, recall that $\{\theta_i\} = \lambda(A_{11}) \cup \lambda(A_{22})$ with A_{11}, A_{22} real symmetric. If $\min_i \theta_i > 0$, then $A_{ii}, i = 1, 2$ are also positive definite. It is well known (see [Bel70] p.134) that

$$\min_i \theta_i \leqslant \min_i a_{ii},$$

where $\{a_{ii}\}$ is the diagonal element set of A_s (or A_z^{ib}). We note that the matrix A_z^{in} of the input balanced form has the same diagonal elements as A_z^{ib}. Clearly, (11.64) then follows from Theorem 11.4 with $\min_i \theta_i \geq \frac{1}{2}$.

12

Generalized transfer function parametrizations for adaptive estimation

12.1 Introduction

All the results so far in this book, and most of the results available in the literature on the minimization of the effects of Finite Word Length (FWL) errors - due to FWL encoding and/or to arithmetic roundoff - have been for applications in filter design or, more recently, in controller design. Although a few other approaches are possible (as we have seen in Section 3.9), for the most part these analyses lead to the formulation and the solution of optimal state space design problems, in which the search is performed over the set of all equivalent (i.e. similar) state space realizations of a filter or a controller.

Less attention has been paid to the minimization of numerical errors in identification or parameter estimation problems. The existing results in this area are for the most part on the selection of algorithms having good numerical properties: [LL85], [VD86] and [Ver89], but not on the issue of parametrization. A notable exception is [Goo88], where Goodwin addressed the numerical issues in parameter estimation and filtering problems, and showed that better accuracy can be obtained if a δ-operator parametrization is used instead of the usual shift operator parametrization.

In this chapter we generalize this idea: we show that there is an equivalence class of transfer function parametrizations, and that the numerical accuracy with which Least Squares (LS) parameter estimation problems are solved are intimately connected to the parametrization that is used to represent the input-output relation. In fact, we first connect the choice of a parametrization with that of a data prefilter. This establishes the link between the choice of parametrization and the classical techniques that are used by numerical analysts to solve LS problems in a numerically efficient way. These techniques come down to a transformation of the information matrix (the matrix made up of the regression vectors) using QR or other transformation techniques:

see e.g. [Bie77], [Ste73]. Therefore these techniques can be seen as both a prefiltering of the regressor vector (and hence of the data processes) and a corresponding transformation of the parameter vector.

An optimal transformation, leading to an information matrix that is optimally conditioned, is therefore possible only if the spectral properties of the input and output process (or, equivalently, of the input process and the system being estimated) are known. This is a reasonable assumption in an off-line parameter estimation application, but not in adaptive identification or filtering applications. Our major contribution in this chapter is to show that some parametrizations, including a particular one that we shall call γ-operator parametrization, have some robustness properties in that with these parametrizations the parameter estimation problem uses an information matrix that has desirable properties irrespective of the data processes and of the unknown system. Thus, this γ-operator parametrization, obtained from the shift operator parametrization by a bilinear transformation, is particularly suited for the case where no prior information is available about the data processes or its spectral properties, such as in adaptive parameter estimation problems.

This chapter is organized as follows. In Section 12.2 we explain the importance of the information matrix in parameter estimation problems, in particular its role in determining the accuracy of the parameter estimate and the convergence speed of an adaptive parameter estimation algorithm. In Section 12.3 we establish the connections between the choice of parametrization, a filtering of the regressor vector and a filtering of the data processes. Optimal and suboptimal parametrization design games are formulated and discussed in Section 12.4, while in Section 12.5 we present our γ-operator parametrization and exhibit the properties it gives to the information matrix in a LS parameter estimation problem. A comparison is also made with the by now more classical δ-operator parametrization. Finally, the advantage of the γ-operator parametrization in adaptive identification and filtering problems is spectacularly evidenced in Section 12.6 by simulations.

12.2 Parameter estimation and the information matrix

Consider that it is desired to estimate the parameters $\{a_i\}$ and $\{b_i\}$ of the transfer function

$$H(z) = \frac{b_0 z^n + b_1 z^{n-1} + \ldots + b_n}{z^n + a_1 z^{n-1} + \ldots + a_i z^{n-i} + \ldots + a_n} \qquad (12.1)$$

from input and output data, $\{u(t)\}$ and $\{y(t)\}$. Assuming that the data and/or the model are not exact, then the following is a commonly used model for this input-output relation:

$$\begin{aligned} y(t) &= -\sum_{i=1}^{n} a_i y(t-i) + \sum_{i=0}^{n} b_i u(t-i) + v(t) \\ &= \theta_z^T \Psi_z(t) + v(t), \end{aligned} \qquad (12.2)$$

where

$$\Psi_z(t) \triangleq \Big(-y(t-1) \quad \cdots \quad -y(t-n) \quad u(t) \quad \cdots \quad u(t-n) \Big)^T \qquad (12.3)$$

is the *regressor vector*,

$$\theta_z \triangleq \begin{pmatrix} a \\ \bar{b} \end{pmatrix} \qquad (12.4)$$

is the parameter vector, a and $\bar{b}$ are defined as

$$\begin{aligned} a &= (a_1 \ \ldots \ a_n)^T \\ \bar{b} &= (b_0 \ b_1 \ \ldots \ b_n)^T, \end{aligned} \qquad (12.5)$$

and $v(t)$ is a noise process. The model (12.2) is an *equation error model*, because an error $v(t)$ is assumed on the input-output equation; it is more commonly called an ARX[1] model or a *Least Squares (LS) regression* model in identification jargon. We note that it is a model in the shift operator, because the regressor contains delayed (i.e. shifted) versions of the input and output signals; hence the notation θ_z. For future use, we also introduce the following expression for $\Psi_z(t)$:

$$\Psi_z(t) = z^{-n} \begin{pmatrix} \nu_n(z) & 0 \\ 0 & \nu_{n+1}(z) \end{pmatrix} \begin{pmatrix} -y(t) \\ u(t) \end{pmatrix}, \qquad (12.6)$$

---

[1] A horrible jargon, coined down by econometricians, and supposed to mean Auto Regressive with eXogeneous inputs.

where $\nu_k(z) \triangleq \begin{pmatrix} z^{k-1} & \cdots & z & 1 \end{pmatrix}^T$.

Perhaps the most classical way of estimating the parameter vector θ_z is by minimizing the sum of the squares of the equation errors, $\sum_t v^2(t)$. The LS estimate of θ_z is then obtained as the solution of the following equations:

$$\left(\sum_t \Psi_z(t) \Psi_z^T(t) \right) \hat{\theta}_z = \sum_t \Psi_z(t) y(t),$$

or equivalently,

$$R_z \hat{\theta}_z = p_z, \qquad (12.7)$$

where

$$R_z \triangleq \sum_t \Psi_z(t) \Psi_z^T(t) \quad \text{and} \quad p_z \triangleq \sum_t \Psi_z(t) y(t). \qquad (12.8)$$

The matrix R_z is called the *information matrix*. The estimate $\hat{\theta}_z$ is the solution of the linear system (12.7): it is unique if R_z is nonsingular.

The information matrix plays a crucial role in estimation and optimal filtering, particularly with respect to the accuracy of the solution. It is constructed from the data (or signals), and hence its properties depend on these data. For example, some 'sufficient richness' is required on the signals to make the matrix R_z nonsingular, an essential requirement to obtain a unique solution for $\hat{\theta}_z$: see e.g. [Lju87]. Here we formulate some observations about the properties of the information matrix, and their connection with the accuracy of the solution. In the next section we show how these properties can be changed by the use of transformations on the regression vector which, as we shall show, correspond to changing the basis for the representation of the numerator and denominator polynomial of $H(z)$.

Observation A:

In practice the elements in R_z and p_z are incorrect because of the FWL encoding of the signals and the arithmetic roundoff in the computation of these elements. When the information matrix R_z and the vector p_z are perturbed, the solution of the equation (12.7) is also perturbed. It can be shown that the relative error of the solution is proportional to the condition number² of the information matrix. This condition number is called the sensitivity of the solution of this linear equation [GL83].

Usually the data are obtained by sampling the corresponding continuous time signals. If fast sampling is used, the corresponding information matrix is poorly conditioned because the input and output signals are both highly correlated. The condition number actually tends towards infinity when the sampling period goes to zero [MG90].

Observation B:

Consider the problem of solving (12.7) by using the off-line Gradient Algorithm,

$$\hat{\theta}_z(k+1) = \hat{\theta}_z(k) + \mu[p_z - R_z\hat{\theta}_z(k)], \qquad (12.9)$$

where $\hat{\theta}_z(k)$ is the estimate of $\hat{\theta}_z$ at the k-th iteration and μ is the adaptation step size, which controls the speed of adaptation. To guarantee the convergence of the algorithm (12.9) to the solution of (12.7), the step size has to be chosen small enough so that all the eigenvalues of $(I - \mu R_z)$ are absolutely smaller than one. For a fixed μ, the speed of convergence of the algorithm can be measured by the maximum of the absolute eigenvalue of $(I - \mu R_z)$. By minimizing this value, the optimal step size μ_{opt} can be shown to be (see [HM84]):

$$\mu_{opt} = 2/(\lambda_{\min} + \lambda_{\max}), \qquad (12.10)$$

where $\lambda_{\min}$ and $\lambda_{\max}$ are the minimum and maximum eigenvalues of the information matrix R_z, respectively. In this case, the maximum of the absolute eigenvalues of $(I - \mu R_z)$ is

$$\max_i |\lambda_i(I - \mu_{opt} R_z)| = \frac{\frac{\lambda_{\max}}{\lambda_{\min}} - 1}{\frac{\lambda_{\max}}{\lambda_{\min}} + 1} \triangleq \frac{\kappa - 1}{\kappa + 1}, \qquad (12.11)$$

where κ is the condition number of R_z. Clearly, the convergence of the algorithm (12.9) will be very slow when the information matrix is badly conditioned. Hence the condition number influences not only the accuracy of the solution, but also the speed of convergence of the iterative

---

[2]The condition number of a matrix A, denoted $\kappa(A)$, is defined as $\kappa(A) = \frac{\sigma_{max}(A)}{\sigma_{min}(A)}$, where σ_{max} and σ_{min} denote the largest and smallest singular values, respectively.

324   Generalized transfer function parametrizations

gradient algorithm.

Observation C:

Now, suppose we compute the LS solution by handling the data one by one using the on-line LMS (Least Mean Square) adaptive algorithm (sometimes called stochastic gradient algorithm):

$$\hat{\theta}_z(t+1) = \hat{\theta}_z(t) + \mu e(t)\Psi_z(t), \qquad (12.12)$$

where

$$e(t) \triangleq y(t) - \hat{y}(t) = y(t) - \hat{\theta}_z^T(t)\Psi_z(t) \qquad (12.13)$$

is the prediction error at time t. A statistical analysis of the convergence of this algorithm can be performed as follows. Denote by

$$E[\Delta\theta_z(t)] \triangleq E[\theta_z - \hat{\theta}_z(t)] \qquad (12.14)$$

the statistical mean of the parameter vector error at time t. If the errors $v(t)$ have zero mean, it then follows that

$$E[\Delta\theta_z(t+1)] = E[\Delta\theta_z(t)] - \mu R_z E[\Delta\theta_z(t)], \qquad (12.15)$$

provided the step size μ is chosen small enough so that $\Delta\theta_z(t)$ and $\Psi_z(t)$ can be considered to be uncorrelated [HM84]. The convergence and the convergence speed of the mean error $\Delta\theta_z(t)$ depend again on the value of μ.

Observation D:

A more subtle phenomenon, related to the convergence issue, involves the "colouration" of the regressor vector (and hence of the data processes). If the regressor vector is white, then R_z is a diagonal matrix. In this case, the negative gradient points towards the minimum of the mean square of the error. However, if the regressor vector is coloured, then R_z has significant off-diagonal elements, and the negative gradient misses the optimal direction [RM87]. In fact, the off-diagonal elements in R_z produce couplings (or correlations) between different parameters in the parameter vector estimate. Consider, for example, the error equation (12.15). The i-th equation is given as

$$E[\Delta\theta_{z,i}(t+1)]$$
$$= E[\Delta\theta_{z,i}(t)] - \mu \sum_{j=1}^{2n+1} R_z(i,j) E[\Delta\theta_{z,j}(t)]. \qquad (12.16)$$

This expression shows that, even if some component of the parameter estimate has already reached its optimal value, it could well be moved away from it if at least one other component has not reached its optimal value, as long as the corresponding correlation coefficient $R_z(i,j)$ between the two components is nonzero. The same analysis holds for the off-line algorithm (12.9), with the same conclusions. This shows the advantage of having an information matrix that is not only well conditioned, but also as near a diagonal matrix as possible.

We conclude that to improve the accuracy of a parameter estimate and to accelerate the convergence of the estimation algorithm, it is desirable to have an information matrix that is

- well conditioned

- of diagonal form, or with many zero off-diagonal elements.

The arguments in our observations A to D have been developed for parametrizations in the shift operator, where the information matrix is made up from regression vectors that contain shifted versions of the signals: see (12.3). The same arguments apply to any Least Squares estimation scheme, and our whole pursuit in this chapter will be to replace the shift operator parametrizations by appropriately chosen polynomial operator parametrizations so that the corresponding information matrices will be better conditioned and closer to diagonal matrices.

The information matrix is constructed from the signal processes and its properties therefore depend on these signals. In particular, when fast sampling is used, the information matrix R_z corresponding to the shift operator parametrization (12.2) is very ill conditioned. An intuitive idea is to change the processes to ones that have better properties by using prefilters, and then to form a new regressor vector using these filtered signals. We shall see in the next section that prefiltering the data is in fact equivalent to changing the basis vectors in the parametrization of the system. This will then lead us to select parametrizations in such a way as to produce well conditioned information matrices for the corresponding regression vectors.

There is an extensive literature on the numerical solution of Least Squares problems, and there exist several methods for computing $\hat{\theta}_z$ without actually inverting R_z (see e.g. [Lju87], [Bie77], [GL83]). These methods use Householder transformations or Cholesky factorizations: they all come down to computing square roots of R_z, but directly from the data, that is without first computing the data products that enter into the elements of R_z.

An alternative way to produce better behaved regression vectors is to filter the data processes through an *adaptive lattice filter*. After convergence of this filter, the backward residuals form an orthogonal process. If these backward residuals are used to form the regression vector, the convergence speed of the parameter estimator is greatly increased: see [Cha71] and [Mak78]. This route effectively requires a succession of two convergences: a convergence of the parameters of the adaptive lattice filter until the backward residuals eventually become orthogonal, and the subsequent convergence of the parameters of the system being identified to their 'true' values. However, a significant improvement can already be obtained if the residuals are nearly orthogonal and thus the parameters of the lattice filter need not have converged to their optimal values.

The methods just described come down to filtering the data with filters that are *data dependent*. In this chapter we show that one way to improve the numerical properties of a LS parameter estimator is to replace the shift operator parametrization of the transfer function by appropriately chosen polynomial-operator parametrizations so that the corresponding information matrices are better conditioned and closer to a diagonal matrix. The use of such polynomial-operator parametrizations *in lieu of* the classical shift operator parametrization is effectively a way of filtering the data (and hence the regression vector), and as such it could be seen as nothing more than a variation on the classical methods that numerical analysts have developed for solving LS problems in a numerically efficient way. Indeed, we shall formally display this link and show how the parametrization can be optimally selected on the basis of spectral information about the data. But, besides the fact that these generalized parametrizations highlight the role of the basis functions in the description of a transfer function, we shall show that, even when no information is available about the statistical properties of the data, *parametrizations can be chosen that have some robustness properties, in that they yield information matrices that have nice properties whatever the data*. One such example is our γ-operator parametrization.

12.3 Connection between parametrization and data filtering

In this section we show how a filtering of the regression vector can improve the properties of the information matrix, and we establish the connection between such regression filtering and the parametrization of the transfer function.

We consider again the model (12.1)-(12.4) for some given parameter vector θ_z. Now let $T \in \mathbb{R}^{(2n+1) \times (2n+1)}$ be any nonsingular matrix and let $p \in \mathbb{R}^n$ and $q \in \mathbb{R}^{n+1}$ be any two known vectors. We can then define a new vector θ_p as the solution of the following set of equations:

$$\theta_z = \begin{pmatrix} a \\ b \end{pmatrix} = \begin{pmatrix} p \\ q \end{pmatrix} + T^T \begin{pmatrix} \alpha \\ \bar{\beta} \end{pmatrix} \triangleq \theta_z^0 + T^T \theta_p \qquad (12.17)$$

where $\theta_z^0 \triangleq \begin{pmatrix} p^T & q^T \end{pmatrix}^T$ and $\theta_p \triangleq \begin{pmatrix} \alpha^T & \bar{\beta}^T \end{pmatrix}^T$ with $\alpha \in \mathbb{R}^n$ and $\bar{\beta} \in \mathbb{R}^{n+1}$. It then follows from (12.2) that

$$y(t) = (\theta_z^0)^T \Psi_z(t) + \theta_p^T T \Psi_z(t) + v(t),$$

or in short

$$y^*(t) = \theta_p^T \Psi_p(t) + v(t), \qquad (12.18)$$

where

$$\begin{aligned} y^*(t) &\triangleq [1 + p_1 z^{-1} + \cdots + p_n z^{-n}] y(t) \\ &\quad - [q_0 + \cdots + q_n z^{-n}] u(t) \\ &\triangleq p(z) z^{-n} y(t) - q(z) z^{-n} u(t) \end{aligned} \qquad (12.19)$$

$$\begin{aligned} \Psi_p(t) &\triangleq T \Psi_z(t) = T \begin{pmatrix} \nu_n(z) & 0 \\ 0 & \nu_{n+1}(z) \end{pmatrix} z^{-n} \begin{pmatrix} -y(t) \\ u(t) \end{pmatrix} \\ &\triangleq F(z) \begin{pmatrix} -y(t) \\ u(t) \end{pmatrix} \end{aligned} \qquad (12.20)$$

with

$$p(z) \triangleq z^n + \sum_{i=1}^{n} p_i z^{n-i} \quad \text{and} \quad q(z) \triangleq \sum_{i=0}^{n} q_i z^{n-i}. \qquad (12.21)$$

We note that $F(z)$ is a $(2n+1) \times 2$ proper rational filter:

$$F(z) \triangleq z^{-n} T \begin{pmatrix} \nu_n(z) & 0 \\ 0 & \nu_{n+1}(z) \end{pmatrix}. \qquad (12.22)$$

For given T and θ_z^0, $y^*(t)$ and $\Psi_p(t)$ can be computed from the data $\{u(t)\}$ and $\{y(t)\}$. The problem of estimating θ_z from $\{u(t)\}$ and $\{y(t)\}$ is then converted into one of estimating θ_p using $\{y^*(t)\}$ and $\{\Psi_p(t)\}$ which are filtered versions of the original signals. The information matrix of $\Psi_p(t)$ will be denoted by R_p:

$$R_p = \sum_t \Psi_p(t) \Psi_p^T(t). \qquad (12.23)$$

It follows immediately from (12.20) that the information matrices of $\Psi_p(t)$ and $\Psi_z(t)$ are related by

$$R_p = T R_z T^T. \qquad (12.24)$$

The manipulations above allow us to make some interesting observations.

1. Filtering the data and the regressor is effectively equivalent to a reparametrization of the initial model: θ_z is replaced by θ_p.

2. The vectors p and q in (12.17), and hence the polynomials $p(z)$ and $q(z)$, are arbitrary. The filter $\frac{q(z)}{p(z)}$ can thus be set as an (a priori) estimate of the transfer function $H(z)$ of the system. The effect of the filtering operation (12.19) is to centre the parameter vector θ_z on $[p^T q^T]^T$.

3. The key observation is that by proper filtering - that is by proper choice of T or, equivalently, of $F(z)$ - the new regression vector $\Psi_p(t)$ can be made to have nice properties, such as a better conditioned information matrix and, possibly, a diagonal or quasi-diagonal structure, leading to decoupled or quasi-decoupled parameters.

We illustrate the latter point with an example.

Example 12.1: Let $H(z)$ be the finite impulse response (FIR) model:

$$H(z) = 1 - 2.1000 z^{-1} + 1.4600 z^{-2} - 0.3360 z^{-3},$$

corresponding to an input-output relation

$$y(t) = \sum_{i=1}^{4} b_i u(t-i) = \theta_z^T \Psi_z(t), \qquad (12.25)$$

with $\Psi_z(t) = (u(t)\ u(t-1)\ u(t-2)\ u(t-3))^T$. We consider the estimation of the parameters b_i from input-output data. With $\{u(t)\}$ obtained from low pass filtered Gaussian noise, the regression vector $\Psi_z(t)$ produces the following information matrix,

$$R_z = \begin{pmatrix} 0.2465 & 0.2235 & 0.1636 & 0.0905 \\ 0.2235 & 0.2465 & 0.2232 & 0.1634 \\ 0.1636 & 0.2232 & 0.2465 & 0.2231 \\ 0.0905 & 0.1634 & 0.2231 & 0.2465 \end{pmatrix}, \quad (12.26)$$

whose condition number is 3.7500×10^3 with $\lambda_{min}(R_z) = 2.1221 \times 10^{-4}$ and $\lambda_{max}(R_z) = 0.7958$.

We now apply a succession of two appropriately chosen[3] transformations $T_1 \in \mathbb{R}^{4\times 4}$ and $T_2 = UT_1$, with U orthogonal, both with $p = 0$ and $q = 0$. The system is then reparametrized as

$$y(t) = \theta_{p1}^T \Psi_{p1}(t) = \theta_{p2}^T \Psi_{p2}(t),$$

with $\Psi_{p1}(t) = T_1 \Psi_z(t)$ and $\Psi_{p2}(t) = UT_1 \Psi_z(t)$. In the new parametrization, the exact parameter vectors are:

$$\theta_{p1}^T \triangleq \theta_z^T T_1^{-1} = \begin{pmatrix} 0.1121 & 0.1815 & 0.1016 & 0.0240 \end{pmatrix}$$

and

$$\theta_{p2}^T \triangleq \theta_z^T T_1^{-1} U^T = \begin{pmatrix} -0.0200 & -0.0151 & 0.0441 & 0.2320 \end{pmatrix}.$$

The corresponding information matrices are given, respectively, by

$$R_{p1} = \begin{pmatrix} 0.3727 & 0.0002 & -0.2079 & -0.0012 \\ 0.0002 & 0.2255 & 0.0011 & -0.1424 \\ -0.2079 & 0.0011 & 0.1921 & 0.0002 \\ -0.0012 & -0.1424 & 0.0002 & 0.2117 \end{pmatrix}$$

and

$$R_{p2} = \begin{pmatrix} 0.0946 & 0.0811 & -0.0289 & -0.0335 \\ 0.0811 & 0.4052 & 0.0973 & -0.0307 \\ -0.0289 & 0.0973 & 0.4067 & 0.0804 \\ -0.0335 & -0.0307 & 0.0804 & 0.0956 \end{pmatrix}.$$

---

[3] We shall explain later in this section how such transformations can be "appropriately chosen".

Because these two information matrices are related by an orthogonal transformation matrix U, they have exactly the same eigenvalues and hence the same condition number: for both of them $\lambda_{max} = 0.5091, \lambda_{min} = 0.0558$. However, the off-diagonal terms in R_{p1} are much smaller than in R_{p2} and we now illustrate the perverse effect of these off-diagonal terms. For both θ_{p1} and θ_{p2}, we have run the iterative LS parameter estimator (12.9) with $\mu = \mu_{opt} = 3.5404$ and with initial conditions that were exact, except for the *second* component which was perturbed by 10 percent. We have then observed the evolution, over the iterations, of the error $\tilde{\theta}_{pi,1}(t) \triangleq \theta_i(1,t) - \hat{\theta}_i(1,t)$ in the *first* parameter estimate for the two vectors, i.e. for $i = 1, 2$. Figure 12.1 shows the simulation results.

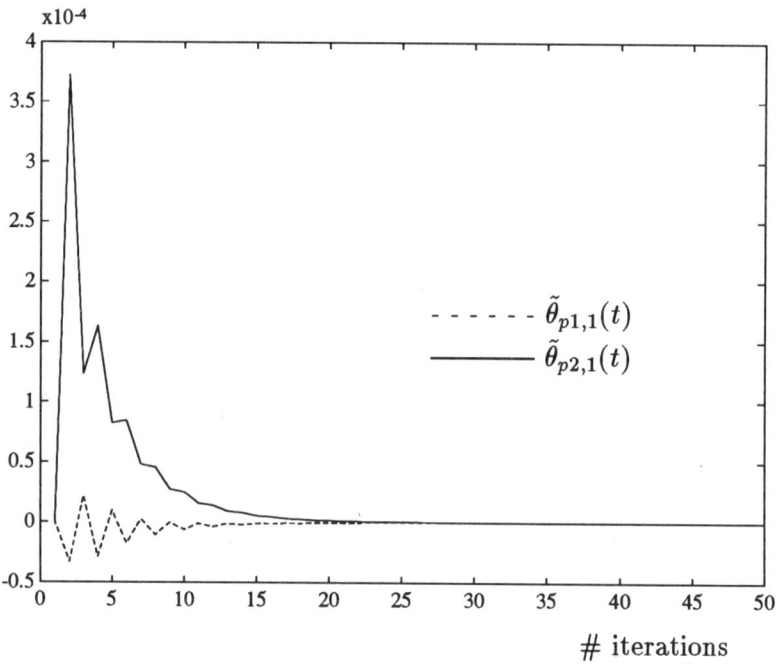

FIGURE 12.1. Effect of a correlated information matrix on parameter errors

Clearly, $\tilde{\theta}_{p1,1}(t)$ is hardly affected by the initial error in $\hat{\theta}_{p1,2}(t)$, while $\tilde{\theta}_{p2,1}(t)$ exhibits a rather large transient error. This is evidently due to the fact that the first and second components of $\hat{\theta}_{p1}$ are much less correlated than those of $\hat{\theta}_{p2}$.

Having illustrated the advantage of a reparametrization of the system, we now establish some further connections between the information matrix of the reparametrized model, the filter applied to the data processes, and the spectral or covariance properties of these data processes. These connections will then enable us to formulate some optimal and suboptimal design problems that can be seen either as filter design problems or as parametrization design problems.

We first briefly recall some basic properties of spectral density functions of filtered signals. Let $\{u(t)\}$ be a stationary discrete time process. Its autocorrelation function $R_u(\tau)$ and spectral density function (or spectrum)[4] $\Phi_u(e^{j\omega})$ are defined as

$$R_u(\tau) \triangleq E[u(t+\tau)u^T(t)], \quad \Phi_u(z) \triangleq \frac{1}{2\pi} \sum_{\tau=-\infty}^{\infty} R_u(\tau)z^{-\tau}. \quad (12.27)$$

$R_u(\tau)$ is obtained from $\Phi_u(z)$ as follows:

$$R_u(\tau) = \int_{-\pi}^{\pi} \Phi_u(e^{j\omega}) e^{j\omega\tau} d\omega. \quad (12.28)$$

In particular:

$$R_u(0) = \int_{-\pi}^{\pi} \Phi_u(e^{j\omega}) d\omega. \quad (12.29)$$

Let $\{x_1(t)\}$ and $\{x_2(t)\}$ be two jointly stationary processes. The cross-correlation function of the two processes and the cross-spectral density function are defined, respectively, as

$$R_{x_1 x_2}(\tau) \triangleq E[x_1(t+\tau)x_2^T(t)], \quad \Phi_{x_1 x_2}(z) \triangleq \frac{1}{2\pi} \sum_{\tau=-\infty}^{\infty} R_{x_1 x_2}(\tau)z^{-\tau}. \quad (12.30)$$

The following lemma presents some well known and useful spectral density function expressions for filtered signals.

Lemma 12.1 : *Let $H(z)$ and $G(z)$ be two filters driven by the same input $u(t)$, with $y(t) = H(z)u(t)$ and $w(t) = G(z)u(t)$. Then*

$$\begin{aligned} \Phi_{uy}(z) &= \Phi_u(z) H^T(z^{-1}) \\ \Phi_{yw}(z) &= H(z) \Phi_u(z) G^T(z^{-1}) \\ \Phi_y(z) &= H(z) \Phi_u(z) H^T(z^{-1}). \end{aligned} \quad (12.31)$$

Proof: see e.g. [Åst70]. ∎

---

[4] We use the terms 'spectral density function' and 'spectrum' to denote the same object.

We now establish two useful expressions relating the spectrum of the filtered regression vector $\Psi_p(t)$ to that of the shift operator regressor $\Psi_z(t)$ and also directly to the spectra of the signal processes $\{u(t)\}$ and $\{y(t)\}$. These expressions lead to similarly useful expressions for the corresponding information matrix of $\Psi_p(t)$.

Theorem 12.1 : Let $\Psi_p(t)$ be the filtered regression vector defined by (12.20), where T and $F(z)$ are connected by (12.22). Then the spectral density matrix of $\Psi_p(t)$ is given by:

$$\Phi_p(z) = F(z) \begin{pmatrix} \Phi_y & -\Phi_{yu} \\ -\Phi_{uy} & \Phi_u \end{pmatrix} F^T(z^{-1}) \qquad (12.32)$$

$$= T\Phi_z(z)T^T. \qquad (12.33)$$

The information matrix R_p corresponding to $\Psi_p(t)$ can be computed by

$$R_p = T R_z T^T \qquad (12.34)$$

$$= \int_{-\pi}^{\pi} F(e^{j\omega}) \begin{pmatrix} \Phi_y(e^{j\omega}) & -\Phi_{yu}(e^{j\omega}) \\ -\Phi_{uy}(e^{j\omega}) & \Phi_u(e^{j\omega}) \end{pmatrix} F^T(e^{-j\omega}) d\omega. \qquad (12.35)$$

Proof: The expressions for $\Phi_p(z)$ follow immediately from Lemma 12.1 using (12.20). The expression for R_p then follows from (12.29). ∎

We have now established the equivalence between a reparametrization and a filtering of the regressor obtained from shift operator models. This, in turn, is equivalent to constructing a new regressor vector, $\Psi_p(t)$, whose components are filtered versions of the signals $u(t)$ and $y(t)$: see (12.20). In addition, we have derived a few formulas connecting the spectra and the information matrices of $\Psi_p(t)$ and $\Psi_z(t)$. We can now turn to optimal and suboptimal design choices for the transformation matrix T or, equivalently, for the matrix filter $F(z)$, given that we know what kind of properties are desirable for the information matrix of $\Psi_p(t)$, in terms of accuracy and convergence speed of parameter estimation algorithms.

12.4 Optimal and suboptimal design choices

In this section, we study the choice of parametrizations with a view to optimizing or improving the numerical properties of parameter estimation algorithms. It is clear from (12.32)-(12.35) that if one wants to give some special properties to the information matrix R_p (for example $R_p = I$), then the filter $F(z)$ that achieves this aim depends on the data through the spectral densities Φ_u, Φ_y and Φ_{yu} or, equivalently, through the spectral density matrix Φ_z. In adaptive filtering applications, for example, such information is not available and one then wants to design a filter that does not explicitly depend on the data.

We therefore examine two different strategies. We first discuss the design of optimal (but data dependent) filters or parametrizations; we then show that some parametrizations yield information matrices with desirable properties independently of the spectrum of the data. This shows that such parametrizations exhibit a robustness property.

12.4.1 Manipulating the information matrix with data information

Suppose that it is desired to design a $(2n+1) \times 2$ filter $F(z)$ such that the filtered regression vector $\Psi_p(t)$ has a prescribed information matrix R_p. If the spectral density functions $\Phi_u(e^{j\omega})$, $\Phi_y(e^{j\omega})$, and $\Phi_{yu}(e^{j\omega})$ are given, then the $(2n+1) \times 2$ optimal filter $F(z)$ can be obtained by solving (12.35).

For example, if $R_p = I$ is desired, then the optimal transformation matrix T is the inverse of any square root of R_z. We note that R_z is uniquely defined by the signal spectra Φ_u, Φ_y and Φ_{yu}:

$$R_z = \int_{-\pi}^{\pi} \begin{pmatrix} \nu_n(e^{j\omega}) & 0 \\ 0 & \nu_{n+1}(e^{j\omega}) \end{pmatrix} \begin{pmatrix} \Phi_y(e^{j\omega}) & -\Phi_{yu}(e^{j\omega}) \\ -\Phi_{uy}(e^{j\omega}) & \Phi_u(e^{j\omega}) \end{pmatrix}$$
$$\times \begin{pmatrix} \nu_n(e^{-j\omega}) & 0 \\ 0 & \nu_{n+1}(e^{-j\omega}) \end{pmatrix}^T d\omega. \quad (12.36)$$

Since R_z is a symmetric positive definite matrix, it is always possible to factor it as

$$R_z = DD^T. \quad (12.37)$$

So, if a unit matrix is desired for R_p, then the optimal transformation matrix T^{opt} is given by $T^{opt} = D^{-1}$. The filter $F(z)$ is then uniquely determined from T by (12.22). The idea of orthogonalizing the regression vector has been used in many estimation and signal processing applications [Bie77], [Mak78] and [Goo88]. As we stated earlier, the orthogonalization can (and should) actually be performed on the data themselves, rather than on the covariance matrix or its estimate. It is interesting to note that the condition number of the optimal matrix T^{opt} above is the square root of the condition number of the original information matrix R_z.

Since the decomposition of (12.37) is non-unique, D can be chosen to be in a sparse form such as in triangular form. D is then the *Cholesky factor* of R_z [Bie77], and the filtering can then be implemented using only $n(n+1)/2$ parameters. This simplifies the implementation burden and speeds up the processing. The number of parameters to be implemented can be further reduced by taking advantages of some special properties of the information matrix.

The optimal filter $F(z)$ depends on the spectral (or covariance) properties of the data processes. When this information is not available (i.e. $\Phi_u(z), \Phi_y(z)$ and $\Phi_{yu}(z)$ are unknown), it is of interest to find a "robust" parametrization that yields interesting numerical properties for the regressor independently of the properties of the signals.

12.4.2 A ROBUSTNESS PROPERTY

We now show that, by proper reparametrization - or, equivalently, by proper data filtering - a new regressor can be obtained whose information matrix can be made to have some desirable properties independently of the spectral properties of the data processes.

We first need the following lemma.

Lemma 12.2 : *Let $H(z)$ and $G(z)$ be two scalar filters driven by the same scalar input $u(t)$, with $y(t) \triangleq H(z)u(t)$ and $w(t) \triangleq G(z)u(t)$. Denote by $\arg(c)$ the argument of a complex number c and let $R_{yw}(\tau) \triangleq E[y(t+\tau)w^T(t)]$. Then*

$$R_{yw}(0) = 0 \tag{12.38}$$

if there exists an integer function $m(\omega)$ such that, for all $\omega \in [0, \pi]$,

$$\phi(\omega) \triangleq \arg(H(e^{j\omega})) - \arg(G(e^{j\omega})) = m(\omega)\pi + \pi/2. \tag{12.39}$$

Proof: By Lemma 12.1,

$$\Phi_{yw}(z) = H(z)\Phi_u(z)G(z^{-1}).$$

It then follows that

$$\begin{aligned} R_{yw}(0) &= \int_{-\pi}^{\pi} \Phi_{yw}(e^{j\omega})d\omega = \int_{-\pi}^{\pi} F(\omega)e^{j\phi(\omega)}d\omega \\ &= 2\int_{0}^{\pi} F(\omega)\cos[\phi(\omega)]d\omega, \end{aligned} \tag{12.40}$$

where $F(\omega) \triangleq |H(e^{j\omega})||G(e^{j\omega})|\Phi_u(e^{j\omega})$. The last equality is due to the fact that $F(\omega)$ and the phase difference are even and odd functions of ω, respectively. The lemma then follows from (12.39). ∎

Comment: An important observation is that the condition (12.39) is independent of the processes $\{y(t)\}$ and $\{u(t)\}$. It depends only on the filters $H(z)$ and $G(z)$. This property will now enable us to produce filtered regression vectors whose information matrices have special structure. Such regression vectors will be obtained by the introduction of a representation of the transfer function not in terms of the usual shift operator z, but in terms of a new operator, based on the bilinear transformation, that we shall call the γ-operator.

12.5 γ-operator parametrizations

Recently, the use of δ-operator parametrizations has been advocated, *in lieu of* the usual shift operator parametrizations, for estimation and control applications [Pet86], [Goo88]. Two major advantages are claimed for the δ-operator parametrization: a theoretically interesting unified formulation of continuous time and discrete time estimation and control problems under fast sampling, and a range of practically interesting numerical advantages connected with finite wordlength effects [MG90]. The unification stems from the fact that, with $z \triangleq e^{jT_s\omega}$, when the sampling period T_s goes to zero, the Delta operator tends towards $j\omega$:

$$\delta \triangleq \frac{z-1}{T_s} \longrightarrow j\omega \left(\Longleftrightarrow \frac{d(\cdot)}{dt} \right) \text{ when } T_s \longrightarrow 0. \qquad (12.41)$$

In other words, the (discrete time) δ-operator converges to the (continuous time) derivative operator $j\omega$ or $d(\cdot)/dt$.

As for the numerical advantages claimed for the δ-operator parametrizations, we have seen in Chapter 11 that these derive from the property that the δ-operator maps the point $(1,0)$ in the z-plane to the point $(0,0)$ in the δ-plane. Now, consider the following operator that we shall call γ-operator for lack of a more imaginative name:

$$\gamma \triangleq \frac{2}{T_s} \frac{z-1}{z+1}. \qquad (12.42)$$

Up to the scalar factor $\frac{2}{T_s}$ it is the familiar bilinear transformation. It is easy to see with the same manipulation as above that

$$\gamma = \frac{2}{T_s} \frac{z-1}{z+1} \longrightarrow j\omega \left(\Longleftrightarrow \frac{d(\cdot)}{dt} \right) \text{ when } T_s \longrightarrow 0. \qquad (12.43)$$

This shows that the γ-operator also converges to the derivative operator. In other words, the γ-operator can also unify the continuous time and discrete time theory. In addition we note that the origin $(0,0)$ in γ-plane is also a mapping of the point $(1,0)$ in z-plane. Thus, we observe a clear similarity between the γ-operator and the δ-operator.

We shall now use the γ-operator as a basis for the representation of transfer functions - just as the δ-operator has been used thus - and we shall demonstrate that, for parameter estimation applications, it yields regression vectors with superior properties. More precisely, we shall show that γ-operator parametrizations applied in estimation problems produce information matrices that have better robustness[5] properties than those of either shift- or δ-operator parametrizations.

12.5.1 A TRANSFER FUNCTION IN γ-OPERATOR

We redefine the γ-operator as follows:

$$\gamma(z) \triangleq \frac{2}{\Delta} \frac{z-1}{z+1}, \qquad (12.44)$$

where Δ is not necessarily the sampling period T_s but any positive number. To stress its dependence on z, we have written $\gamma(z)$, and we note that it is a proper and inverse proper function of z. We can now rewrite the transfer function (12.1) as a function of this γ-operator,

$$H(z) = \frac{\beta_0 \gamma^n + \beta_1 \gamma^{n-1} + \ldots + \beta_n \gamma^0}{\gamma^n + \alpha_1 \gamma^{n-1} + \ldots + \alpha_n \gamma^0} \triangleq H_\gamma(\gamma). \qquad (12.45)$$

The relationship between the coefficients $\{a_i, b_i\}$ in (12.1) and $\{\alpha_i, \beta_i\}$ in (12.45) can be obtained by rewriting (12.45) in terms of the following polynomial parametrization in z:

$$H(z) = \frac{\beta_0 p_{\gamma_0}(z) + \beta_1 p_{\gamma_1}(z) + \cdots + \beta_n p_{\gamma_n}(z)}{p_{\gamma_0}(z) + \alpha_1 p_{\gamma_1}(z) + \cdots + \alpha_n p_{\gamma_n}(z)}, \qquad (12.46)$$

---

[5] By 'robustness' we mean that these properties of the information matrix can be achieved without knowledge of the data processes.

where
$$p_{\gamma_i}(z) \triangleq \left(\frac{z+1}{2}\right)^n \gamma^{n-i} \quad i = 0, 1, \ldots, n. \tag{12.47}$$

The difference between (12.45) and (12.46) is that the rational transfer functions γ^i have been replaced by polynomials in z, the $p_{\gamma_i}(z)$. The coefficients $\{a_i, b_i\}$ in (12.1) and $\{\alpha_i, \beta_i\}$ are related by:

$$\begin{pmatrix} 1 \\ a_1 \\ \cdot \\ \cdot \\ \cdot \\ a_n \end{pmatrix} = k_\gamma T_\gamma^T \begin{pmatrix} 1 \\ \alpha_1 \\ \cdot \\ \cdot \\ \cdot \\ \alpha_n \end{pmatrix}, \quad \begin{pmatrix} b_0 \\ b_1 \\ \cdot \\ \cdot \\ \cdot \\ \beta_n \end{pmatrix} = k_\gamma T_\gamma^T \begin{pmatrix} \beta_0 \\ \beta_1 \\ \cdot \\ \cdot \\ \cdot \\ \beta_n \end{pmatrix},$$

where $T_\gamma \in \mathbb{R}^{(n+1)\times(n+1)}$ is given by

$$P_\gamma(z) \triangleq \begin{pmatrix} p_{\gamma_0}(z) \\ p_{\gamma_1}(z) \\ \cdot \\ \cdot \\ \cdot \\ p_{\gamma_n}(z) \end{pmatrix} = T_\gamma \begin{pmatrix} z^n \\ z^{n-1} \\ \cdot \\ \cdot \\ \cdot \\ 1 \end{pmatrix} = T_\gamma \nu_{n+1}(z), \tag{12.48}$$

and where k_γ is a normalizing constant that makes $a_0 = \alpha_0 = 1$. This is always possible as long as the system has no pole at $z = -1$. Since T_γ is uniquely determined by the $\{p_{\gamma_i}(z)\}$ defined in (12.47), the relationship between the two parameter sets $\{a_i, b_i\}$ and $\{\alpha_i, \beta_i\}$ is a one-to-one mapping. We then call both (12.45) and (12.46)-(12.47) the γ-operator parametrization of the system.

The γ-operator parametrizations have many interesting properties, but in this chapter we limit our discussion to those properties that are of interest for parameter estimation purposes.

12.5.2 Computing the information matrix

Using the expression (12.46) of $H(z)$, the input-output relation can be rewritten as:

$$p_{\gamma_0}(z)z^{-n}y(t) = -\sum_{i=1}^{n} \alpha_i p_{\gamma_i}(z)z^{-n}y(t) + \sum_{i=0}^{n} \beta_i p_{\gamma_i}(z)z^{-n}u(t). \tag{12.49}$$

We observe that the polynomials $p_{\gamma_i}(z)$ are all of degree n. Comparing with (12.18)-(12.22), one can see now that the γ-operator parametrization (12.45) corresponds to a special choice of data filtering with the following choices:

$$p(z) = p_{\gamma_0}(z) = (\frac{z+1}{2})^n \gamma^n = (\frac{z-1}{\Delta})^n,$$
$$q(z) = 0$$
$$F_\gamma(z) = \begin{pmatrix} F_{\gamma_1}(z) & 0 \\ 0 & F_{\gamma_2}(z) \end{pmatrix} \quad (12.50)$$

$$F_{\gamma_1}(z) \triangleq z^{-n} \begin{pmatrix} p_{\gamma_1}(z) & \cdots & p_{\gamma_n}(z) \end{pmatrix}^T \triangleq z^{-n} T_{\gamma_1} \nu_{n+1}(z)$$
$$F_{\gamma_2}(z) \triangleq z^{-n} \begin{pmatrix} p_{\gamma_0}(z) & \cdots & p_{\gamma_n}(z) \end{pmatrix}^T \triangleq z^{-n} T_\gamma \nu_{n+1}(z)$$
$$(12.51)$$

where T_γ was defined in (12.48) and where $T_{\gamma_1} \in \mathbb{R}^{n \times (n+1)}$ is made up of the last n rows of T_γ.

The Least Squares regression equation is then written as

$$y^*(t) = \theta_\gamma^T \Psi_\gamma(t) + v(t), \quad (12.52)$$

where

$$\theta_\gamma \triangleq \begin{pmatrix} \alpha_1 & \cdots & \alpha_n & \beta_0 & \cdots & \beta_n \end{pmatrix}^T. \quad (12.53)$$

Here the new process $y^*(t)$ and the new regression vector $\Psi_\gamma(t)$ are computed as (see (12.19) and (12.20)):

$$y^*(t) \triangleq p(z) z^{-n} y(t) = (\frac{1 - z^{-1}}{\Delta})^n y(t)$$
$$\Psi_\gamma(t) \triangleq F_\gamma(z) \begin{pmatrix} -y(t) \\ u(t) \end{pmatrix} = \begin{pmatrix} T_{\gamma_1} & 0 \\ 0 & T_\gamma \end{pmatrix} \begin{pmatrix} -y(t) \\ \Psi_z(t) \end{pmatrix} \quad (12.54)$$

It is worth noting that $\Psi_\gamma(t)$ cannot be written as a linear transformation of $\Psi_z(t)$ because the first n components of the regressor $\Psi_\gamma(t)$ actually contain the signal $y(t)$ at time t. This clearly follows from the expression of $F(z)$ (see (12.50) and (12.51)) and our earlier remark that all polynomials $p_{\gamma_i}(z)$ have degree n. In fact, it will prove useful to rewrite $\Psi_\gamma(t)$ in the following alternative form, which makes the dependence on $y(t)$ very clear:

$$\Psi_\gamma(t) = \left(\frac{1+z^{-1}}{2}\right)^n \begin{pmatrix} -\gamma^{n-1} y(t) \\ \cdot \\ \cdot \\ \cdot \\ -\gamma^0 y(t) \\ \gamma^n u(t) \\ \gamma^{n-1} u(t) \\ \cdot \\ \cdot \\ \cdot \\ \gamma^0 u(t) \end{pmatrix} \quad (12.55)$$

This expression follows directly from (12.54) and from the expressions (12.50), (12.51) and (12.47).

We have established the relationship between the γ- and z- operator parametrizations, and we have shown how to compute the regression vector $\Psi_\gamma(t)$ in this γ-operator parametrization. We now explore the properties of its information matrix.

To do so, we first prove a key orthogonality property of the operator γ, considered as a filter, $\gamma(z)$.

Lemma 12.3 : *Let* $y(t) \triangleq \gamma^i(z) u(t)$ *and* $w(t) \triangleq \gamma^k(z) u(t)$. *Then*

$$R_{yw}(0) \triangleq \int_{-\pi}^{\pi} \gamma^i(e^{j\omega}) \Phi_u(e^{j\omega}) \gamma^k(e^{-j\omega}) d\omega = 0 \quad (12.56)$$

for all Φ_u *if* $|i - k|$ *is an odd number. In particular,*

$$\int_{-\pi}^{\pi} \gamma(e^{j\omega}) \Phi_u(e^{j\omega}) d\omega = 0 \quad \text{for all } \Phi_u. \quad (12.57)$$

Proof: It follows from elementary high school level calculus[6] that $\gamma(e^{j\omega}) = j\frac{2}{\Delta} tg(\omega/2)$ and hence

$$\arg\{\gamma(e^{j\omega})\} = \frac{\pi}{2}, \quad \forall \omega \in [0, \pi]$$

and

$$\arg\{\gamma^k(e^{j\omega})\} = \frac{\pi}{2} k, \quad \forall \omega \in [0, \pi].$$

---

[6] It took one of the authors as much time to verify this piece of high school level calculus as it took to understand the rest of the proof; so, don't despair if it does not appear obvious.

Therefore,
$$arg\{\gamma^i(e^{j\omega})\} - arg\{\gamma^k(e^{j\omega})\} = \frac{\pi}{2}(i-k) \quad \forall \omega \in [0,\pi].$$

The result then follows from Lemma 12.2. ∎

The structure of the information matrix corresponding to $\Psi_\gamma(t)$ follows immediately from this lemma and from the expression (12.55) of the regression vector.

Theorem 12.2 : *Let $\Psi_\gamma(t)$ be the regression vector (12.55) obtained from the γ-operator parametrization. Then the corresponding information matrix has the form:*

$$R_\gamma = \begin{pmatrix} R_{\gamma_1} & X \\ X^T & R_{\gamma_2} \end{pmatrix}, \qquad (12.58)$$

where X has no particular structure while $R_{\gamma_1} \in \mathbb{R}^{n\times n}$ and $R_{\gamma_2} \in \mathbb{R}^{(n+1)\times(n+1)}$ are of the following form: if n is even, then

$$R_{\gamma_1} = \begin{pmatrix} x & 0 & x & 0 & \cdot & \cdot & x & 0 \\ 0 & x & 0 & x & \cdot & \cdot & 0 & x \\ x & 0 & x & 0 & \cdot & \cdot & x & 0 \\ \cdot & \cdot & \cdot & \cdot & \cdot & \cdot & \cdot & \cdot \\ \cdot & \cdot & \cdot & \cdot & \cdot & \cdot & \cdot & \cdot \\ \cdot & \cdot & \cdot & \cdot & \cdot & \cdot & \cdot & \cdot \\ x & 0 & x & 0 & \cdot & \cdot & x & 0 \\ 0 & x & 0 & x & \cdot & \cdot & 0 & x \end{pmatrix} \qquad (12.59)$$

where the elements denoted x are arbitrary, while $R_{\gamma 2}$ has the same structure with one more row and column.

Proof: The proof follows directly from the expression (12.55) of $\Psi_\gamma(t)$ and from the previous lemma. ∎

Thus, using a γ-operator parametrization yields a regression vector whose information matrix has the nice structure (12.58)-(12.59). The important point is that this property is totally independent of the data processes $\{u(t)\}$ and $\{y(t)\}$.

We have seen in Section 12.2 that the presence of off-diagonal coupling terms in the information matrix has the effect of slowing down the convergence of parameter estimation schemes. We shall contrariwise illustrate the beneficial effects of the zero off-diagonal elements in R_γ through several simulations. These beneficial effects are in fact due to the orthogonalizing properties of the γ-operator displayed in Lemma 12.3. But first, we provide an analysis of the condition number of R_γ under a simplifying assumption of narrow band signals, and we show how this condition number can be made small by a normalization that is again independent of the spectral properties of the signals.

12.5.3 ANALYSIS OF THE INFORMATION MATRIX

A low condition number for the information matrix is probably as important for the convergence speed of the adaptive estimation algorithms as the presence of zero off-diagonal elements. The computation of the condition number of R_z and R_γ, with R_γ in the form (12.58)-(12.59), appears to be very hard in general. Here we provide an analysis under two simplifying assumptions:

- we first consider that the model is either an FIR model (that is, $\{a_i = 0\}$ in (12.2) or $\{\alpha_i = 0\}$ in (12.52)-(12.53)), or an AR model (that is, $\{b_i = 0\}$ in (12.2) or $\{\beta_i = 0\}$ in (12.52)-(12.53)). At the end of this subsection, we shall give a few elements for the case of a general ARX model.

- we assume that the spectrum of the signals $\{u(t)\}$ and $\{y(t)\}$ is narrow compared to the sampling frequency.

Under these assumptions, we give a number of arguments that all point towards the direction that the condition number of R_γ can be made much smaller than that of R_z, particularly if a further normalization is introduced on the regressor $\Psi_\gamma(t)$ - normalization that is again data independent.

Thus, consider for example an AR model, in which case $\Psi_z(t)$ and $\Psi_\gamma(t)$ contain only filters of $y(t)$. The analysis for an FIR model is identical. First we compute the expressions for an arbitrary element of the corresponding information matrices R_z and R_γ. Recall that:

$$F_{z,i}(z) = z^{-i}, \quad F_{\gamma,i}(z) = (\frac{1+z^{-1}}{2})^n \gamma^{n-i}. \qquad (12.60)$$

Therefore,

$$R_z(i,k) = \int_{-\pi}^{\pi} e^{j(k-i)\omega} \Phi_y(e^{j\omega}) d\omega$$

$$\begin{aligned}
&= 2\int_0^\pi \cos[(k-i)\omega]\Phi_y(e^{j\omega})d\omega\\
R_\gamma(i,k) &= \int_{-\pi}^\pi F_{\gamma,i}(e^{j\omega})F_{\gamma,k}(e^{-j\omega})\Phi_y(e^{j\omega})d\omega\\
&= 2\cos(\frac{\pi(k-i)}{2})\int_0^\pi (\cos\frac{\omega}{2})^{2n}(\frac{2}{\Delta}tg\frac{\omega}{2})^{2n-i-k}\Phi_y(e^{j\omega})d\omega.
\end{aligned}$$
(12.61)

Clearly, the exact value of R_ρ, for $\rho = z$ or γ, depends on the power spectral density function $\Phi_y(z)$, and it is difficult to give more precise expressions for a general $\Phi_y(z)$. To proceed further, we make the following narrow bandwidth assumption about the process $\{y(t)\}$:

$$\Phi_y(e^{j\omega}) = \begin{cases} C_y & |\omega| \leq \epsilon \ll \pi \\ 0 & \text{otherwise} \end{cases}$$

By taking $\Delta = \epsilon$ in (12.44) we then get:

$$\begin{aligned}
R_z(i,k) &\approx 2\epsilon C_y \triangleq \sigma_y^2\\
R_\gamma(i,k) &\approx \frac{\sigma_y^2}{2n+1-i-k}\cos(\frac{\pi(k-i)}{2}),
\end{aligned}$$
(12.62)

where $\cos(\frac{\pi(k-i)}{2})$ is zero for $|i-k|$ odd and ± 1 for $|i-k|$ even. In particular,

$$R_\gamma(i,i) \approx \frac{\sigma_y^2}{2(n-i)+1}.$$
(12.63)

Comments:

1. When fast sampling is used, the data sequence is strongly correlated and the corresponding information matrix R_z goes towards

$$\begin{pmatrix} 1 & 1 & \cdots & 1 \\ 1 & 1 & \cdots & 1 \\ \cdot & \cdot & \cdots & 1 \\ 1 & 1 & \cdots & 1 \end{pmatrix} \sigma_y^2.$$

The condition number of such R_z is nearly infinite. This is the key explanation for the poor numerical properties of shift operator regressors when fast sampling is used.

2. The expression (12.62) of $R_\gamma(i,k)$ is zero for $|i-k|$ odd, as should be. In addition (12.63) shows that the diagonal elements of R_γ range from $\frac{\sigma_y^2}{2n-1}$ to σ_y^2. Thus, by multiplying the i-th component of the regressor, $\Psi_{\gamma,i}(t)$, by $d_i \triangleq \sqrt{2(n-i)+1}$, all diagonal elements of R_γ can be made identical. We note that such normalization can be performed without any knowledge of the spectral properties of $\{y(t)\}$. We now indicate that this normalization improves the condition number of R_γ. We give a proof for $n=3$; this proof can probably be extended to higher order models, although we have not done it.

For $n=3$, the information matrix R_γ has the following general form:

$$R_\gamma = \begin{pmatrix} r_1 & 0 & x \\ 0 & r_2 & 0 \\ x & 0 & r_3 \end{pmatrix} > 0. \tag{12.64}$$

We then have the following result.

Lemma 12.4: *Let R_γ be a symmetric strictly positive definite matrix of the form (12.64). Let $D \triangleq \mathrm{diag}\left(\sqrt{\frac{r_2}{r_1}}, 1, \sqrt{\frac{r_2}{r_3}}\right)$, and define*

$$\bar{R}_\gamma \triangleq D R_\gamma D = \begin{pmatrix} r_2 & 0 & y \\ 0 & r_2 & 0 \\ y & 0 & r_2 \end{pmatrix},$$

where $y \triangleq d_3 d_1 x = r_2 x / \sqrt{r_1 r_3} \triangleq r_2 k_{13}$. Then

$$\mathrm{cond}(R_\gamma) > \mathrm{cond}(\bar{R}_\gamma). \tag{12.65}$$

Proof: It is easy to show that

$$\lambda(R_\gamma) = \left\{r_2, \frac{r_1+r_3}{2} + \sqrt{x^2 + \left(\frac{r_1-r_3}{2}\right)^2}, \frac{r_1+r_3}{2} - \sqrt{x^2 + \left(\frac{r_1-r_3}{2}\right)^2}\right\}.$$

Therefore,

$$\lambda(\bar{R}_\gamma) = \{r_2, r_2+|y|, r_2-|y|\}$$

and hence

$$\mathrm{cond}(\bar{R}_\gamma) = \frac{r_2+|y|}{r_2-|y|} = \frac{1+|k_{13}|}{1-|k_{13}|}. \tag{12.66}$$

Clearly,

$$\text{cond}(R_\gamma) = \frac{\lambda_{max}(R_\gamma)}{\lambda_{min}(R_\gamma)} \geq \frac{\frac{r_1+r_3}{2} + \sqrt{x^2 + (\frac{r_1-r_3}{2})^2}}{\frac{r_1+r_3}{2} - \sqrt{x^2 + (\frac{r_1-r_3}{2})^2}}$$

$$\triangleq \frac{1+\eta}{1-\eta}, \qquad (12.67)$$

where $\eta \triangleq \sqrt{x^2 + (\frac{r_1-r_3}{2})^2}/\frac{r_1+r_3}{2}$. Since $R_\gamma > 0$, it follows that $|k_{13}| < 1$. A few algebraic manipulations then show that $\eta > |k_{13}|$. The result then follows from (12.66) and (12.67). ∎

The result can be extended to $R_\gamma \in \mathbb{R}^{4 \times 4}$, but we have not proved it for arbitrary n. However, another argument that pleads for making the diagonal elements of R_γ all identical or nearly identical is the Gershgorin disc theory on the location and perturbation of eigenvalues [SS90]. It states that when a matrix is diagonally dominant (meaning that the sum of the absolute values of the off-diagonal elements of every row is smaller than the corresponding diagonal element), then the eigenvalues of that matrix are essentially determined by the diagonal elements. For such matrices, the condition number can thus be reduced by making the diagonal elements identical. We recall from the analysis above that, at least for narrow band signals $\{y(t)\}$, the information matrix R_γ has nearly identical diagonal elements if the regressor $\Psi_\gamma(t)$ is transformed by the diagonal matrix

$$D \triangleq \text{diag}(d_1, \ldots, d_n),$$

where

$$d_i = \sqrt{2(n-i)+1} \quad \forall i. \qquad (12.68)$$

It then follows from (12.63) that for narrow band signals $\bar{R}_\gamma(i,i) \approx \sigma_y^2 \ \forall i$.

From now on, we consider that the γ-operator regressor $\Psi_\gamma(t)$ has been normalized, that is the regressor $\bar{\Psi}_\gamma(t)$ is used,

$$\bar{\Psi}_\gamma(t) \triangleq D\Psi_\gamma(t), \qquad (12.69)$$

for which the information matrix is $\bar{R}_\gamma = D R_\gamma D$, with D defined in (12.68). We now examine the nonzero off-diagonal elements of $\bar{R}_\gamma$ and compare them with those of R_z. It follows from (12.62) and the normalization that, for narrow band Φ_y,

$$|\bar{R}_\gamma(i,k)| \approx \sigma_y^2 \sqrt{1 - (\frac{k-i}{2n+1-i-k})^2} |\cos\frac{\pi(k-i)}{2}|. \qquad (12.70)$$

We then observe that

$$\bar{R}_\gamma(i,k) \leq \bar{R}_\gamma(1,3) = \sigma_y^2 \sqrt{1 - (\frac{2}{2n-3})^2}$$
$$< \sigma_y^2 \quad \forall i,k. \qquad (12.71)$$

By comparison, recall that $R_z(i,k) = \sigma_y^2 \ \forall i,k$.

Comments:

1. We first note that, even the nonzero off-diagonal elements of $\bar{R}_\gamma$ are smaller than those of R_z; thus the corresponding elements of the parameter vector estimate $\hat{\theta}_\gamma$, with the normalization by D, are less correlated than those of $\hat{\theta}_z$.

2. All nonzero off-diagonal elements of $\bar{R}_\gamma$ are smaller than $\bar{R}_\gamma(1,3)$, which is itself smaller than the (normalized) diagonal elements. Thus, even though this is not sufficient to prove that $\bar{R}_\gamma$ is diagonally dominant, it certainly points in the right direction. It can further be verified from (12.70) that the nonzero elements that are further away from the diagonal are smaller than those that are closer. This means that, for any even m, $\hat{\theta}_\gamma(i)$ is less correlated with $\hat{\theta}_\gamma(i+m+2)$ than with $\hat{\theta}_\gamma(i+m)$.

We conclude from these various analyses that, not only is $\bar{R}_\gamma$ better conditioned than R_z, but in addition the estimates of the components of $\hat{\theta}_\gamma$ are much less correlated than those of $\hat{\theta}_z$.

We have performed our analysis for the case where the underlying model is purely autoregressive, that is the regressor contains only filters of $y(t)$. For the more general case of an ARX model such as (12.1) or (12.45), the diagonal elements of R_ρ^u and R_ρ^y, for $\rho = z$ or γ, can be made identical if only the variances σ_u and σ_y are known. Indeed, in such case it suffices to pre-scale the output $y(t)$ by σ_u/σ_y, that is to replace the output $y(t)$ by $\bar{y}(t) \triangleq \sigma_u/\sigma_y y(t)$. The identification can then be performed with $u(t)$ and $\bar{y}(t)$. In the case where a γ-operator parametrization is used, the components of $\Psi_\gamma^u(t)$ can be scaled in the same way as shown above by d_i defined in (12.68) to obtain identical diagonal elements. Again, all elements of the analysis point towards a superiority of the γ-operator parametrization.

We conclude by saying that, even without the suggested normalization but with the choice $\Delta = \epsilon$, the simulations have shown that the information matrix R_γ has its diagonal elements of the same order of magnitude, and that it has a much better condition number than R_z does.

12.5.4 COMPARISON WITH δ-OPERATOR PARAMETRIZATIONS

An alternative parametrization, much acclaimed for its numerical properties, is the δ-operator parametrization which we have extensively discussed in Chapter 11. In [Goo88] this parametrization was studied in the context of parameter estimation, and it was claimed that estimating parameters obtained from a δ-operator transfer function leads to regressors whose information matrices have approximately zero elements in the odd off-diagonals. We now examine this claim in some detail.

Consider a system represented in a δ-operator transfer function:

$$H(z) = \frac{\beta_0 \delta^n + \beta_1 \delta^{n-1} + \ldots + \beta_n \delta^0}{\delta^n + \alpha_1 \delta^{n-1} + \ldots + \alpha_n \delta^0} \triangleq H_\delta(\delta). \qquad (12.72)$$

The input-output relation can be written as

$$\delta^n y(t) = -\sum_{i=1}^{n} \alpha_i \delta^{n-i} y(t) + \sum_{i=0}^{n} \beta_i \delta^{n-i} u(t).$$

It is easy to see that the corresponding regression vector has the form:

$$\Psi_\delta(t) = z^{-n} \begin{pmatrix} -\delta^{n-1} y(t) \\ \cdot \\ \cdot \\ -\delta^0 y(t) \\ \delta^n u(t) \\ \delta^{n-1} u(t) \\ \cdot \\ \cdot \\ \delta^0 u(t) \end{pmatrix}. \qquad (12.73)$$

The properties of the information matrix corresponding to $\Psi_\delta(t)$ are an immediate consequence of the following lemma.

Lemma 12.5 : Let $y(t) \triangleq \delta^i(z)u(t)$ and $w(t) \triangleq \delta^k(z)u(t)$. Then

$$R_{yw}(0) \triangleq \int_{-\pi}^{\pi} \delta^i(e^{j\omega})\Phi_u(e^{j\omega})\delta^k(e^{-j\omega})d\omega \approx 0 \qquad (12.74)$$

if $|i-k|$ is an odd number and if u has a narrow band low pass spectrum.

Proof: It follows from even more elementary high school level calculus that

$$\delta(e^{j\omega}) = \frac{e^{j\omega}-1}{\Delta} \approx \left(j\frac{\omega}{\Delta}\right) \text{ for } |\omega| \ll 1, \qquad (12.75)$$

which leads to $\arg\{\delta(e^{j\omega})\} \approx \frac{\pi}{2}$ for small ω. Therefore:

$$\arg\{\delta^i(e^{j\omega})\} - \arg\{\delta^k(e^{j\omega})\} \approx \frac{\pi}{2}(i-k)$$
$$\text{for } |\omega| \ll 1 \text{ and for } i,k = 1,\ldots,n. \qquad (12.76)$$

It then follows from Lemma 12.2 that the information matrix R_δ formed from the corresponding regression vector $\Psi_\delta(t)$ is approximately of the form (12.58)-(12.59), but only for small ω. ∎

Comments:

1. Comparing the results for δ-operator and γ-operator parametrizations, we notice that LS parameter estimation problems solved using a δ-operator parametrization lead to an information matrix that has the same interesting properties as those achieved with a γ-operator parametrization, but only for narrow band input signals.

2. On the other hand, one advantage of the δ-operator parametrization is that the transformation matrices T_{δ_1} and T_δ, corresponding to T_{γ_1} and T_γ in (12.51) for the γ-operator parametrization, are upper triangular while T_{γ_1} and T_{γ_2} are fully parametrized. This means that the δ-operator parametrization requires less computational work than its corresponding γ-operator equivalent if the latter is implemented directly from (12.54).

In Section 12.4 we have shown how to reparametrize this model - or, equivalently, how to filter the data - in order to obtain a regressor with an optimally conditioned information matrix when data information is available, while in the present section we have shown that some suboptimal properties can be obtained for this information matrix by using a γ-operator parametrization or, to a lesser extent, a δ-operator parametrization. The time has come now to illustrate our claims with some applications.

12.6 Applications in estimation and adaptive filtering

In this section we illustrate several of the theoretical results of this chapter, as well as some of the numerical aspects of parameter estimation algorithms, with a particular emphasis on the connection between the properties of the information matrix, the accuracy of the parameter estimates and the convergence speed of the adaptive algorithms. We first present an identification application, then an application to adaptive noise cancellation.

Parameter Estimation

Consider again the FIR model of Example 12.1, which is represented in shift operator form as follows:

$$H(z) = 1 - 2.1000z^{-1} + 1.4600z^{-2} - 0.3360z^{-3}. \tag{12.77}$$

We consider the problem of identifying the parameters of this model from input-output data. The input $\{u(t)\}$ is obtained by feeding white Gaussian noise through a low pass filter with a bandwidth of about $\pi/4$. With 950 data points, the regression vector $\Psi_z(t)$ produces the information matrix R_z represented in (12.26). Recall that the condition number of R_z is $\kappa = 3.7500 \times 10^3$.

In the same way, we compute the information matrix corresponding to the normalized regressor $\bar{\Psi}_\gamma(t)$ of (12.69). The scaling constant Δ in the γ-operator (see (12.44)) has been selected according to the design rules of Subsection 12.5.3 on the basis of a passband of $\pi/4$, such as to produce diagonal elements of the same order of magnitude. The corresponding information matrix, computed again from 950 data, is:

$$\bar{R}_\gamma = \begin{pmatrix} 0.3377 & -0.0000 & -0.1982 & -0.0000 \\ -0.0000 & 0.2163 & 0.0001 & -0.1351 \\ -0.1982 & 0.0001 & 0.1816 & 0.0002 \\ -0.0000 & -0.1351 & 0.0002 & 0.2151 \end{pmatrix}.$$

For $\bar{R}_\gamma$ we get: $\lambda_{min}(\bar{R}_\gamma) = 0.0466$, $\lambda_{max}(\bar{R}_\gamma) = 0.4727$, $\kappa = 10.1410$.

Comments:

1. We observe that the condition number of $\bar{R}_\gamma$ is much smaller than that of R_z. In addition, the computations confirm that the γ-operator parametrization yields an information matrix of zero odd-off-diagonal elements, which guarantees the nice decoupling property of this parametrization.

2. An examination of the matrices R_z and $\bar{R}_\gamma$ confirms our theoretical analysis of Subsection 12.5.3 that the elements of R_z are all very close to one another, while in $\bar{R}_\gamma$ the diagonal elements dominate the off-diagonal elements reasonably well.

To confirm our theoretical results, some simulations have been performed in which the parameters have been estimated using the following Least Mean Square (LMS) adaptive algorithm:

$$\begin{aligned}\hat{\theta}_\rho(t+1) &= \hat{\theta}_\rho(t) + \mu \Psi_\rho(t) e_\rho(t) \\ e_\rho(t) &= y(t) - \hat{\theta}_\rho^T(t) \Psi_\rho(t)\end{aligned} \quad (12.78)$$

where ρ stands for either z or γ, $\hat{\theta}_\rho(t)$ and $\Psi_\rho(t)$ are the parameter vector estimate and the regressor vector at time t in their particular parametrization, respectively. The signal $e_\rho(t)$ is the prediction error, while μ is a constant adaptation step size, which must be chosen small enough in order to guarantee the stability of the algorithm. One possible choice for μ, which requires an estimate of $E[\Psi_\rho^T(t)\Psi_\rho(t)]$, the power of the regressor vector, is to set

$$\mu \approx \frac{1}{E[\Psi_\rho^T(t)\Psi_\rho(t)]}. \quad (12.79)$$

In practice, μ is chosen smaller than this value. The adaptive identifier (12.78) is represented schematically in Figure 12.2.

The results of our simulations are shown in Figures 12.3 and 12.4. They present, respectively, the evolution over time of the prediction error processes $e_z(t)$ and $e_\gamma(t)$ for the shift and γ-operator parametrizations. For the sake of comparison, the step size μ has been set at 0.1 and the initial condition for the parameter vector has been set at zero for both cases.

Generalized transfer function parametrizations

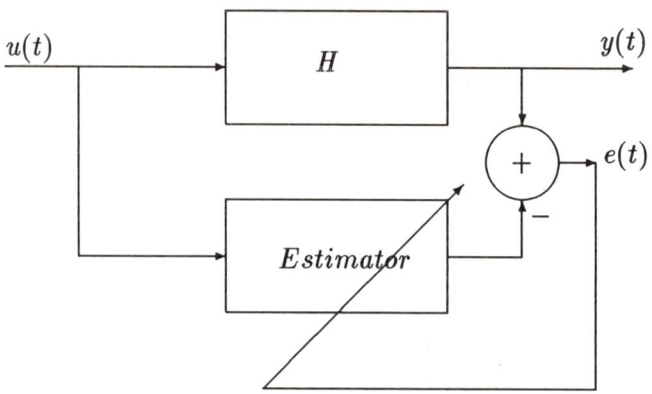

FIGURE 12.2. A scheme for adaptive identification

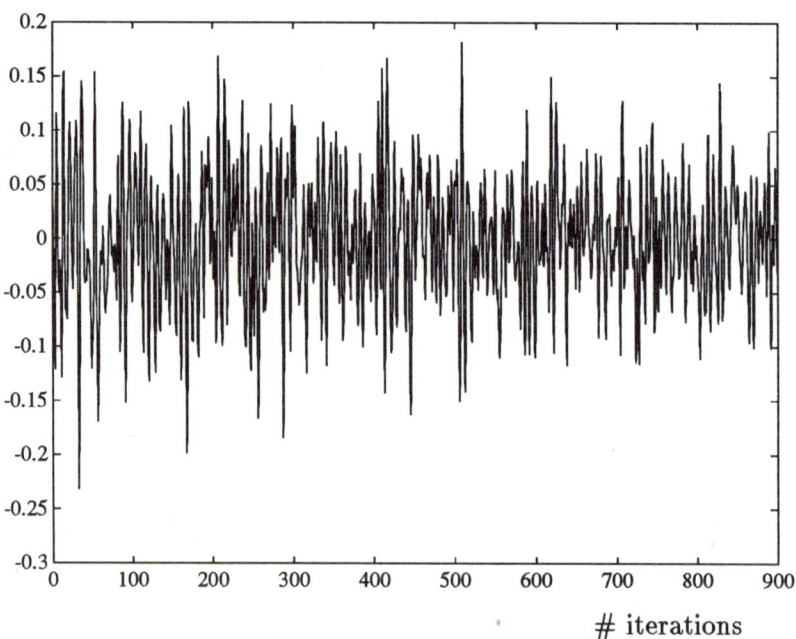

FIGURE 12.3. Prediction error sequence $e_z(t)$

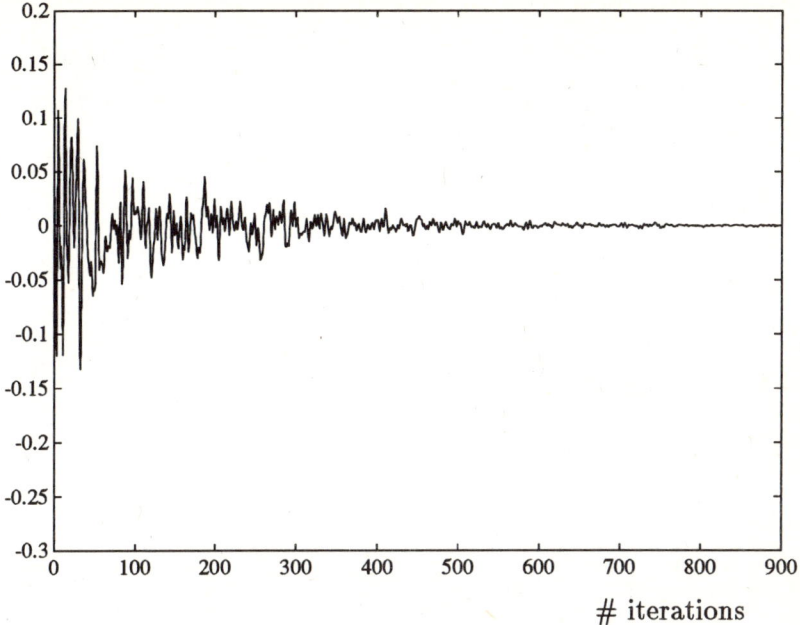

FIGURE 12.4. Prediction error sequence $e_\gamma(t)$

The superiority of the γ-operator parametrization over the classical shift operator parametrization in terms of convergence speed is so obvious that no further comment should be required. But since 99.5% of our readers probably still use the shift operator parametrization, let us further comment that, after 900 iterations, the parameter estimate $\hat\theta_z(t)$ still shows no tendency to converge to its true value $[1 \quad -2.1000 \quad 1.4600 \quad -0.3360]^T$. In contrast, for the γ-operator parametrization the estimate $\hat\theta_\gamma(450)$ is very close to its true value $[0.1121 \quad 0.1815 \quad 0.1016 \quad 0.0240]^T$, while the estimate is almost exact after 900 iterations: $\hat\theta_\gamma(900) = [0.1110 \quad 0.1813 \quad 0.0999 \quad 0.0238]^T$.

In order to give an idea of how the parameter estimates evolve over the iterations, we have plotted the first components $\hat\theta_{z,1}(t)$ and $\hat\theta_{\gamma,1}(t)$ for the two parameter vector estimates in Figure 12.5. This figure shows clearly the superiority of the γ-operator parametrization again.

The comparison just made in terms of convergence speed of the parameter estimates is somewhat unfair to the shift operator estimate $\hat\theta_z(t)$, because the zero initial condition is closer to the exact value of θ_γ than it is to the exact value of θ_z. We have therefore run another simulation with an initial condition $[1.0000 \quad -1.5000 \quad 0.7500 \quad -0.1250]^T$

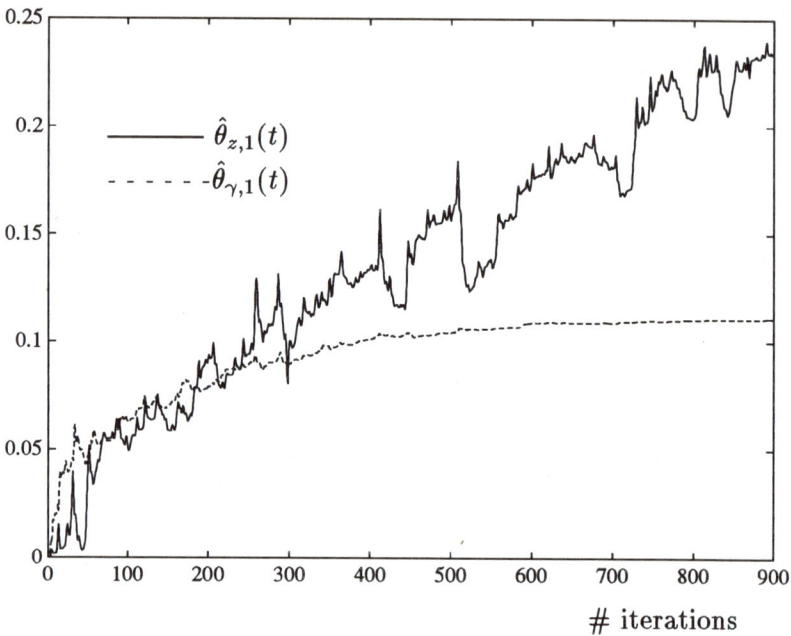

FIGURE 12.5. Evolution of $\hat{\theta}_{\rho,1}(t)$ with zero initial condition

for both parametrizations. This choice now favours the shift operator parametrization for the same reason as mentioned above. The evolution of the estimates of the first component of $\hat{\theta}_z(t)$ and $\hat{\theta}_\gamma(t)$ are shown in Figure 12.6.

The values of the parameter vector estimates after 900 iterations are:

$$\begin{aligned}\hat{\theta}_z(900) &= (0.8343 \quad -1.5567 \quad 0.8164 \quad -0.0346)^T \\ \hat{\theta}_\gamma(900) &= (0.1195 \quad 0.1817 \quad 0.1124 \quad 0.0233)^T.\end{aligned}$$

Note that after 900 iterations $\hat{\theta}_{z,1}(900)$ is far from its true value of 1, even though the initial condition $\hat{\theta}_{z,1}(0)$ is correct. These simulations clearly exhibit the perverse effect of the off-diagonal terms of the information matrices in the convergence behaviour of the estimation algorithm, since even when the initial condition of a component of $\hat{\theta}_z$ is exact, it is perturbed by the other non exact terms through the coupling effects.

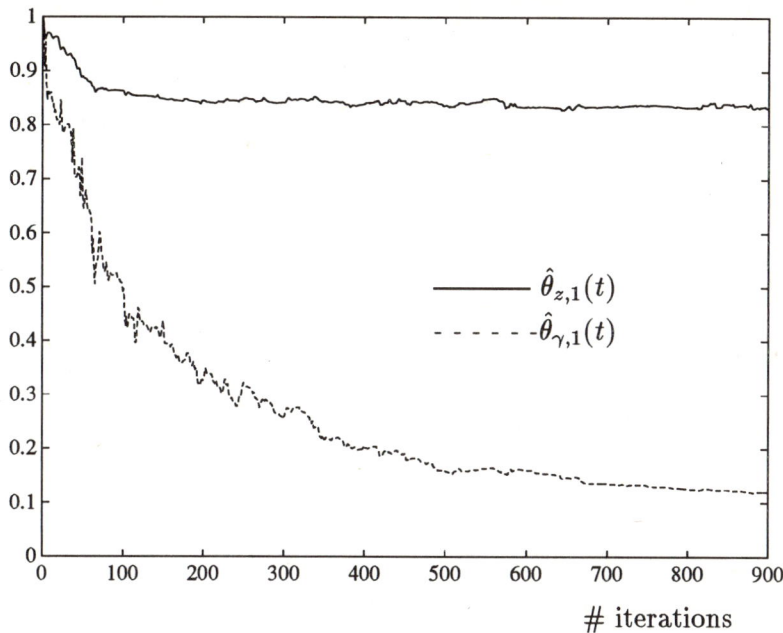

FIGURE 12.6. Evolution of $\hat{\theta}_{\rho,1}(t)$ with non-zero initial condition

Adaptive noise cancellation

We now consider another application in which it is desired to extract a useful signal that is buried in noise, in a situation where the spectral densities of the signal and the noise overlap. A noise source, correlated with the noisy signal that perturbs the useful signal, is available. A typical application is in echo cancellation, where an adaptive filter is often used to extract the useful signal from the noise. A schematic diagram of such an adaptive filter is represented in Figure 12.7.

The measured signal, $d(t)$, (called primary input) consists of a desired signal $s(t)$ and a noise $n(t)$. They are assumed to be uncorrelated with one another. The input of the adaptive filter, $u(t)$, (called reference input) is a noise source that is correlated with the noise signal $n(t)$ in the primary input $d(t)$.

The adaptive filter (AF) is tuned so as to produce an output $\hat{n}(t)$, which is an estimate of the noise $n(t)$. This estimate is subtracted from the primary signal to produce an estimate, $\hat{s}(t)$, of the useful signal $s(t)$. This estimate is used to drive the adaptation of the filter parameters.

We now consider a simulation where the noise source $u(t)$ and the

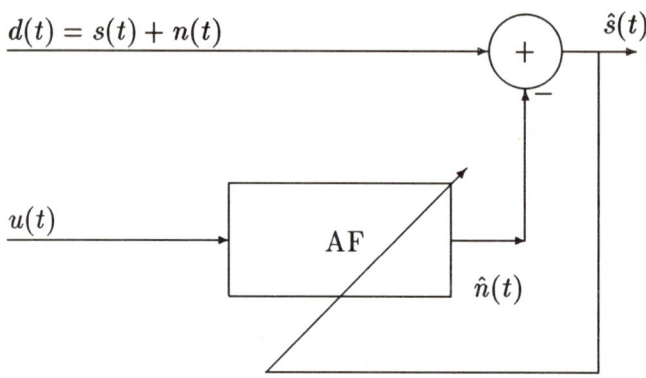

FIGURE 12.7. Block diagram of an adaptive noise canceller

perturbation $n(t)$ are, respectively, the input and output of the FIR model $H(z)$ of (12.77) used in the previous simulations, that is

$$n(t) = u(t) - 2.1000u(t-1) + 1.4600u(t-2) - 0.3360u(t-3).$$

The input $u(t)$ is the same low pass filtered Gaussian noise as in the previous simulation, whereas the useful signal $s(t)$ is a pure sinusoid: $s(t) = 0.05 \times sin(2\pi 100t)$. The Fast Fourier Transforms (FFT) of $s(t)$ and $d(t)$ based on 500 data points with a sampling frequency of $1000Hz$ are shown, respectively, in Figures 12.8 and 12.9. These were computed with MATLAB.

It is clear from Figure 12.9 that the useful signal $s(t)$ is buried in noise. In fact, the variance of the signal $s(t)$ and of the noise $n(t)$ are, respectively, 1.2500×10^{-3} and 7.6541×10^{-3}, that is the signal to noise ratio is 0.1633, which is small by signal processing standards.

We now perform a simulation for the estimation of $\hat{s}(t)$ using the LMS adaptive filter (12.78) with the following definitions:

$$y(t) = d(t), \quad e_\rho(t) = d(t) - \hat{n}(t) = d(t) - \hat{\theta}_\rho^T(t)\Psi_\rho(t) = \hat{s}(t).$$

For $\rho = z$, we take

$$\Psi_z(t) = (u(t) \;\; u(t-1) \;\; u(t-2) \;\; u(t-3))^T,$$

and for $\rho = \gamma$ we have

$$\Psi_\gamma(t) = (\frac{1+z^{-1}}{2})^n \left(d_0\gamma^n \;\; d_1\gamma^{n-1} \;\; d_2\gamma^{n-2} \;\; d_3\gamma^{n-3} \right)^T,$$

Generalized transfer function parametrizations 355

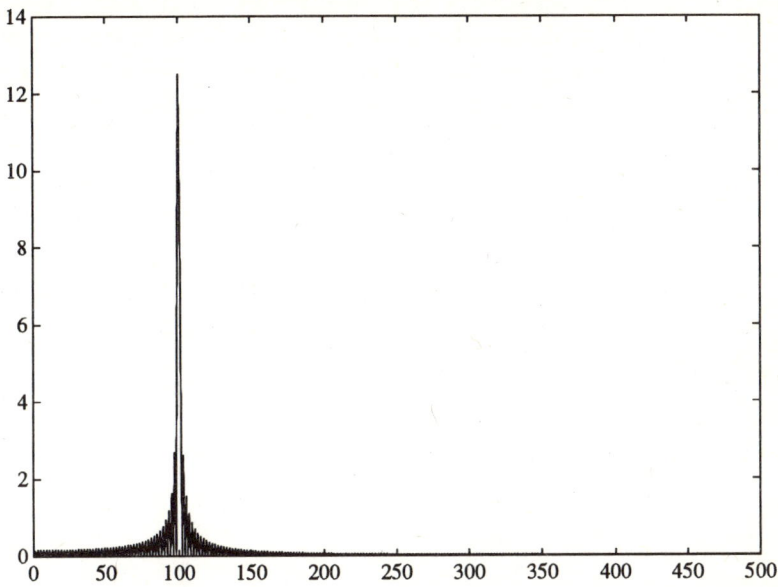

FIGURE 12.8. Fast Fourier Transform of $s(t)$

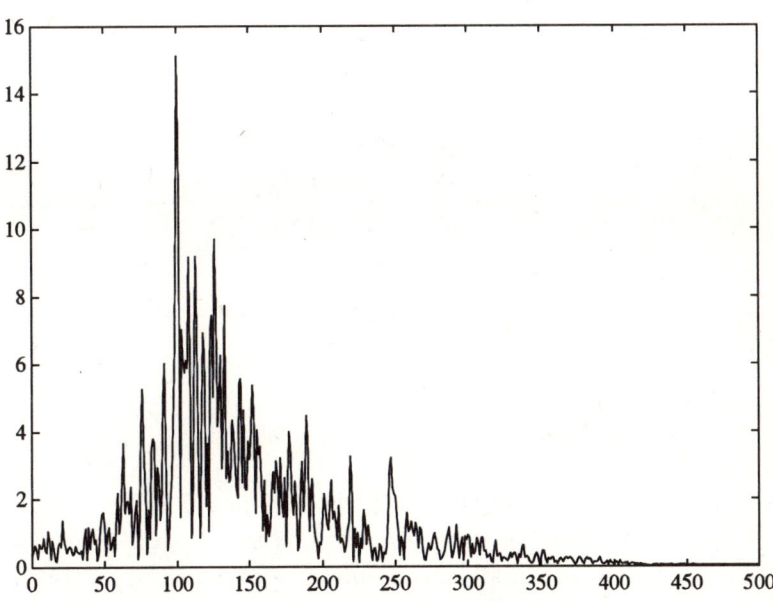

FIGURE 12.9. Fast Fourier Transform of $d(t) = s(t) + n(t)$

where the normalizing constants d_i are defined in (12.68). We have taken a step size $\mu = 0.1$ for both cases, and initial conditions $\hat{\theta}_z(0) = \hat{\theta}_\gamma(0) = 0$. The parameter adaptation has been performed over a data set of 900 data. In order to compare the convergence speeds of both parametrizations, we have plotted in Figures 12.10 and 12.11 the FFT's of the estimates $\hat{s}_z(t)$ and $\hat{s}_\gamma(t)$ computed on the last 450 of these 900 data points. The figures show once again that the adaptive noise canceller using the γ-operator parametrization arrives much more quickly than the shift operator parametrization at an efficient noise cancellation, and hence at a good extraction of the useful signal.

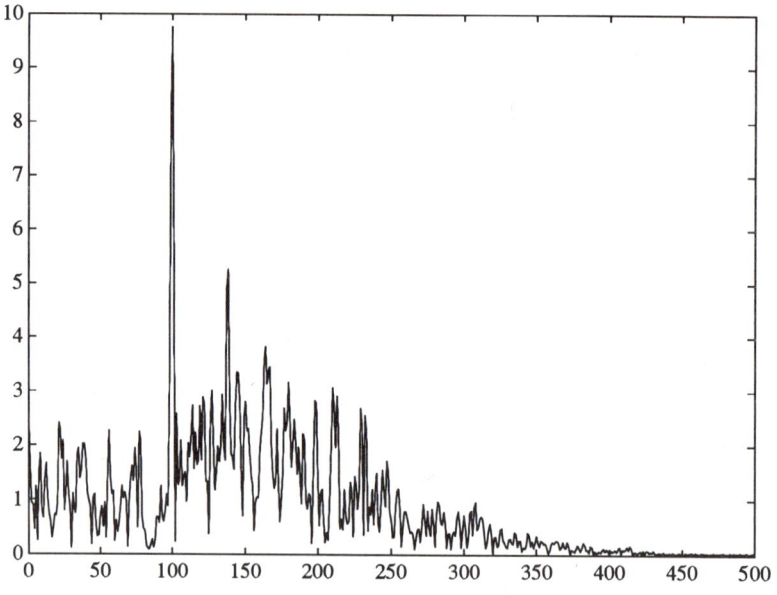

FIGURE 12.10. Fast Fourier Transform of $\hat{s}_z(t)$

12.7 Conclusions

In the previous chapters of this book we have posed and solved a number of optimal parametrization problems, where the optimization was performed over the set of all equivalent (also called similar) state space realizations of a system. The objective was to compute realizations of filters or controllers that had optimal numerical properties, where the optimality was defined by one of a range of possible criteria.

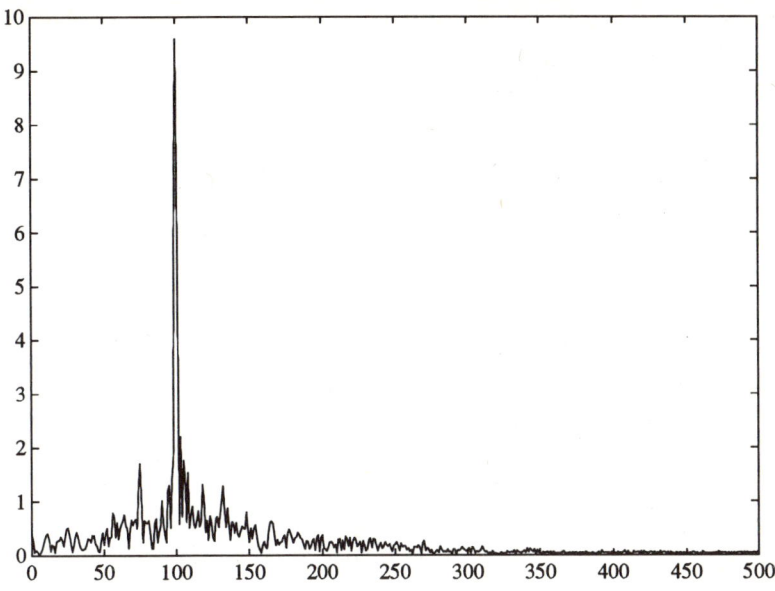

FIGURE 12.11. Fast Fourier Transform of $\hat{s}_\gamma(t)$

This chapter has dealt with another equivalence class, namely the class of equivalent transfer functions, where the choice is in terms of the polynomial basis functions in which the numerator and denominator polynomials of the transfer function are expressed. Changing the basis functions corresponds to changing the coordinate basis of the parameter vector and, correspondingly, of the regression vector. It can thus also be seen as a filtering of the input and output data. Such a coordinate transformation of the parameter vector is therefore relevant for parameter estimation problems. Indeed, as we have shown in Section 12.2, it corresponds to a transformation operated on the information matrix, which has a direct bearing on the numerical properties of parameter estimation algorithms, such as accuracy of the solution and convergence speed of the algorithm.

The idea of changing the parametrization of a transfer function within the context of parameter estimation was addressed by Goodwin [Goo88] who showed that by estimating the parameters of a linear transfer function model using a δ-operator model of this transfer function, rather than the classical shift operator model, advantages could be gained in both the accuracy of the estimated parameters and the convergence speed of the estimator. This superior numerical behaviour of the δ-operator parametrization is essentially due to an orthogonalizing property of the δ function (seen as a transfer function operator), at least when applied to narrow band signals.

Here, we have formally established the connection between parametrization of the transfer function and data filtering or, equivalently, regression filtering. With these formal links firmly established, one can then play some optimal data filtering (or optimal parametrization) games in order to ameliorate the properties of the ensuing information matrix. This is effectively what the properly chosen algorithms for solving LS regression problems do. However, to perform *optimal* transformations, one need to know either the spectral densities of the input and output data, or the spectral density of the input process and the parameters of the unknown system that one is identifying.

In off-line identification, the spectral information can indeed be estimated from the data, but in adaptive applications this information is not available. It is then of interest to develop transformations of the data processes, or equivalently reparametrizations, that yield nicely conditioned and almost diagonal information matrices *independently of the data processes.* The δ-operator parametrization yields an information matrix with zero odd-off-diagonal terms, but this property evanesces if the signals are wide band. We have shown that an alternative parametrization, called γ-operator parametrization achieves the same robust (i.e. data independent) orthogonalization over all frequencies. In addition, we have shown that the condition number of the information matrix obtained with γ-operator representations can be improved further by a normalization that is again data independent.

Simulations have shown the dramatic improvements that can be achieved using the γ-operator parametrization in adaptive parameter estimation. Applications to identification and adaptive noise cancellation have been presented.

We conclude by saying that the formalism introduced by the dual idea of reparametrizing the transfer function or filtering the data opens up many alternative ways of representing transfer functions using basis functions other than the powers of z. The recently introduced Laguerre models can also be seen to fit into this framework [Wah91]; they have also been shown to have interesting numerical properties in the context of the approximation of transfer functions.

References[1]

[AB75] R.C. Agarwal and C.S. Burrus, 'New Recursive Digital Filter Structures Having Very Low Sensitivity and Roundoff Noise,' *IEEE Trans. on Circuits and Systems*, Vol. CAS-22, No. 12, pp. 921-927, Dec. 1975.

[AGG92] B.D.O. Anderson, Y. Genin and M. Gevers, 'Sensitivity minimization of digital controllers for sampled data systems,' in preparation.

[AM79] B.D.O. Anderson and J.B. Moore, *Optimal Filtering*, Prentice-Hall, Inc. Englewood Cliffs, New Jersey, 1979.

[AM89] B.D.O. Anderson and J.B. Moore, *Optimal Control*, Prentice-Hall, Englewood Cliffs, New Jersey, 1989.

[AP89] A.I. Abu-El-Haija and A.M. Peterson, 'An approach to eliminate roundoff errors in digital filters,' *IEEE Trans. on Acoustics, Speech, and Signal Processing*, Vol. ASSP-27, pp. 195-198, April 1979.

[AS88] G. Amit and U. Shaked, 'Small Roundoff Realization of Fixed-Point Digital Filters and Controllers,' *IEEE Trans. on Acoustics, Speech, and Signal Processing*, Vol. ASSP-36, No. 6, pp. 880-891, June 1988.

[Ave72] E. Avenhaus, 'On the design of digital filters with coefficients of limited wordlength,' *IEEE Trans. on Audio Electroacoust.*, Vol. AU-20, August 1972.

[Bel70] R. Bellman, *Introduction to Matrix Analysis*, New York : McGraw-Hill, Second Edition, 1970.

[BGW90] R.R. Bitmead, M. Gevers and V. Wertz, *Adaptive Optimal Control: The Thinking Man's GPC*, Prentice Hall, New York, 1990.

[BH84] B. W. Bomar and J. C. Hung, 'Minimum Roundoff Noise Digital Filters with Some Power-of-Two Coefficients,' *IEEE Trans. on Circuits and Systems*, Vol. CAS-31, pp. 833-840, October 1984.

[Bie77] G. Bierman, *Factorization Methods for Discrete Sequential Estimation*, Academic Press, New York, San Francisco, London, 1977.

[Bod45] H.W. Bode, *Network Analysis and Feedback Amplifier Design*, New York, D. Van Nostrand Co., 1945.

---

[1] A note to our readership. The references starting with the Swedish letter Å, all by our colleague K.J. Åström, have correctly been put at the end of this reference list, as required by the Swedish alphabet.

[BS72] R.H. Bartels and G.W. Stewart, 'Solution to The Matrix Equation AX + XB = C,' *Comm. ACM*, Vol. 15, pp. 820-826, September 1972.

[CF91a] T. Chen and B.A. Francis, 'Linear Time-varying H_2-optimal Control of Sampled-data Systems,' *Automatica*, Vol. 27, No. 6, pp. 963-974, November 1991.

[CF91b] T. Chen and B.A. Francis, 'H_2-optimal Control,' *IEEE Trans. on Automatic Control*, Vol. AC-36, No. 4, pp. 387-397, April 1991.

[CGS84] S.W. Chan, G.C. Goodwin and K.S. Sin, 'Convergence Properties of The Riccati Difference Equation in Optimal Filtering of Non-stabilizable Systems,' *IEEE Trans. on Automatic Control*, Vol. AC-29, pp. 110-119, February 1984.

[Cha71] R.W. Chang, 'A New Equalizer Structure for Fast Start-up Digital Communication,' *Bell Syst. Tech. J.*, Vol. 50, pp. 1969-2014, 1971.

[Cha78] T.L. Chang, 'A Low Roundoff Noise Digital Filter Structure,' in *Proc. IEEE Int. Symp. on Circuits and Syst.*, New York, pp. 1004-1008, May 1978.

[Cha79] D. S. K. Chan, 'Constrained minimization of roundoff noise in fixed-point digital filters,' *Proc. 1979 IEEE Int. Conf. on Acoust., Speech, and Signal Processing*, pp. 335-339, April 1979.

[Cro74] R.E. Crochiere, *Digital Network Theory and Its Application to The Analysis and Design of Digital Filters*, Ph.D. dissertation, Dept. of Elec. Eng., M.I.T., April 1974.

[Cro75] R.E. Crochiere, 'A New Statistical Approach to The Coefficient Word Length Problem for Digital Filters,' *IEEE Trans. on Circuits and Systems*, Vol. CAS-22, pp. 190-196, March 1975.

[Cru72] J.B. Cruz, *Feedback Systems*, New York: McGraw-Hill, 1972.

[Cur67] E.E. Curry, 'The Analysis of Roundoff and Truncation Errors in a Hybrid System', *IEEE Trans. on Automatic Control*, Vol. AC-13, pp. 601-604, October 1967.

[DD81] P. Van Dooren and P. Dewilde, 'Minimal Cascade Factorization of Real and Complex Rational Transfer Matrices,' *IEEE Trans. on Circuits and Systems*, Vol. CAS-28, pp. 390-400, May 1981.

[Doo81] P. Van Dooren, 'Numerical Linear Algebra: An Increasing Interest in Linear System Theory', in *Proc. European Conf. Theory Design*, The Hague, Netherlands, pp. 243-251, 1981.

[Doo89] P. Van Dooren, 'Numerical Aspects of Systems and Control Algorithms,' *Journal A*, Vol. 1, pp. 25-32, 1989.

[DV85] P. Van Dooren and M. Verhaegen, 'On The Use of Unitary State-Space Transformations,' *Contemporary Mathematics*, Vol. 47, pp. 447-463, 1985.

[DW78] E. J. Davison and S. H. Wang, 'An Algorithm for the Calculation of Transmission Zeros of the System (C,A,B,D) Using High Gain Output Feedback', *IEEE Trans. on Automatic Control*, Vol. AC-23, No.4, pp. 738-741, August 1978.

[ED82] A. Emani-Naeini and P. Van Dooren, 'Computation of Zeros of Linear Multivariable Systems,' *Automatica*, Vol. 18, No. 4, pp. 415-430, July 1982.

[Fet72] A. Fettweis, 'On the Connection Between Multiplier Word Length Limitation and Roundoff Noise in Digital Filters,' *IEEE Trans. on Circuit Theory*, Vol. 19, pp. 486-491, September 1972.

[Fet74] A. Fettweis, 'On Properties of Floating-Point Roundoff Noise,' *IEEE Trans. on Acoustics, Speech and Signal Processing*, Vol. ASSP-22, pp. 149-151, April 1974.

[FN83] K.V. Fernando and H. Nicholson, 'On the Cauchy index of linear systems,' *IEEE Trans. on Automatic Control*, Vol. AC-28, pp. 222-224, February 1983.

[Gan77] F.R. Gantmacher, *The Theory of Matrices*, Vol. 1 and 2, Chelsea, New York, 1977.

[Gaw80] P. J. Gawthrop, 'Hybrid self tuning control,' *Proc.* vol.127, pt. D.5, 1980.

[GL83] G.H. Golub and C.F. Van Loan, *Matrix Computations*, North Oxford Academic, 1983.

[Goo88] G.C. Goodwin, 'Some Observations on Robust Stochastic Estimation,' in *Proc. of 8th IFAC/IFORS Symposium on Identification and System Parameter Estimation*, Beijing, pp. 22-32, 1988.

[Gra89] W.S. Gray, *A Geometric Approach to The Parametric Sensitivity of Dynamical Systems*, Ph.D. dissertation, Georgia Institute of Technology, Georgia, USA, 1989.

[GS84] G.C. Goodwin and K.J. Sin, *Adaptive Filtering, Prediction and Control*, Prentice-Hall, Englewood Cliffs, NJ, USA, 1984.

[Hen82] P. Henrici, *Essentials of Numerical Analysis*, Wiley, 1982.

[HJ85] R. A. Horn and C. A. Johnson, *Matrix Analysis*, Cambridge University Press, 1985.

[HM84] M.L. Honig and D.G. Messerschmitt, *Adaptive Filters: Structures, Algorithms, and Applications*, Bell Communications Research Inc., 1984.

[HM91] U. Helmke and J.B. Moore, 'L^2 sensitivity minimization of linear system representations via gradient flows,' submitted to *J. of Mathematical Systems, Estimation and Control*, 1991.

[HP79] A.I. Abu-El-Haija and A.M. Petersen, 'An approach to eliminate roundoff errors in digital filters,' *IEEE Trans. on Acoustics, Speech, and Signal Processing*, Vol. ASSP-27, pp. 195-198, April 1979.

[Hwa77] S.Y. Hwang, 'Minimum Uncorrelated Unit Noise in State-Space Digital Filtering,' *IEEE Trans. on Acoust., Speech, and Signal Processing*, Vol. ASSP-25, No. 4, pp. 273-281, August 1977.

[IKH90] M. Iwatsuki, M. Kawamata and T. Higuchi, 'Statistical Sensitivity and Minimum Sensitivity Structures with Fewer Coefficients in Discrete Time Linear Systems,' *IEEE Trans. on Circuits and Systems*, Vol. 37, No. 1, pp. 72-80, January 1990.

[Jac76] L.B. Jackson, 'Roundoff noise bounds derived from coefficient sensitivities for digital filters,' *IEEE Trans. on Circuits and Systems*, Vol. CAS-23, pp. 481-485, August 1976.

[JLK79] L.B. Jackson, A.G. Lindgren, and Y. Kim, 'Optimal synthesis of second-order state-space structures for digital filters,' *IEEE Trans. on Circuits and Systems*, Vol. CAS-26, pp. 148-153, March 1979.

[KA92] J.P. Keller and B.D.O. Anderson, 'A New Approach to the Discretization of Continuous-Time Controllers,' *IEEE Trans. on Automatic Control*, Vol. AC-37, No. 2, pp. 214-223, February 1992.

[Kai66] F. J. Kaiser, *Digital Filters*, in 'System Analysis by Digital Computers,' F. F. Kuo and J. F. Kaiser Eds., New York: Wiley, 1966, pp. 218-285.

[Kai80] T. Kailath, *Linear Systems*, Prentice-Hall, Englewood Cliffs, New Jersey, 1980.

[Kar79] R. J. Karwoski, 'Introduction to the Z transformation and its derivation,' *TRW Psi Products, El Segundo*, CA, Tutorial paper, Sept. 1979.

[KE65] J.B. Knowles and R. Edwards, 'Effects of A Finite-Word-Length Computer in Sampled-Data Feedback Systems,' *Proc. Inst. Elec. Eng.*, Vol. 112, pp. 1194-1207, June 1965.

[KH85] M. Kawamata and T. Higuchi, 'A Unified Approach to the Optimal Synthesis of Fixed-Point State-Space Digital Filters,' *IEEE Trans. on Acoust., Speech, and Signal Processing*, Vol. ASSP-33, No. 4, pp. 911-920, August 1985.

[KO68] J.B. Knowles and E.M. Olcayto, 'Coefficient accuracy and digital filter response,' *IEEE Trans. on Circuit Theory*, Vol. CT-15, pp. 31-41, March 1968.

[KS72] H. Kwakernaak and R. Sivan, *Linear Optimal Control Systems*, Wiley-Interscience, New York, 1972.

[Kun78] S.Y. Kung, 'A new identification and model reduction algorithm via SVD,' in *IEEE Proc. 12th Asilomar Conf. on Circuits, Systems and Computers*, Pacific Grove, CA, pp. 705-714, 1978.

[LAGP92] G. Li, B.D.O. Anderson, M. Gevers and J.E. Perkins, 'Optimal FWL Design of State-Space Digital Systems with Weighted Sensitivity Minimization and Sparseness Consideration,' *IEEE Trans. on Circuits and Systems*, Vol., CAS-39, pp. 365-377, May 1992.

[Lau80] A.J. Laub, 'Computation of Balancing Transformations,' in *Proc. JACC*, San Francisco, CA, Vol. 1, paper FA 8-E, 1980.

[LFM82] D.T.L. Lee, B. Friedlander and M. Morf, 'Recusive Ladder Algorithms for ARMA Modeling,' *IEEE Trans. on Automatic Control*, Vol. AC-27, No. 4, pp. 753-764, August 1982.

[LG90a] G. Li and M. Gevers, 'Sensitivity and Roundoff Noise Optimization of a State-Estimate Feedback Controller,' *Proc 11th IFAC World Congress*, Tallinn, URSS, Vol. 4, pp. 303-310, August 1990.

[LG90b] G. Li and M. Gevers, 'Optimal Finite Precision Implementation of a State-Estimate Feedback Controller,' *IEEE Trans. on Circuits and Systems*, Vol. CAS-38, No. 12, pp. 1487-1498, December 1990.

[LG90c] G. Li and M. Gevers, 'Comparative Study of Finite Wordlength Effects in Shift and Delta Operator Parametrizations,' *Proceedings of 29th IEEE Conf. on Decision and Control*, Honolulu, Hawaii, pp. 954-959, December 1990.

[LG93] G. Li and M. Gevers, 'Roundoff Noise Minimization Using Delta-Operator Realizations,' to appear in *IEEE Trans. on Acoustics, Speech, and Signal Processing*, February 1993.

[LG91] G. Li and M. Gevers, 'Minimization of Finite Wordlength Effects in Compensator Design,' *Proceedings of the First European Control Conference*, Grenoble, France, pp. 544-549, July 1991.

[LH88] W.J. Lutz and S. Louis Hakimi, 'Design of Multi-Input Multi-Output Systems with Minimum Sensitivity,' *IEEE Trans. on Circuits and Systems*, Vol. 35, No. 9, pp. 1114-1122, September 1988.

[Li90] G. Li, 'Finite precision aspects in the parametrizations of control, estimation and filtering problems', Ph.D. dissertation, Louvain University, Belgium, 1990.

[Liu71] B. Liu, 'Effect of Finite Word Length on The Accuracy of Digital Filters - A Review,' *IEEE Trans. on Circuit Theory*, Vol. CT-18, pp. 670-677, November 1971.

[Lju87] L. Ljung, *System Identification: Theory for the User*, Prentice-Hall Inc., Englewood Cliffs, New Jersey, 1987.

[LL85] S. Ljung and L. Ljung, 'Error Propagation Properties of Recursive Least Square Adaptive Algorithms,' *Automatica*, Vol. 21, No. 2, pp. 157-167, March 1985.

[LL86] A.J. Laub and A. Linnemann, 'Hessenberg and Hessenberg/Triangular Forms in Linear System Theory,' *Int. J. Control*, Vol. 44, No. 6, pp. 1523-1547, 1986.

[LS90] K. Liu and R. Skelton, 'Optimal Controllers for Finite Wordlength Implementation,' *in Proc. of the American Control Conference*, San Diego, 1990.

[LSG92] K. Liu, R. Skelton and K. Grigoriadis, 'Optimal Controllers for Finite Wordlength Implementation', submitted for publication.

[Mak78] J. Makhoul, 'A Class of All-Zero Lattice Digital Filters: Properties and Applications,' *IEEE Trans. on Acoustics, Speech, and Signal Processing*, Vol. ASSP-26, No. 4, pp. 304-314, August 1978.

[Man68] P.E. Mantey, 'Eigenvalue Sensitivity and State Variable Selection,' *IEEE Trans. on Automatic Control*, Vol. AC-13, pp. 263-269, June 1968.

[MG75] J.D. Markel and A.H. Gray, 'Roundoff noise characteristics of a class of orthogonal polynomial structures,' *IEEE Trans. on Acoustics, Speech, and Signal Processing*, Vol. ASSP-23, No. 5, pp. 473-486, October 1975.

[MG86] R.H. Middleton and G.C. Goodwin, 'Improved Finite Word Length Characteristics in Digital Control Using Delta Operators,' *IEEE Trans. on Automatic Control*, Vol. 31, No. 11, pp. 1015-1021, November 1986.

[MG90] R.H. Middleton and G.C. Goodwin, *Digital Control and Estimation : A Unified Approach*, Prentice Hall, Englewood Cliffs, New Jersey, 1990.

[MH82] A. G. J. MacFarlane and Y. S. Hung, 'A Quasi-Classical Approach to Multivariable Feedback System Design', *Proc. of the Second IFAC Symposium on Computer Aided Design of Multivariable Technological Systems*, West Lafayette, Ind., pp. 38-48, September 1982.

[Moo81] B.C. Moore, 'Principal Component Analysis in Linear Systems: Controllability, Observability and Model Reduction,' *IEEE Trans. on Automatic Control*, Vol. AC-26, No. 1, pp. 17-32, February 1981.

[MR76a] C.T. Mullis and R.A. Roberts, 'Synthesis of Minimum Roundoff Noise Fixed-Point Digital Filters,' *IEEE Trans. on Circuits and Systems*, Vol. CAS-23, pp. 551-562, September 1976.

[MR76b] C.T. Mullis and R.A. Roberts, 'Filter structures which minimize roundoff noise in fixed-point digital filters,' *Proc. IEEE Int. Conf. Acoustics, Speech and Signal Processing*, pp. 505-508, 1976.

[MR76c] C.T. Mullis and R.A. Roberts, 'Roundoff noise in digital filters: Frequency transformations and invariants,' *IEEE Trans. on Acoustics, Speech and Signal Processing*, Vol. ASSP-24, pp. 538-550, December 1976.

[MWH80] P. Moroney, A.S. Willsky and P.K. Houpt, 'The Digital Implementation of Control Compensators: The Coefficient Wordlength Issue,' *IEEE Trans. on Automatic Control*, Vol. AC-25, No. 4, pp. 621-630, August 1980.

[Nei67] F. Neiss, *Determinanten und Matrizen*, Springer-Verlag, Berlin, 1967.

[OM84] G. Orlandi and G. Martinelli, 'Low Sensitivity Recursive Digital Filters Obtained Via the Delay Replacement', *IEEE Trans. on Circuits and Systems*, Vol. CAS 31, No. 7, pp. 453-460, July 1984.

[OS89] A.V. Oppenheim and R.W. Schafer, *Discrete-time Signal Processing*, Prentice Hall, Signal Processing Series, Englewood-Cliffs, N.J., 1989.

[Pet86] V. Peterka, 'Control of Uncertain Processes: Applied Theory and Algorithms,' *Kybernetika*, Vol. 22, pp. 1-102, 1986.

[PHM90] J.E. Perkins, U. Helmke and J.B. Moore, 'Balanced Realizations via Gradient Flow Techniques', *Systems and Control Letters*, Vol. 14, pp. 369-380, 1990.

[PT82] R. V. Patel and M. Toda, 'Quantitive Measures of Robustness for Multivariable Systems', *Proc. of the Second IFAC Symposium on Computer Aided Design of Multivariable Technological Systems*, West Lafayette, Ind., pp. 153-158, September 1982.

[RC79] R.E. Rink and H.Y. Chong, 'Performance of State Regulator Systems with Floating-Point Computation,' *IEEE Trans. on Automatic Control*, Vol. AC-24, pp. 411-420, June 1979.

[RG75] L.R. Rabiner and B. Gold, *Theory and Application of Digital Signal Processing*, Englewood Cliffs, New Jersey, Prentice-Hall, 1975.

[RM87] R.A. Roberts and C.T. Mullis, *Digital Signal Processing*, Addison-Wesley, 1987.

[RR72] L.R. Rabiner and C.M. Rader, *Digital Signal Processing*, IEEE Press, The Institute of Electrical and Electronics Engineers Inc., New York, 1972.

[SB88] L.M. Smith and B.W. Bomar, 'An Algorithm for Constrained Round-off Noise Minimization in Digital Filters with Application to Two-Dimensional Filters,' *IEEE Trans. on Circuits and Systems*, Vol. CAS-35, No. 11, pp. 1359-1368, November 1988.

[SDD84] W.D. Stanley, G. R. Dougherty and Dougherty, *Digital Signal Processing*, Englewood Cliffs, New Jersey, Prentice-Hall, 1984.

[Smi68] R.A. Smith, 'Matrix Equation $AX + XB = C$,' *J. Appl. Math.*, Vol. 16, pp. 198-201, 1968.

[Spi65] M. Spivak, *Calculus on Manifolds*, W.A. Benjamin Inc, New York, NY, 1965.

[Sri78] A.B. Sripad, *Models for Finite Precision Arithmetic, with Applications to the Digital Implementations of Kalman Filters*, Sc.D. dissertation, Washington University, Sever Institute, January 1978.

[Sri81] A.B. Sripad, 'Performance Degradation in Digitally Implemented Kalman Filters,' *IEEE Trans. on Aerospace Electron. Syst.*, Vol. AES-17, pp. 626-634, September 1981.

[SS77] A. Sripad and D.L. Snyder, 'A necessary and sufficient condition for quantization errors to be uniform and white,' *IEEE Trans. on Acoustics, Speech, and Signal Processing*, Vol. ASSP-25, pp. 442-448, October 1977.

[SS90] G.W. Stewart and Ji-guan Sun, *Matrix Perturbation Theory*, Academic Press, 1990.

[Ste73] G.W. Stewart, *Introduction to Matrix Computations*, Academic Press, 1973.

[SW84] R. E. Skelton and D. A. Wagie, 'Minimal root sensitivity in linear systems,' *J. Guidance Contr.*, Vol. 7, pp. 570-574, September-October 1984.

[Thi84] L. Thiele, 'Design of Sensitivity and Roundoff Noise Optimal State-Space Discrete Systems,' *Int. J. Circuit Theory Appl.*, Vol. 12, pp. 39-46, 1984.

[Thi86] L. Thiele, 'On The Sensitivity of Linear State-Space Systems,' *IEEE Trans. on Circuits and Systems*, Vol. CAS-33, pp. 502-510, May 1986.

[TT84] V. Tavsanoglu and L. Thiele, 'Optimal Design of State-Space Digital Filters by Simultaneous Minimization of Sensitivity and Roundoff Noise,' *IEEE Trans. on Circuits and Systems*, Vol. CAS-31, No. 10, pp. 884-888, October 1984.

[VB89] B.D. Van Veen and R. Baraniuk, 'Matrix Based Computation of Floating Roundoff Noise,' *IEEE Trans. on Acoustics, Speech and Signal Processing*, Vol. ASSP-37, No. 12, pp. 1995-1998, December 1989.

[VD86] M. Verhaegen and P. Van Dooren, 'Numerical Aspects of Different Kalman Filter Implementations,' *IEEE Trans. on Automatic Control* Vol. AC-31, No. 10, pp. 907-917, October 1986.

[Ver84] E.I. Verriest, 'Gain Correction in Optimal Filtering Using Floating-Point Arithmetic,' *Proc. 23rd IEEE Conference on Decision and Control*, Las Vegas, NV, December 1984.

[Ver89] M. Verhaegen , 'Roundoff Error Propagation in Four Generally-applicable, Recursive, Least-squares Estimation Schemes,' *Automatica*, Vol. 25, No. 3, pp. 437-444, May 1989.

[VK83] E.I. Verriest and T. Kailath, 'On Generalized Balanced Realizations,' *IEEE Trans. on Automatic Control*, Vol. AC-28, No. 8, pp. 833-844, August 1983.

[VLG90] E.I. Verriest, G. Li and M. Gevers, *Computationally Optimal System Implementations via Polynomial Operators*, presented at MTNS-91, Japan, 1990.

[Wah91] B. Wahlberg, 'System Identification Using Laguerre Models,' *IEEE Trans. on Automatic Control*, Vol. 36, No. 5, pp. 551-562, May 1991.

[Wil63] J.H. Wilkinson, *Roundoff Errors in Algebraic Processes*, Prentice-Hall Inc., Englewood Cliffs, New Jersey, 1963.

[Wil79] A.S. Willsky, *Digital Signal Processing in Control and Estimation Theory - Points of Tangency, Areas of Intersection, and Parallel Directions*, Cambridge, MA, M.I.T. Press, 1979.

[Wil85] D. Williamson, 'Finite Wordlength Design of Digital Kalman Filters for State Estimation,' *IEEE Trans. on Automatic Control*, Vol. AC-30, No. 10, pp. 930-939, October 1985.

[Wil86] D. Williamson , 'Roundoff Noise Minimization and Pole-Zero Sensitivity in Fixed-Point Digital Filters Using Residue Feedback,' *IEEE Trans. on Acoustics, Speech, and Signal Processing*, Vol. ASSP-34, No. 5, pp. 1210-1220, October 1986.

[Wil88] D. Williamson, 'Delay Replacement in Direct Form Structures,' *IEEE Trans. on Acoustics, Speech, and Signal Processing*, Vol. 36, No. 4, pp. 453-460, April 1988.

[Wil91] D. Williamson, *Digital Control and Implementation-Finite Wordlength Considerations*, Prentice Hall Int., London, 1991.

[WK83] D.A. Wilson and A. Kumar, 'Symmetry Properties of Balanced Systems,' *IEEE Trans. on Automatic Control*, Vol. AC-28, No. 9, pp. 927-929, September 1983.

[WK89] D. Williamson and K. Kadiman, 'Optimal Finite Wordlength Linear Quadratic Regulation,' *IEEE Trans. on Automatic Control*, Vol. AC-34, No. 12, pp. 1218-1228, December 1989.

[Won90] P.W. Wong, 'Quantization noise, fixed-point multiplicative round-off noise, and dithering,' *IEEE Trans. on Acoustics, Speech, and Signal Processing*, Vol. ASSP-38, No. 2, pp. 286-300, February 1990.

[WS85] D. Williamson and S. Sriharan, 'Residue Feedback in Digital Filters Using Fractional Feedback Coefficients,' *IEEE Trans. on Acoustics, Speech, and Signal Processing*, Vol. ASSP-33, No. 2, pp. 477-483, April 1985.

[ZN91] B. Zeng and Y. Neuvo, 'Analysis of Floating Point Errors Using Dummy Multiplier Coefficient Sensitivities', *IEEE Trans. on Circuits and Systems*, Vol. 38, No. 6, pp. 602-612, June 1991.

[ÅSH84] K.J. Åström, J. Sternby and P. Hagander, 'Zeros of Sampled Systems,' *Automatica*, Vol. 20, No. 1, pp. 31-38, January 1984.

[Åst70] K.J. Åström, *Introduction to Stochastic Control Theory*, Academic Press Inc., New York, 1970.

[ÅW84] K.J. Åström and B. Wittenmark, *Computer Controlled Systems: Theory and Design*, Prentice-Hall Inc., Englewood Cliffs, New Jersey, 1984.

Index

The page numbers in bold refer to the page where a definition can be found.

A

Accumulator, 36, 42
Adaptive filter, 4, 353
Adder, 45
Adjoint matrix, **133**
Analog to digital converter, 23, 217, 252, 266
ARX model, **321**
Autocorrelation function, **331**

B

Balanced realization, 7, 14, 100, 118, 185, 198, 240, 302
    input balanced, **311**, 318
    internally balanced, **57**, 58, 98, 192, 311, 317
Baliverne, 21
Basis functions, 4, 79, 325, 326, 357
    generalized, 4
B_c, 33
Bias, 27
Binary code, 25, 28
Binary point, 26
Bit, 25
Bitlength, 262, 275, 276, 284, 286
Bomar-Hung algorithm, 203
B_s, 33

C

Canonical form, 7, 45, 66, 202
Cascade realization, 77
Cauchy index, 192
Cauchy-Schwartz inequality, 53, 90, 98, 137, 147, 299
Cholesky factorization, 57, 325, 334
Coefficient error
    finite wordlength, 50

Computational cost, 77, 211, 347
Computational speed, 183, 192, 194, 202, 218, 220
Condition number, 137, 323, 341, 346
Controllable canonical form, 122, 283
Controller design, 14, 217
    model reference, 236
    synthetic measure, 285
Controller implementation, 218
Controller realization
    sparse, 220
Controller reduction, 220
Convergence speed, 16, 323, 324, 348, 356, 357

D

Degrees of freedom, 186, 192, 201
Delay replacement, 80, 289
Delta operator, 3, 15, 20, 79, 80, 289, **291**, 335
    implementation, 295
    motivation, 290
Digital controller, 218
    FWL effects, 217
Digital to analog converter, 217, 252
Direct form, **4**, 45, 76, 78, 81, 127, 151, 175, 178, 183, 292
Dynamic range, 24–26, 39, 75, 302
    constraint, 63, 66, 161, 168, 202, 306
    δ form, 298
    fixed point, 28, 31
    floating point, 28

E

Eigenvalue sensitivity, 19, 131, 134, 136
    measure, 144
    multiple eigenvalues, 135
    overall measure, **138**

simple eigenvalue, 131
Estimation, 4, 16
Exponent, 28

F
Fast sampling, 3, 15, 20, 79, 81, 218, 291, 302, 323, 335, 342
Finite precision, 23
Finite wordlength, 1, 24, 50
    coefficient error
        closed loop, 268, 271
        controller design, 221, 255, 261
            LQG, 221
            model reference, 222, 223
        controller realization, 255, 261
        error, 25
        representation, 24, 25
    roundoff error
        closed loop, 269
FIR, 16
Fixed point, 25, 29, 33, 161, 296
    scaled, 26
    unscaled, 26
Floating point, 25, 28, 29, 296
    coprocessor, 29
Fractional part, 26, 33
Frequency transformation, 65
FWL, *see* Finite wordlength

G
Gamma operator, 16, 21, **336**
G_{cl}, **276**, 281
$G_{cl,T}$, **275**, 280, 282
Givens rotation, **65**
Gradient algorithm, 98, 117, 150, 323
Gradient flow, 117
Gramian, 113, 192, 203, 300
    controllability, 7, **42**, 48, 54, 63, 91, 168, 235, 276
    lower bound, 56
    observability, 7, 48, **54**, 91, 165, 233, 273, 274, 307
    similarity transformation, 55, 57
    weighted, 89, **90**, 106, 230, 273
G_T, 19, 160, 166, 169

H
Hankel singular values, **56**, 65, 300, 305, 309, 310, 312
Hessenberg form, 77, 185, **187**
High school, 339
Householder transformation, **188**, 325

I
Identification, 4, 16, 319, 348
Implicit function theorem, 132
Impulse response, 40
Infinite precision, 1
Information matrix, 21, **322**, 325, 327, 332, 337, 341, 348, 357
    δ form, 347
    γ-operator form, 340
Integer part, 26
Invariant, 65

K
Kalman filter, 13, 16, 225

L
Lattice filter, 78, 326
Least Squares, 319
Least Squares estimation, 322
LMS algorithm, 17, 324, 349
Low pass, 99, 134, 198, 313
LQG control, 13, 20, 217, 219, 220, **245**
    roundoff noise analysis, 248
LS, *see* Least Squares
Lyapunov equation, **43**, 54, 106, 165, 247, 249

M
Mantissa, 28
Matrix
    diagonally dominant, 344

involutory, **188**
orthogonal, **188**
normal, *see* Normal matrix
$M_{cl,L_{12}}$, **229**
M_{cl,L_2}, **276**, 280
$M_{\delta,L_{12}}$, **297**
$\bar{M}_{\delta,L_{12}}$, **299**
M_{δ,L_2}, **298**
Memory register, 26, 45
Microprocessors, 3, 218
MIMO system, 7
Minor of a matrix, **133**
MISO system, **162**, 265
$M_{L_{12}}$, **52**
$M_{L_{12}}^*$, **88**
$\bar{M}_{L_{12}}$, **54**, 167
M_{L_2}, **110**, 113, 165, 166
Model reference control, 218, 219, 264
M_{pz}, **129**, 144, 148, 209
MRH structure, 6, **64**, 65, 202, 209
l_2-scaled, 243, 282
Multiplier, 45

N

Narrow band, 66, 79, 81, 99, 134, 198, 313, 341, 342, 347
Noise cancellation, 353
Non-defective matrix, **141**, 146
Norm
  Frobenius, **52**
  L_1, 52
  L_2, 52
  l_2, **42**
  L_p, **52**
Normal matrix, **10**, 19, 32, 120, 138–141, 146, 185, 194
Normalization, 39

O

Observer, 224, 236
Optimal control, 218
Optimal controller design, 219
  L_1/L_2, 230, 237
  L_2

l_2 scaled, 280
LQG
  l_2 scaled, 252, 253
  roundoff noise gain, 222, 231, 234, 242, 245, 261
  l_2 scaled, 235
  l_2 scaled, 281
  LQG control, 250
  sensitivity, 222, 226, 261
  synthetic measure, 222, 261
  l_2 scaled, 276
Optimal filter design, 50, 219
  block optimal, **214**
  classical results, 67
  frequency weighted L_1/L_2 measure, 89, 99, 230
  L_1/L_2 measure, 56, 59, 187
    δ form, 299
  L_2 measure, 113, 170, 187
    δ form, 299
  pole measure, 139, 151
  pole-zero measure, 148, 150, 155, 175, 187, 209
  roundoff noise gain, 62, 63, 185, 211
    δ form, 305, 308
  sectional optimal, **214**
  synthetic measure, 111
  total noise gain, 168, 170, 174
  zero measure, 146, 150, 152
Orthogonal filter, 78
Overflow, 23, 24, 29, 39
  oscillations, 32
  saturation characteristic, 32
  two's complement characteristic, 31

P

Parallel realization, 77
Parameter estimation, 319
  accuracy, 325, 348
  adaptive, 320
Parametrization, 1, 24, 49
  δ-operator, 346, 347, 358
  γ-operator, 336, 347, 358

transfer function, 327
Performance criterion, 50
Perturbation of eigenvalues,
    see Eigenvalue sensitivity
Pole placement, 14, 20
Pole sensitivity, 120, 127
Polynomial parametrization, 20, 79, 326
Power-of-two, 203, 206, 211
Prefiltering, 325

Q
Quantization, 23, 161
    step size, 26
Quantization error, 33

R
RAM, see Roundoff after multiplication
RBM, see Roundoff before multiplication
Realization, 1, 24
    $0-1$, 209
    block balanced, 192
    block optimal, 185
    cascade, 77, 184, 212, 213
    δ model, 294
    minimal, 224
    parallel, 77, 184, 211, 213
    similar, **55**
    sparse, 20, 77, 184, 190, 194, 200, 215
        suboptimal, 201
    state space, **46**, 49
Realization set, **49**, 295
    controller, **226**, 266
Regressor, 321
    δ form, 346
    γ-operator, 339
Regulation, 219
Representation, 1
Residue feedback, 80, 305, 306
Residue modes, 305, 309, 310
Resolution, 28
Roundoff, 1

Roundoff after multiplication, 34, 37, 61
Roundoff before multiplication, 34, 36, 61, 80, 161, 231, 248, 269, 305
Roundoff error, 27, 38, 60, 232
    δ-operator, 296
Roundoff noise, **1**, 24, 34, 60, 159, 219, 220
Roundoff noise gain, 18, 39, 47, **62**, 66, 159, 202
    actual, 203, 209
    closed loop, 14, 225, 227, **234**, 243, 263, 276, 281
    δ form, 307
    LQG regulator, 250
    minimum, 64
    theoretical, 209

S
Sampled data system, 13, 217, 222
Sampling period, 3, 289, 290, 292
S_c, **226**
Scaling, 24, 26, 39
    dynamical, 43
    l_2, **41**, 42, 63, 168, 209, 214, 276
        controller states, 235, 280
Schur form, 77, 185, **190**, 213
    normal, 197
Schur realization, **190**, 199
    pole sensitivity of, 195, 197
S_δ, **79**, 292
S_δ^{opt}, **300**
Second order modes, **56**
Sensitivity, 2, 39, 220
Sensitivity function, **51**, 111
    closed loop, 229, **257**, 267, 272
    pole, 6, 9, 10, 129, 291
    root, 290
    transfer function, 47, 164
    weighted, **88**
    zero, 9, 129, 142
Sensitivity measure
    actual, 193, 203

closed loop, 14, 225, 227, 268
  L_1/L_2, **229**
  L_2, 262, **276**, 281
  upper bound, 230, 243
eigenvalue, **136**
frequency weighted, 8, 18
L_1/L_2, 7, **52**, 85, 110, 118, 159, 167, 192
  actual, 200
  δ form, **297**
  lower bound, 59
  upper bound, 54, 57, 90
  weighted, 87, **88**
L_2, 19, 109, **110**, 118, 160, 165, 210
  δ form, **298**
pole, 120, 139, 141, 144, 185, 194
  actual, 197
  lower bound, 138
  overall measure, **130**, 139
  theoretical, 197
pole-zero, 19, 128, **129**, 144, 185, 209
theoretical, 193
transfer function, 6, 51
zero, 142, **143**, 145
  lower bound, 146
  overall measure, 130, 143
zero measure, 13
Separation principle, 228
S_H, 49
Shift operator, 3, 20, 49, 78, 289
Sign bit, 25
Signal processing
  analog, 4
  digital, 4, 23
Signature matrix, 59, 118, 120, 192
Similar realizations, 6
Similarity transformation, 6, **43**, 46, 50, 90, 140, 144, 186
  closed loop, 226
  constrained, 202
  orthogonal, 186
  time varying, 203

Singular value decomposition, 57, 171, 207
Singular value decomposition, 115, 120
Sparse, **2**, 20, 183, 186, 211, 314
Spectral density function, **331**
Spoon-feed, 47
S_ρ, **295**
Statistical approach, 163
Step size, 323, 349
Stochastic gradient algorithm, 324
Sufficient richness, 322
SVD, *see* Singular value decomposition
Synthetic measure, 14, 20, 125
System Hessenberg form, **187**, 188, 189, 198

T

Total Noise Gain, 19, 160, 166, 178, 210
  closed loop, **275**, 277
Tracking, 219
Tradeoff
  accuracy/speed, 77, 184, 201
  overflow/roundoff, 33, 40
  range/precision, 28, 218
Transfer function, 49
  δ form, 294, 346
Trivial parameters, **183**
Truncation, 26
Truncation error, 27
Two's complement, 25
Two-degree-of-freedom controller, 14, 160, 262, 263, 279

W

White noise, 33, 34, 162
Wordlength, 25, 167

Z

Zero sensitivity, 127

Printing: Druckhaus Beltz, Hemsbach
Binding: Buchbinderei Kränkl, Heppenheim